Mathematical Modeling of Physical Systems

AN INTRODUCTION

Diran Basmadjian

Department of Chemical Engineering and Applied Chemistry, University of Toronto

New York Oxford
OXFORD UNIVERSITY PRESS
2003

Oxford University Press

Oxford New York
Auckland Bangkok Buenos Aires Chennai
Dar es Salaam Delhi Hong Kong Istanbul Karachi Kolkata
Kuala Lumpur Madrid Melbourne Mexico City Mumbai
Nairobi São Paulo Shanghai Taipei Tokyo Toronto

Published by Oxford University Press, Inc.
198 Madison Avenue, New York, New York, 10016
http://www.oup-usa.org

Library of Congress Cataloging-in-Publication Data

Basmadjian, Diran.
Mathematical modeling of physical systems: an introduction/Diran Basmadjian.
p. cm.
Includes bibliographical references and index.
ISBN 0-19-515314-6
1. Mathematical models. I. Title.

QA401 .B382 2002
511′.8–dc21

2002070373

Printing number: 9 8 7 6 5 4 3 2 1

Printed in the United States of America
on acid-free paper

CONTENTS

PREFACE

This book is intended to provide the beginning student and the professional with an easily understood introduction to the topic of mathematical modeling. Modeling, which consists of assembling mathematical expressions as a description of a system or process, now permeates the daily lives of professionals, and an early introduction to the subject at the university level has become desirable, even mandatory.

The requirements for successful modeling are numerous, onerous, and complex. They comprise, first and foremost, an understanding of the underlying physics of the systems and the laws that govern them. Next in importance is the ability to make suitable simplifying assumptions, which will reduce the complexities of a system to manageable proportions while retaining a valid and realistic description of its behavior. In many instances this will mean that we shall have to content ourselves with upper and lower bounds to the solution. We term this "bracketing the solution," and it is surprising how often such bounds are all one needs to arrive at satisfactory answers. Choosing appropriate assumptions, however, is no easy task and requires a good "feel" for the system under consideration. It is as much an art as it is science, and it involves skills that can be acquired by only sustained practice. It is one of the aims of this text to provide that practice with numerous illustrations drawn from a wide range of disciplines. This policy of choosing examples from diverse fields is a deliberate one on our part. The route we have chosen is therefore a somewhat difficult one, but drawing on a wide range of concepts and physical laws ultimately proves more fruitful than a narrowly focused approach.

There are three sound reasons for this. First, it is rare in modeling that one does not have to reach out into at least some unfamiliar territory, be it mathematical or physical. This is particularly so in the present-day climate of interdisciplinary activity. Indeed, one must often apply new concepts without fully understanding them or having the time to study them in depth. This necessity may be unpalatable on pedagogical grounds, but it is a stark fact of real-world modeling. The present text attempts to ease the pain in two ways. In the first instance, we provide our audience with numerous convenient tables of physical laws and mathematical formulae. Second, we lead the reader through an examination of new concepts by asking whether they make physical or mathematical sense, then by showing that they do. This provides a strong sense of reassurance of being on the right track and allows us to proceed to the solution with some confidence.

A second reason for choosing the present format is that studying modeling techniques used in other disciplines can immensely enhance one's own modeling skills and power. "There is gold in them-there hills," and one would be foolish not to seek it out. One must, however, be aware that it sometimes turns out to be fools' gold.

Third, different physical processes often lead to similar, if not identical, models. We show this in the present text on a number of occasions and urge the reader to use the insights provided in other contexts to enhance the quality of the solution being sought.

Setting up the model is followed by its solution, which can be obtained by analytical or numerical methods. The latter are very powerful, and their application is eased by a plethora of available numerical packages. We shall employ both methods, but the emphasis will be on analytical solutions, which are unsurpassed in providing insight into the system behavior. The analysis of solutions, both numerical and analytical, for important and unusual features is, in fact, an indispensable part of fruitful modeling.

The text is confined to deterministic models at the level of algebraic and ordinary differential equations. To keep length within reasonable bounds, stochastic processes are not addressed. Otherwise, the text is broad in its approach. Both discrete and continuous systems are considered. Optimization methods are taken up, and the occurrence of maxima and minima is brought out in a number of carefully chosen examples. Chapter 3 is entirely devoted to graphical and geometrical techniques, some of which are of an unusual character. We hope to induce the reader to apply these techniques in other contexts. The examples are accompanied by a detailed commentary highlighting important features of each problem and its solution and drawing the reader's attention to the lessons to be learned. We term this process "closure," and consider it to be an indispensable part of modeling.

The first chapter, Getting Started and Beyond, provides the reader with some useful tools to overcome what is often the greatest hurdle, that of initiating the modeling process. We tabulate a number of the more important underlying fundamental laws and provide some useful guidelines for choosing simplifying assumptions. The methodology is illustrated with several examples and practice problems drawn from different disciplines.

Chapter 2 addresses, at a simple and understandable level, some of the mathematical tools that will be required in subsequent chapters. They comprise short treatments, with examples, of vector algebra, matrices, and ordinary differential equations. This is followed, in Chapter 3, by examples whose principal feature is the use of a variety of geometrical concepts. These are applied to navigational problems, including the use of the global positioning system (GPS), problems of management and finance, the elucidation of the structure of the DNA double helix, and several other cases drawn form diverse fields. We have, as well, a first encounter with optimization procedures in which the reader is introduced to the concept of linear programming.

Chapter 4 deals with the effect of forces, a topic with wide ramifications. We examine forces due to electrical and magnetic fields and explore Thomson's famous experiment for the determination of e/m, the ratio of the charge of an electron to its mass. Mechanical and gravitational forces are taken up in the context of the flow of fluids, the bending of beams (which includes Euler's celebrated formula for the buckling of a strut), and the path of projectiles and satellites. The hot air balloon provides an interesting example of the combination of gravity and lift obtained from chemical energy. Several additional examples drawn from different fields round off the chapter.

Chapter 5 introduces the reader to the notion of compartmental models, which arise in a variety of contexts and assume that the state variables change at most with time and not at all with distance. The examples here are drawn largely from the fields of biomedical and environmental engineering and should have a particular appeal to students and professionals in the biological and environmental sciences, as well as in chemical engineering. An example of interest to all disciplines concerns the dynamics of the human immunodeficiency virus (HIV), a topic that has only recently come to the fore.

In Chapter 6 we consider unidirectional distributed systems. Here there is no variation with time; instead, changes in the state variables occur along one principal direction. We examine distributions that arise in processes involving heat and mass transfer, fluid flow, and chemical reaction, and we show how these primary distributions can often be converted to characteristic efficiency indices for engineering use. Chapters 5 and 6 involve algebraic as well as ordinary differential equations.

Chapter 7 deals with simple networks, a term we use to describe systems with several state variables, giving rise to sets of equation. Such sets are encountered in a wide range of disciplines, for example, as electrical circuits, and we demonstrate their occurrence with numerous examples. Readers who have often questioned the usefulness of matrices will find some fruitful applications here.

In the final chapter we take up the twin topics of dimensional and numerical analysis, which are indispensable in the solution of the more complex problems. Some interesting examples are presented, among them the design of a depth charge and the effect of drag on the trajectory of an artillery piece.

The treatment is throughout kept at an elementary level and does not go beyond simple applications of vector algebra, matrices, and ordinary differential equations. Partial differential equations are studiously avoided. This does not mean that the examples and practice problems presented are trivial. Most of the examples are important classical and contemporary problems, which illuminate the disciplines from which they are drawn. These include the fields of mechanics, electricity, magnetism, chemistry, materials science, biology, economics, and finance, as well as the civil, mechanical, chemical, electrical, and biomedical engineering disciplines.

The material is suitable as a text in modeling or applied mathematics for students of mathematics, the sciences, and engineering at the third-year level. It will serve as a reference text for students and professionals alike who are looking for an introduction to modeling that is easy to follow without being trivial.

The text has been scaled to allow its coverage in one term. It has also been designed so that any number of the examples or even entire chapters can be omitted without disturbing continuity. We emphasize that our primary concerns are to address the novice and to provide a simple and lucid account of what mathematical modeling can accomplish. In the process we hope to open up for the reader the physical world that has benefited from modeling.

As on other occasions, I have the pleasant task of expressing my gratitude to Arlene Fillatre, who miraculously converted handwritten notes to camera-ready format, and to Linda Staats, who worked her usual magic to produce the illustrations seen in this text. I am immensely grateful to S. Vijayakumar, who provided some inspired examples for the text and helped in many important ways. The constant encouragement provided by my wife, Janet, was an essential ingredient in the completion of this work.

NOTATION

a	Acceleration, m/s^2
a	Specific surface area, m^2/m^3 packing
A	Area, m^2
A_C	Cross-sectional area, m^2
B	Magnetic field, N/A m = T
c	Particle speed, m/s
c	Velocity of sound, m/s
C	Capacitance, C/V
C	Concentration, mol/m^3
C_D	Drag coefficient, dimensionless
C_p	Specific heat at constant pressure, J/mol K or J/kg K
C_v	Specific heat at constant volume, J/mol K or J/kg K
d	Diameter, m
d	Distance, m
D	Diffusivity, m^2/s
D	Dilution rate, 1/s
D	Rate of evaporation, mol/s
D_{AB}	Distance (surveying), m
e	Charge of an electron
E	Catalyst pellet effectiveness factor, dimensionless
E	Extruder uniformity index, dimensionless
E	Heat exchanger fin efficiency, dimensionless
E	Residence time distribution, 1/s
E	Young's modulus, Pa
E	Electrical field, N/C
E_A	Activation energy, J/mol
E_K	Kinetic energy, J/kg or J/mol
E_p	Potential energy, J/kg or J/mol
f	Friction factor, dimensionless
f	Self-purification rate $k_L a/k_r$, dimensionless
F	Force, N
F	Mass flow rate, kg/s
F_D	Drag force, N
Fr	Froude number, $v^2/l\,g$, dimensionless

g	Gravitational acceleration, 9.81 m/s^2
G	Mass velocity, F/A, kg/m^2 s
G	Universal gravitational constant, 6.6720×10^{-11} N m^2/kg^2
h	Heat transfer coefficient, J/m^2 s K
h	Height, m
H	Enthalpy, J/kg or J/mol
H	Height, m
H	Henry's constant, various units
Δh_f	Friction head, $\Delta p/\rho$, N m/kg
ΔH_s	Heat (enthalpy) of solidification, J/kg
i	Current, A
I	Moment of inertia, N m^2
I	Second area moment, m^4
k	Thermal conductivity, J/m s K
k_c	Mass transfer coefficient, various units
k_e	Elimination rate constant, 1/s
k_L	Liquid phase mass transfer coefficient, m/s
k_r	Reaction rate constant, various units
K	Bulk modulus, Pa
K	Cooling constant, 1/s
K	Overall mass transfer coefficient, various units
K	Partition coefficient, dimensionless
K	Permeability, m^2
K_m	Michaelis–Menten constant, mol/m^3 s
K_p	Equilibrium constant, Pa^n
K_s	Monod constant, mol/m^3
l	Length, m
L	Amount of liquid, kg
L	Inductance, V s/A
L	Length, m
Lk	Link number, dimensionless
m	Equilibrium constant, Y*/X, dimensionless
m	Mass, kg
M	Molar mass, kg/mol
M	Moment, N m
Ma	Mach number, v/c, dimensionless
n	Number of moles
N	Mass transfer rate, mol/s
N	Number of particles
N_{Av}	Avogadro's number, 6.02×10^{23} 1/mol
Nu	Nusselt number, hd/k, dimensionless
p	Population
p	Pressure, Pa
P	Momentum, kg m/s
P	Perimeter, m
P	Power, J/s = W

P^0 Vapor pressure, Pa
Pr Prandtl number, $C_p\mu/k$, dimensionless
q Electrical charge, C
q Heat transfer rate, J/s
Q Volumetric flow rate, m^3/s
r Radial distance, m
r Reaction rate, mol/m^3 s
r_C Core radius, m
r_{max} Maximum reaction rate, mol/m^3s
R Radius, m
R Resistance, Ω
R Universal gas constant, Pa m^3/mol K
Re Reynolds number, $lv\rho/\mu$, dimensionless
s Arc length, rad
s Laplace transform parameter
S Amount of adsorbent, kg
S Fractional area covered, dimensionless
S Rate of heat generation, J/m^3 s
S Specific resistivity, Ω m
S Substrate concentration, mol/m^3
Sc Schmidt number, $\mu/\rho D$, dimensionless
Sh Sherwood number, k_cl/D, dimensionless
St Stanton number, $h/v\rho C_p$, dimensionless
t Time, s
T Period of revolution, 1/s
T Tangent, m
T Temperature, °C or K
T_C Critical temperature, °C or K
Tw Twist number
u Velocity, m/s
U Overall heat transfer coefficient, J/m^2 s K
U_s Stored strain energy, J
v Velocity, m/s
V Voltage, V
V Volume, m^3
W Residue, mol
W Width, m
W Work, J
Wr Writhing number, dimensionless
x Axial distance, m
x Mole fraction, mol/total moles
x_i Variable i
x_s Solubility mole fraction, mol/total moles
X Conversion, moles reacted/moles fed
X Mass ratio, kg/kg
y Mole fraction, mol/total moles

Y Mass ratio, kg/kg
Y_{CS} Mass of cell produced/mass of substrate consumed
Y_{PC} Mass of product formed/mass of cells
z Axial distance, m

Greek Letters
α Bearing (surveying), degrees
γ Ratio of specific heats, C_p/C_v
δ Boundary layer thickness, m
δ Displacement, m
ε Emissivity, dimensionless
ε Strain, m/m
ε Void fraction, m^3/m^3
ε_0 Permittivity constant, $8.8 \times 10^{-12}\ C^2/N\,m^2$
θ Direction (surveying), degrees
μ Rate of growth, 1/s
μ Viscosity, Pa s
π Dimensionless group
ρ Density, kg/m^3
σ Normal stress, Pa
τ Residence time, s
τ Shear stress, Pa
τ Tensile stress, Pa
φ Angle, rad
ω Angular velocity, rad/s

Subscripts
a ambient
a arterial
b bulk
d dye
e earth
f feed
f final
f fluid
h heart
H hydrodynamic
i infected
i initial
L liquid
m membrane
m moon
n nozzle
o initial
p particle
p plasma

s sun
S solid
t tangential
t tank
u uninfected
v venous
v virus

Superscripts
$*$ Equilibrium
$-$ Laplace transform

CHAPTER 1

Getting Started and Beyond

A mathematical model, as it is understood here, consists of the assembled equations that describe a physical system or process. The act of setting up these equations and, by extension, solving them, is referred to as mathematical modeling. It is practiced with uncommon frequency in all physical and engineering sciences, including biology and medicine, and beyond those areas in economics and finance. Modeling has even reached into the so-called soft disciplines such as the social and behavioral sciences. Its aim in all these subject areas is to quantify a system in mathematical language and to provide quantitative answers to underlying problems. Modeling can thus be viewed, at its most elementary level, as a problem-solving tool. Although it also has much more profound uses, such as analysis and prediction, and although we shall also give analysis its due, our principal concern here will be with the elementary problem-solving aspects of modeling.

Students who have struggled with problem solving going back to their high school days will recall the uneasiness, even dread, upon being confronted with physical problems, particularly those that fall into the category of "word problems." They will recall, in particular, the difficulty of getting started, of overcoming that first hurdle of defining underlying principles and relations that will serve as building blocks for the model. This process is often an art, which requires physical insight as well as an understanding of physical principles, and the skill, usually attributed to engineers, of making suitable simplifying assumptions.

Being at least part art, the initial steps of modeling do not easily yield to fixed recipes with a guaranteed successful outcome. There are, however, a number of useful steps the student can take which, while not carrying an ironclad guarantee, may serve to ease the initial pain. It is these beginning steps we wish to address here in the first instance and to demonstrate their application by means of some simple examples. This will be followed by a description of what we term "closure," that is, the final and indispensable act of analyzing the results obtained.

1.1 When Not to Model

This is a somewhat startling beginning to a text devoted to the opposite goal of modeling and how to model. Yet the preamble is needed to caution the reader about unnecessary, futile, and other inappropriate uses of modeling. We have for this purpose compiled a list of problems with various modeling requirements. The first nine items comprise problems that either require or would at least benefit from mathematical modeling. The underlying models, or physical principles on which the models are to be based, are not indicated and require some thought and preparation. These are in fact the "initial steps" alluded to earlier, to which we shall return later. In the case of item 1 (see Table 1.1), for example, an initial examination will reveal that the projectile will be subjected to both gravitational and drag forces that must be incorporated in a force balance. In other words, we have to make use of Newton's law:

$$\sum \text{Forces} = \text{Mass} \times \text{Acceleration} \tag{1.1}$$

A similar force balance is needed in item 5 (pendulum), and it will be noted that in both cases the desired information (distance or time) lurks in the acceleration term. Hence there is reason to believe that modeling is appropriate and promises to provide the required answers. We do not take the reader beyond this first peek at the hurdles to be encountered but reiterate that the *very first decision to be made* is whether modeling should be undertaken at all. We reinforce this aspect by turning to the question of when *not* to model. Here we are able to provide some fairly precise answers, which we take up in turn. We ask the indulgence

Table 1.1 When to Model

Information Sought	Model?
1. Distance traveled by a projectile	Yes
2. Time required for a fuel droplet to burn	Yes
3. Deflection of a loaded beam	Yes
4. Acceleration of a charged particle in a magnetic field	Yes
5. Relation between length and period of a pendulum	Yes
6. Resolution of a "spy in the sky" satellite	Yes
7. Size of a water purification unit	Yes
8. Prediction of population growth	Yes
9. Weather prognosis	Yes
10. Contaminant levels in a polluted lake	Yes and no
11. Rupture of a water pipe due to freezing	Yes and no
12. Cause of a fire	Yes and no
13. Cause of an explosion	Yes and no
14. Drag force on an object moving through a fluid at high speed	Dimensional analysis + experiment
15. Cause of a leaky valve	No
16. Cause of corrosion failure of a vessel	No
17. Reaction rate constants	No
18. Heat and mass transfer coefficients in turbulent flow	Dimensional analysis + experiment

of the reader if some of the answers appear to be self-evident. They need nevertheless to be stated.

1. The answer is required within hours or no more than a day. This case arises frequently, particularly in industrial contexts. It is surprising how often a solution is required the next day or not at all. The recommendation here is to attempt to compose a very simple model that may at least provide a guideline, or a lower or upper bound, to the answer being sought. If this is not forthcoming, no modeling should be undertaken.

2. The answer may be obtained by a relatively simple and inexpensive experiment. Here an assessment is needed of the speed and cost of experimentation, as well as the degree of uncertainty that may reside in the model. If laboratory facilities and manpower are conveniently available and the experiment can be rapidly performed, modeling should be set aside in favor of experiment.

3. The client is suspicious of "theory" and prefers the certainty of experiment, even if it should be costly and time-consuming. This situation arises with some frequency in court cases involving insurance claims related to fires and explosions. The audience here (judge, jury) is one that will tend to be more easily swayed by physical facts than by theory. Modeling, if it is undertaken at all, should be at an elementary level easily conveyed to laymen and should in most cases be used as an adjunct to experiment. We have indicated this in Table 1.1 by answering "Yes and no" to items 12 and 13.

4. The system is too complex to be modeled in a meaningful way. The determination of drag coefficients and transport coefficients in complex systems (items 14 and 18 in Table 1.1) provides two cases in point. Although advances in the theoretical treatment of these systems are being made with the aid of powerful computational methods, one is still, at this stage, forced to resort to experimentation in conjunction with dimensional analysis to obtain satisfactory answers (see Chapter 8). In the case of reaction rate constants, one has to rely entirely on experimental data.

5. The answer is self-evident, in some cases making a model unnecessary, whereas at other times some benefits may be derived from modeling. An example of this case is provided by item 11 of Table 1.1 — rupture of a water pipe due to freezing. A pipe containing still water because of pump failure and exposed to subzero temperatures for several weeks will evidently rupture. Suppose, on the other hand, that the temperature drops only a few degrees below zero, and then only for a brief period of time, say 12 hours. The question then arises whether one should provide expensive insulation or use costly heavy gauge specialty steels to forestall rupture. A heat transfer model would in this case provide some fairly precise answers on the temperature drop experienced by the water and the maximum internal pressures that would result from ice formation. If the results indicate that the water will remain above freezing or will freeze only marginally, the costly precautions we had indicated can be avoided altogether. If, on the other hand, the predictions are for a temperature drop below zero, the model will enable us to minimize the cost of the countermeasures to be taken. Here, as in several other items of Table 1.1, the answer must therefore be "Yes and no."

6. Sterile or post-facto modeling. With these two phrases we wish to draw the reader's attention to a class of modeling that is objectionable on the grounds that it fails to provide answers of substance or illuminate the underlying process in a meaningful way. This type of activity occurs in areas that have been thoroughly studied experimentally, where the underlying physics of the process are well understood and the pertinent variables have been quantitatively

interrelated. Applied to such systems, a mathematical model would merely play a confirmatory role rather than being put to use as an investigative or problem-solving tool. Similar objections can be raised against elaborate models that fail to provide significant new information or cannot be implemented in practice. Yet another objectionable category consists of models that are confined to stating the self-evident or trivial. It is surprising that these rather obvious misuses or underuses of mathematical modeling still find their advocates and still persist in the literature. We reinforce the foregoing points with two concrete examples.

◆ EXAMPLE 1.1 The *Challenger* Space Shuttle Disaster

In this well-publicized disaster in the U.S. space program, the *Challenger* craft exploded in flight several minutes after launch. The elaborate investigation that followed revealed the failure of a crucial gasket seal because of embrittlement as one possible cause of the explosion. That embrittlement was brought about by a drop in the temperature near the gasket and an attendant loss of its elasticity and ability to provide a smooth seal. Nobel laureate Richard Feynman, a member of the investigative panel, demonstrated embrittlement in a strikingly simple experiment by placing some gasket material in ice water and allowing his audience to observe its loss of elasticity. This deterioration at what was a relatively benign temperature had not been completely anticipated by the manufacturers. Feynman also scrutinized some model predictions made during construction of the craft and found them wanting. He expressed his misgivings in the statement "When using a mathematical model, careful attention must be given to the uncertainties in the model."

COMMENTS

- Clearly, an enterprise as important as the construction of a manned spacecraft requires both modeling and experimental testing of sensitive components. There is no doubt here that it is necessary to model.
- The reference by Feynman to model uncertainties points to a difficulty that bedevils many, if not most, mathematical models. Removing all uncertainties is often impossible or would lead to expressions of unmanageable complexity. One remedy is to use asymptotic analysis, or what we had termed "bracketing the solution." In this exercise one looks for upper and lower bounds to the solution being sought. One of these bounds should reflect a "worst-case scenario" and will help establish safety factors that will prevent the unbelievable scenario from unfolding.

◆ EXAMPLE 1.2 Loss of Blood Vessel Patency

The deposition of blood components onto the blood vessel wall is a well-known pathological phenomenon that may ultimately lead to complete clogging of the vessel, or "loss of patency." The consequences are loss of mobility of a limb or, in more serious instances, stroke or heart failure. The author examined the process of vessel closure some time ago in a paper that considered the fluid forces on a wall protuberance exerted by the flowing blood. If flow is large enough and the deposit protrudes significantly into the flow field, the fluid forces will tend to disengage the deposit, thus preventing vessel closure. If flow velocity is low, on the other hand, and the vessel diameter is small, deposition may continue unhindered until complete closure has occurred. The model was able to predict the range of flow rates and vessel diameters that would either maintain patency or result in vessel failure.

COMMENTS

- The modeling of life-threatening situations holds a special lure and is on the surface at least of considerable significance. In this particular instance, however, insufficient attention was given to the existing body of knowledge. A wealth of data was available from accumulated medical and surgical experience capable of relating the likelihood of vessel failure to size and blood flow. One implant surgeon commented: "Give me a garden hose and I can guarantee patency." The size of implant that was likely to stay patent or, conversely, would be prone to failure in various locations of the body, was thus well-known.
- Given the facts just stated, one would have to categorize the exercise as "post-facto" modeling and one that failed to go beyond confirming established wisdom. It registered few citations and elicited little interest. The author learned from this early experience and henceforth tried to confine himself to worthier causes.

1.2 Some Initial Tools

To commence the modeling process in a fruitful manner, we must have some initial tools available. In addition to physical insight, these will generally comprise the underlying equations that describe the system or process. Evidently there is a plethora of such relations, and part of the art of modeling consists of choosing the appropriate and correct equations. In physical systems or processes, which need not be limited to the sciences or engineering, one can resort to a preliminary classification or tabulation of the pertinent equations, which provide an aid in the difficult task of getting off to a good start. We give two such short tabulations designed to assist the reader in this endeavor. They are based on the recognition that many systems and processes are based on fundamental laws, which include the important category of conservation laws. A limited number of such laws are listed in Table 1.2 and are supplemented in Table 1.3 by a short compilation of what we term "auxiliary relations."

We note that Table 1.2 contains several conservation laws, among which the laws of conservation of mass and energy (items 1 and 2) stand out. Several subsidiary laws of conservation spring from the law of conservation of energy (items 2a, b, and c and Kirchhoff's second rule, item 7) and from Newton's law (items 3a, b, and c). The latter is sometimes referred to as a momentum balance, keeping in mind that the rate of change of momentum $(d/dt)(m\mathbf{v}) = (d/dt)\mathbf{P}$ is equivalent to a force $\mathbf{F}$. Another conservation law is that of electrical charge, which we have not stated explicitly but have instead presented as Kirchhoff's first rule (item 6), which derives from it. It states that electrical charge is neither created nor destroyed, a fact that is nowadays accepted as self-evident. The gravitational force acting between two masses (item 4) has its electrical counterpart in Coulomb's law (item 8). Fick's law and Fourier's law (items 9 and 10) express the rate of diffusional flux of mass N (mol/s) and of thermal energy q (J/s). Finally we cite two additional laws due to Newton, the law of cooling (item 11) and the viscosity law (item 12). The former expresses the convective (i.e., nondiffusive) heat loss of a hot body to a cooler medium. Such convective heat transfer occurs particularly when the object is immersed in a flowing medium. Equation (1.18) in Table 1.3 derives from Newton's law of cooling and

Table 1.2 Fundamental Laws

1. *Law of Conservation of Mass*

$$\begin{matrix}\text{Rate of} \\ \text{mass in}\end{matrix} - \begin{matrix}\text{Rate of} \\ \text{mass out}\end{matrix} = \begin{matrix}\text{Rate of change} \\ \text{of mass content}\end{matrix}$$

2. *Law of Conservation of Energy*

$$\begin{matrix}\text{Rate of} \\ \text{energy in}\end{matrix} - \begin{matrix}\text{Rate of} \\ \text{energy out}\end{matrix} = \begin{matrix}\text{Rate of change} \\ \text{of energy content}\end{matrix}$$

   (a) Conservation of thermal energy
   (b) Conservation of mechanical energy
   (c) Conservation of kinetic and potential energies

3. *Newton's Law*

$$\mathbf{F} = m\frac{d\mathbf{v}}{dt} = \frac{d\mathbf{P}}{dt} \tag{1.1}$$

   (a) Conservation of linear momentum

$$\sum \mathbf{P} = \mathbf{0} \text{ (no external force)} \tag{1.2}$$

   (b) Conservation of angular momentum

$$\sum I\omega = 0 \text{ (no external torque)} \tag{1.3}$$

   (c) Conservation of moment

$$\sum \mathbf{r} \times \mathbf{F} = 0 \text{ (no external torque)} \tag{1.4}$$

4. *Law of Universal Gravitation*

$$F = G\frac{m_1 m_2}{r^2} \tag{1.5}$$

   G = gravitational constant = 6.67×10^{-11} N m^2/kg^2

5. *Ohm's Law*

$$V = \frac{i}{R} \tag{1.6}$$

6. *Kirchhoff's First Rule (conservation of charge)*

$$\left(\sum i\right)_{\text{junction}} = 0 \tag{1.7}$$

7. *Kirchhoff's Second Rule (conservation of energy)*

$$\left(\sum V\right)_{\text{loop}} = 0 \tag{1.8}$$

8. *Coulomb's Law*

$$F = \frac{1}{4\pi\varepsilon_0}\frac{q_1 q_2}{r^2} \tag{1.9}$$

   where ε_0 = permittivity constant = 8.85×10^{-12} C^2/N m^2

9. *Fick's Law*

$$N = -DA\frac{dC}{dx} \tag{1.10}$$

10. *Fourier's Law*

$$q = -kA\frac{dT}{dx} \tag{1.11}$$

11. *Newton's Law of Cooling*

$$\frac{d}{dt}\Delta T = -K\Delta T \tag{1.12}$$

12. *Newton's Viscosity Law*

$$\tau = -\mu\frac{dv}{dx} \tag{1.13}$$

Table 1.3 Auxiliary Relations

1. *Chemical Reaction Rates*		
First order	$r = k_r C$	(1.14)
Second order	$r = k_r C^2$ or $k_r C_A C_B$	(1.15)
Michaelis–Menten	$r = \frac{r_{max} S}{K_m + S}$	(1.16)
2. *Interphase Transport*		
Mass transfer	$N = k_C A \Delta C$	(1.17)
Heat transfer	$q = hA\Delta T$	(1.18)
3. *Friction and Drag*		
Frictional pressure drop in a pipe	$\Delta p_f = 4f\rho \frac{v^2}{2} \frac{L}{d}$	(1.19)
Drag in flow around a submerged body	$F_D = C_D \rho \frac{v^2}{2} A_C$	(1.20)
4. *Enthalpy per Unit Mass*		
Thermal enthalpy change	$\Delta H = C_p \Delta T$	(1.21a)
Enthalpy of reaction	$H_{products} - H_{reactants}$	(1.21b)
Enthalpy of vaporization	$H_{vapor} - H_{liquid}$	(1.21c)
Enthalpy of solidification	$H_{liquid} - H_{solid}$	(1.21d)

generally applies to a system having one or both of the hot and cold media in convective flow. Both K and h are empirical coefficients that depend on system properties, including velocity, and are termed heat transfer coefficients. Newton's viscosity law, finally, relates the shear force τ (N/m^2) that results (or must be applied) when two adjacent fluid layers in viscous flow move at different velocities. Here again μ (Pa s) is an empirical coefficient and a property of the fluid termed viscosity.

The relations listed in Table 1.3 are termed auxiliary because they enter the fundamental conservation laws listed in Table 1.1 as *adjuncts*. Thus the chemical reaction rates, item 1, enter the conservation of mass statement either under the "Rate in" or the "Rate out" column, depending on whether the rate refers to a species being produced or consumed. The interphase transport equations (item 2), convey, as already mentioned, the rate of mass and heat transfer between two media, one or both of which are in flow. On rarer occasions they are used to describe transport between two stationary media, such as the heat loss of a hot medium to stagnant air. We note that both equations (1.17) and (1.18) are composed of a driving force, ΔC and ΔT, and a coefficient k_C (m/s) and h (J/m^2 s K), which may be viewed as a conductivity, or the reciprocal of a resistance to transport. In this they resemble Ohm's law, equation (1.6), in which voltage, V, plays the role of a driving force and R a resistance, which results in the current flow i. We also note that these relations enter the "Rate in" or "Rate out" columns of the conservation of mass and energy statements, depending on their direction. The relations in item 3 in essence represent the friction forces that arise in flow through a pipe and when an object is in relative motion to a surrounding fluid. We encounter again certain empirical coefficients, the friction factor f and the drag coefficient C_D, both of which depend on fluid velocity, incorporated in the dimensionless Reynolds number $Re = lv\rho/\mu$, where l is some linear dimension such as diameter. Pipe friction varies directly with length L and inversely with pipe diameter d, while the drag force F_D is proportional to the cross-sectional area A_C of the object exposed to the fluid. Finally, in item 4, we relate enthalpy change ΔH to the corresponding temperature change ΔT. This relation is needed to express the energy change of a medium, often in a state of flow, when it receives or loses energy as represented by equation (1.18). It has the advantage of converting energy, which is often not of direct importance, to temperature, which tends to be the variable of interest. Enthalpy changes that occur during reactions, vaporization, and solidification are also tabulated. These, then, are the tools, or typical of the tools, we have available to initiate the modeling process.

We next turn to the consideration of certain initial steps that can be taken, which, together with the tools we have presented in Tables 1.2 and 1.3, will help in a first formulation of the model. We list these steps, some of which may appear self-evident, in the order in which they are to be taken.

1. *Make a Sketch.* Start by diagramming the system to be considered, inserting where possible the important given variables and the desired result or results. This seemingly obvious step is often omitted by the novice, who instead scrutinizes of the problem mentally. This may be appropriate in certain simple situations but rapidly loses its effectiveness with increasing complexity of the system or process. A case can be made for using it under all circumstances.

2. *Draw an Envelope.* If a conservation law is to be invoked, draw an envelope around the segment of the sketch to which it is to be applied, and indicate ingoing and exiting quantities. This is not the easy matter it appears to be. The envelope may have to be drawn around a differential segment, or alternatively, the geometry is a finite one such as a tank. In the case of systems composed of several phases, more than one balance is usually called for.

Differential segments are used if it is planned to track the change with distance of a particular variable. Typically this will be the case if the question asked is: How long must a certain unit be to give a prescribed value of the variable at the exit? That information can be extracted only by first considering the inputs and outputs to a differential entity, which we term a "differential balance," and then integrating the result to the desired length. We will have much more to say about this in Chapter 6, where such "distributed systems" are taken up in more detail. Application of conservation laws to a finite entity, on the other hand, calls for the use of what we term "integral balances," to distinguish them from the differential balances applicable to distributed systems. Typically, such balances provide a relation between ingoing and exiting quantities, or their variation with time. Such time-dependent processes usually result in a differential equation that must be integrated to provide the information sought. More about such systems will appear in Chapter 5.

3. *Introduce Simplifying Assumptions.* Whether to simplify and how to simplify is quite often the crucial question which needs to be answered to ensure success. Simplification can make or break a model. It is also an art that needs to be learned by persistence and repeated application.

Let us see what it is we wish to accomplish by introducing simplifications. There are basically five goals to be attained, all of them mathematical, which can be stated with a fair degree of precision.

(a) *Reduce the Number of Unknown Dependent Variables*. Since this number generally equals the number of equations, the net effect is to reduce the size of the model, a feature often referred to as "model reduction." This reduction is usually accomplished by declaring one of the variables to be of little or no consequence to the outcome. Suppose, for example, that we are considering a chemical reaction taking place in a reactor. Such reactions often occur with an evolution of heat, which in turn affects the rate of reaction. The changes in mass that take place are accommodated by at least one mass balance, while the corresponding heat effects call for the use of the law of conservation of energy. We will thus be dealing with a minimum of two coupled equations, a mass balance and an energy balance. Suppose, however, that (a) the reactor is cooled, or (b) we are dealing with a liquid system that has a high heat capacity capable of absorbing heat with little change in temperature. We may then be in a position to assume that the operation is nearly isothermal and that as a consequence the energy balance becomes redundant. Let us next consider a system in which two species react, rather than a single one; that is, we are dealing with equation (1.15) of Table 1.3. This will call for the use of two mass balances plus possibly an energy balance. However, by carrying out the reaction with a large excess of one of the species, say A, we can ensure that C_A will experience only minor variations, thus eliminating the need to formulate a mass balance for A. This is yet another example of model reduction. Whether these simplifications are justified within acceptable error limits is, of course, a judgment call, the "art" of which we spoke earlier.

(b) *Reduce the Number of Independent Variables*. This has the effect of lowering the dimensionality of the model. Thus a partial differential equation (PDE) in two independent variables would, upon removal of one of them, become an ordinary differential equation (ODE).

There are two frequently used ways of accomplishing this reduction. The first assumes that the dependent variable, such as temperature, concentration, or pressure, is *uniform in space* and depends at most on time. This leads to the concept of a "stirred tank" or "compartment," which we have depicted in Fig. 1.1A. It consists of a finite entity, the tank or compartment, which has an inflow and outflow and can also accommodate not only the chemical reactions taking place within it but also an exchange of mass or energy with the surroundings. Such stirred tanks or compartments are the principal tool of modeling environmental and biological systems, as well as a host of other processes that arise in the sciences and engineering. A typical environmental example is a lake that takes in a contaminant with its inflow. That contaminant may then undergo a reaction within the lake (e.g., by biodegradation), interact with the sediment or suspended matter (e.g., by adsorption), evaporate into the atmosphere, and ultimately exit with the outgoing flow. The concentration within the lake, which equals that in the outflowing water, is assumed to be completely uniform. It does not reflect local variations in concentrations, and hence represents a smoothed average of the levels seen in the lake and in the outflow. Its value will initially vary with time and if contamination persists will ultimately rise to a constant, steady-state result. This is shown in Fig. 1.2.

A second important way of reducing dimensionality is to eliminate all but one distance variable from consideration and assume that there is no variation with time. This leads to what we term the "one-dimensional pipe," depicted in Fig. 1.1B. In this device, a stream carrying mass, energy, or

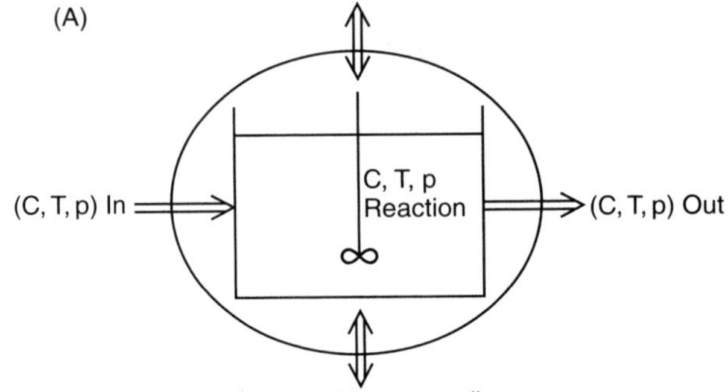

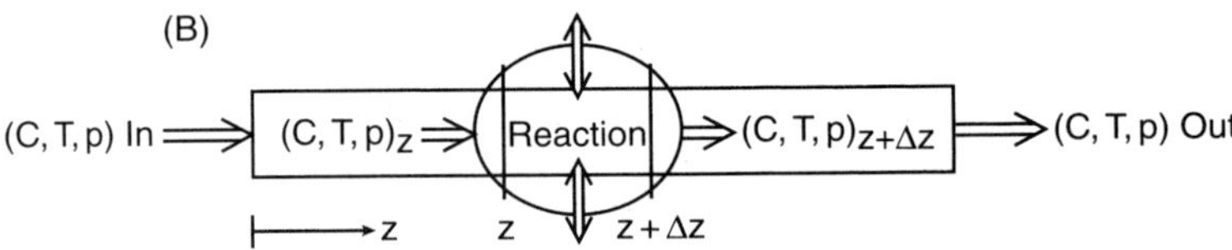

Figure 1.1 Two basic physical models: (A) the stirred tank or compartment and (B) the one-dimensional pipe.

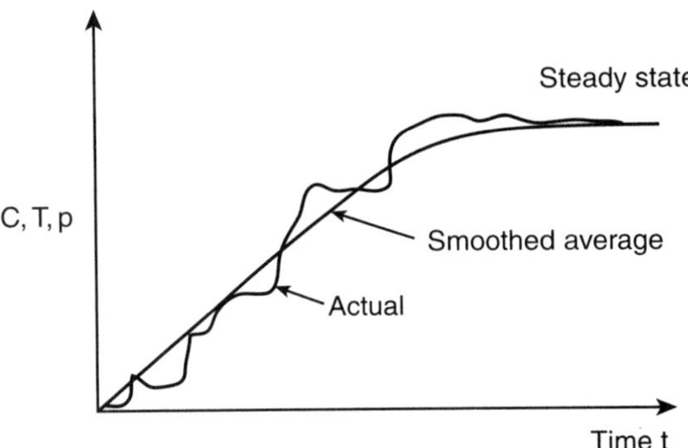

Figure 1.2 Outflow from a stirred tank or compartment.

momentum enters a conduit, and the associated variables (temperature, concentration, pressure) undergo a continuous variation in the direction of flow. Any changes in the radial direction are lumped into an "effective film" near the wall. In this film, concentration and temperature gradients are assumed to be linear, so that any heat or mass transport can be represented by equations (1.17) and (1.18). We repeat them here for completeness:

$$N\,(\text{mol/s}) = k_c A \Delta C \tag{1.17}$$

$$q\,(\text{J/s}) = hA\Delta T \tag{1.18}$$

The situation is depicted in Fig. 1.3 for the case of heat transfer from the bulk of the fluid, which carries a temperature T_b, to the wall, which is at a constant temperature T_w. Then ΔT in

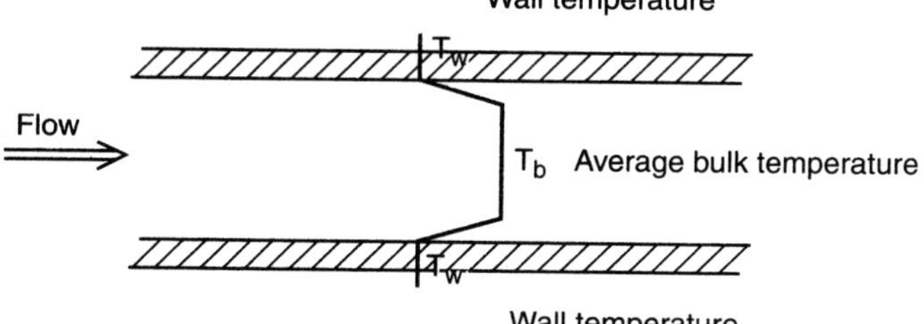

Figure 1.3 Temperature profile in a one-dimensional pipe (turbulent flow).

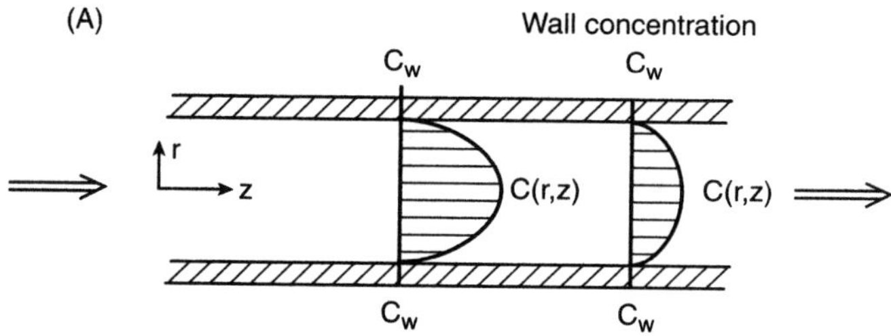

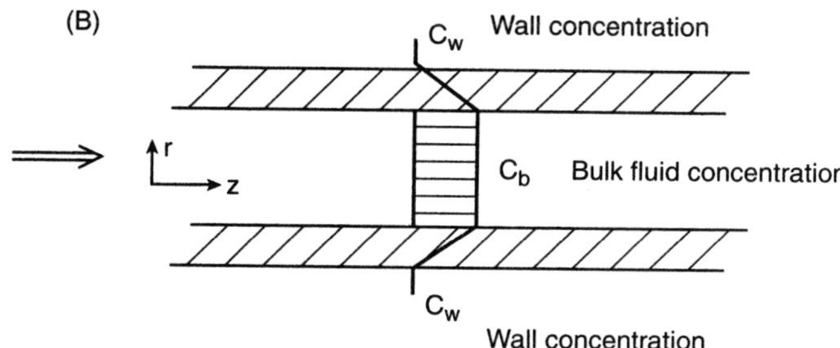

Figure 1.4 Mass transfer in tubular flow: (A) fluid resistance controlling and (B) wall resistance controlling.

expression (1.18) is given by

$$\Delta T = T_b - T_w \tag{1.22}$$

To describe the analogous mass transfer case, we draw on an example from the biomedical field. In the process of blood dialysis, the blood of patients with kidney disease is circulated through a bundle of hollow fibers in which toxins pass from the blood and through the permeable wall of the fibers into an external fluid devoid of toxins. Since flow is usually relatively slow and in the laminar regime, concentration variations will in principle arise in both the radial and axial directions. The model would then have to accommodate two independent variables leading to a partial differential equation (Fig. 1.4A). If, however, wall thickness is substantial, the main resistance will reside in the wall, and radial mass transport will be sufficiently fast to result in a constant cross-sectional concentration (Fig. 1.4B). This situation, which is analogous to the heat transfer case shown in Fig. 1.3, results in an ordinary differential equation.

The two cases we have described here, the stirred tank or compartment, and the one-dimensional pipe, are important starting tools in modeling transport of mass and energy. They reduce what would otherwise be a PDE model to the level of an ODE. We note, however, that this simplification requires good judgment on the part of the analyst to ensure that the underlying assumptions are valid.

Finally, a reduction on the number of independent variables can also be effected by means of what is termed dimensional analysis. This method is listed separately under subsection (f) and discussed in greater detail in Chapter 8.

(c) *Reduce the Number of Terms in an Equation*. We have so far considered the simplifications that result from a reduction in the number of dependent or independent variables. Further gains can be made by giving consideration to a reduction in the number of *terms* that appear in an equation. We illustrate this with two examples.

Suppose a flowing liquid that is being heated from 25 °C to 100 °C, simultaneously undergoes changes in kinetic energy, for example, by entering a constricted or expanded section. This has the effect of increasing or reducing the velocity of the fluid, thus changing its kinetic energy. That change must in principle be incorporated in an overall energy balance of the system, which comprises both thermal and mechanical forms of energy. Let us, however, consider the relative magnitude of the two terms by assuming that the liquid undergoes a change in velocity of 1 m/s, an extreme case. The thermal equivalent of this change is given by the enthalpy change, equation (1.21a), so that for unit mass of liquid flowing, we obtain

$$\underset{\text{Enthalpy}}{C_p \Delta T} = \underset{\substack{\text{Kinetic}\\ \text{energy}}}{\frac{\Delta v^2}{2}} \tag{1.23}$$

Taking water as an example, with a specific heat of 4.2 J/g K, we obtain

$$\Delta T = \frac{\Delta v^2}{2C_p} = \frac{1^2}{2 \times 4.2} = 0.12\,^\circ\text{C} \tag{1.24}$$

Thus the kinetic energy change considered, a rather substantial one, leads to a temperature change of only 0.12 °C, a negligible amount compared with the 75 °C total change undergone by the water. Thus it is safe in this instance to omit kinetic energy from the balance and limit ourselves to thermal energy only.

Consider next the discharge from a cylindrical tank through a nozzle installed at the bottom. This is in essence a process in which potential energy is converted into kinetic energy, and the question arises as to whether friction should be included in composing an appropriate mechanical energy balance. Let us assume the height of the tank equals its diameter and that the tank and nozzle diameters are in the ratio 10:1, so that velocity v_n in the nozzle is 100 times that in the tank, v_t. The reader will already intuitively sense that the change in kinetic energy will outweigh the energy loss due to friction. This can be shown quantitatively by drawing on equation (1.19) for friction and comparing it with the kinetic energy change. We have, per unit mass fluid,

$$\frac{v_n^2 - v_t^2}{2} \quad \text{and} \quad 4f\frac{v_t^2}{2}\frac{L}{d}$$

where the "friction factor" f is at most of the order one, and usually less. Transposing v_t^2 we obtain, for the left side,

$$\frac{1}{2}\left(\frac{v_n^2}{v_t^2} - 1\right) = \frac{1}{2}(10^4 - 1) \approx 10^4$$

while the right side becomes

$$4f\frac{1}{2}v_t^2\frac{L}{d} = 4 \times 1 \times \frac{1}{2} \times 1 \times \frac{1}{1} = 2 \ll 10^4$$

This confirms that the frictional energy losses are minuscule in comparison to the change in kinetic energy undergone by the liquid. We note that the celebrated Bernoulli equation, to be taken up in Chapter 4 (Example 4.7), is based on precisely the same concept of neglecting frictional losses occasioned by flow compared with changes in kinetic, potential, and pressure energies.

(d) *Simplify the Terms in an Equation.* The principal tool here that we wish to draw to the reader's attention is linearization. We can do this in two ways. We can draw on a host of established linear relations and assume their validity in the pertinent range of operation. We have summarized some of the more frequently used linear relations for the reader's convenience in Table 1.4. Some of these relations have already appeared in Tables 1.1 and 1.2 but are repeated for completeness.

A second way of linearizing is to expand the nonlinear term and to discard all higher order terms that may be nonlinear. Suppose it is desired to linearize a second-order term, for example, $r = kC^2$ in a second-order reaction (see equation 1.15 of Table 1.3). Then, setting

$$C = C_0 + \varepsilon\widetilde{C}, \qquad r = r_0 + \varepsilon\tilde{r} \tag{1.25}$$

and equating terms that are linear in ε, we obtain

$$\tilde{r} = k2C_0\widetilde{C} \tag{1.26}$$

Table 1.4 Short Table of Linear Relations

1. Ohm's law	$V = \frac{i}{R}$	(1.6)
2. Fick's law	$N = -DA\frac{dC}{dx}$	(1.10)
3. Fourier's law	$q = -kA\frac{dT}{dx}$	(1.11)
4. Newton's viscosity law	$\tau = -\mu\frac{dv}{dx}$	(1.13)
5. First-order chemical reaction	$r = k_r C$	(1.14)
6. Convective heat transfer	$q = hA\Delta T$	(1.18)
7. Extension of a spring (Hooke's law)	$F = k_s x$	(1.23)
8. Drag on a sphere in viscous flow (Stokes' law)	$F_D = 3\pi\mu dv$	(1.24)

(e) *Bracket the Solution*. This device, a favorite one of ours, avoids meeting a complex problem head on and instead provides lower and upper bounds to the solution being sought. The procedure, often referred to as asymptotic analysis, is a perfectly legitimate tool much used by mathematicians. It can, among other things, provide us with a best-, or worst-case scenario, a very useful piece of information to have. Suppose, for example, that a stored liquid emits toxic fumes by evaporation. This is a complex process that requires consideration of the external temperature variations and transfer of heat to the storage tank, leading to two PDEs (mass and energy balance) in time and distance. If instead we assume that the liquid is at the maximum external temperature and that the air above it is fully saturated with vapor, we will obtain the *maximum* possible concentration in the air. If that value falls within prescribed regulatory limits, we can breathe a sigh of relief. If not, we will at least know that countermeasures will likely have to be taken.

(f) *Dimensional Analysis*. It frequently happens, particularly in industry, that a process is too complex to be modeled in a quantitative way. Heat and mass transfer in turbulent flow, or the drag on a body such as a car exposed to turbulent flow conditions, are examples of processes that cannot at present be satisfactorily described by quantitative modeling. It is convenient in these instances to use dimensional analysis to arrive at a *qualitative* description of the process in terms of *dimensionless ratios*; that relation can then be quantified by means of experimentation. This dual use of theory and experiment has proved to be of great value in unravelling complex phenomena, particularly in the fields of fluid mechanics, heat and mass transfer, and elasticity. A description of the method, accompanied by a number of examples, appears in Chapter 8.

We have in this section attempted to provide the reader with some initial tools to help in initiating the modeling process and to carry it one or two steps further. We did this first by listing a number of important fundamental laws and auxiliary relations that aid in composing the model equations. We followed this up by suggesting three useful steps to accompany this process of composition. Two of these were rather mundane, consisting of a sketch of the system and an envelope for the application of certain conservation laws. They should nevertheless be used. The third tool, that of introducing simplifications while composing the model, was elaborated on in some detail. We made six suggestions in this direction: (a) reducing the number of equations, (b) reducing the number of independent variables, (c) reducing the number of terms, (d) linearization, (e) bracketing the solution, and (f) dimensional analysis. Together, these six avenues play an important, often crucial role in the success of a model. The story does not, of course, end there. The equations still have to be solved, either analytically or numerically. Although numerical methods are also addressed, our own preference is for the analytical approach, which poses new challenges and has its (often severe) limitations. It does, however, provide greater insight and often leads to more elegant results than are possible with numerical solutions. We will address some of the available analytical techniques in Chapter 2. Numerical solutions, which are nowadays standard fare with innumerable solver packages available, are taken up in Chapter 8.

1.3 Closure

With the term "closure" we wish to introduce the notion that the results of a model should be scrutinized carefully before being put aside. In particular, the following questions should be addressed.

1. Is the order of magnitude of the result "reasonable," or is it outlandishly low or high? There is nothing more embarrassing in modeling than to present results that are patently absurd. To avoid this, one should check for conceptual and numerical errors and for dimensional consistency. On occasion, unexpectedly high or low results may in fact constitute the correct answer. An example of this kind is found in Example 5.8. Evaporation of a Pollutant into the Atmosphere. The numerical calculations made there yield the surprising result that fully 87% of mercury pollutant contained in a body of water will be transferred to the atmosphere with the evaporation of only 0.01% of the solution. This result can be justified on physical grounds because mercury has an extremely low solubility in water and consequently exhibits an inordinately high fugacity or "escaping tendency." This is the reason for its high rate of transfer into the atmosphere. The lesson for the reader here is that results that appear to be completely out of the ordinary should not call for automatic disbelief and dismissal. Instead, a careful review of the solution should be undertaken and if the result persists, the underlying physics of the process should be closely scrutinized. If no immediate rationale can be found, one should, with some caution, accept the result and hope that further investigation, perhaps in a changed context, will confirm the answer obtained. One of the thrills of modeling lies in the discovery of the unexpected, and its ultimate explanation and acceptance by the peer community. The currently fashionable topic of chaos owes its discovery in part to unexpectedly erratic behavior found in the solution of nonlinear algebraic and ordinary differential equations. The heroes here are the individuals who did not dismiss the results as numerical aberrations but rather persisted in their quest for an explanation. From this sprang a host of new theories, which became the foundations of the topic of chaos.

2. What is the behavior of the result for large or small values of the variables or parameters? This is akin to the asymptotic analysis mentioned before and often provides upper and lower bounds to the solution being sought. A special case of this procedure is to let the independent variables, time and distance, go to infinity. In the former case, this leads to what is termed the steady-state solution of the system. An example of its use is found in Example 7.4. An Electrical Network: Hitting a Brick Wall and Going Around It. The brick wall in question is the full transient solution, which is not amenable to analytical treatment. No explicit expression in terms of the system variables can therefore be obtained. We circumvent this difficulty by seeking out instead the steady-state solution, which is obtained by setting the time derivatives in the system equations equal to zero and solving the resulting algebraic equations. This yields only a partial answer of the full model, but it is a very important partial answer, since the ultimate steady behavior of a system is oftentimes of greater significance than the intervening transient period.

3. Does the result show unusual behavior for certain values of the variables or parameters? Suppose, for example, that the solution has the form

$$f(x) \propto \frac{1}{b - x} \tag{1.27}$$

and we let x approach b. The function f(x) will then experience an enormous increase as x nears b and will go to infinity when the two values are equal. This behavior is sometimes referred to as "runaway" and may signal either a beneficial or a catastrophic event, for example, an unbounded rise in temperature. In general, whenever the result is made up of fractions that contain *differences* in the denominator, an immediate scrutiny is called for to establish the conditions under which the solution goes to infinity. This is the case in Example 2.12.

The appearance of the exponential term e^{bx}, where b is a positive number, similarly signals an unbounded increase in the dependent variable. We will show how this works in Example 5.13, dealing with the onset of AIDS. Trigonometric terms such as sin bt, on the other hand, imply oscillatory behavior of the solution. This, too, will be of interest to the analyst and may be termed unusual behavior.

4. Is the solution highly sensitive to small changes in one of the parameters or variables? As an example, let us revert to the form of equation (1.27) and write

$$f(x) \propto \frac{1}{x - 1} \tag{1.28}$$

Suppose that initially $x = 1.005$ and its value is raised by a small increment, 0.005, to $x = 1.01$. This seemingly insignificant increase in x of 0.5% will increase f(x) by 100% (i.e., double it). This is again a highly interesting result, which should be sought out by the analyst. Another example of extreme sensitivity to parameter values occurs in the important contemporary subject of chaos. Here the parameters in question are the initial conditions of a set of nonlinear differential equations. Minuscule changes in these conditions lead to enormous changes in the solution and ultimately result in chaotic behavior. The phenomenon was first identified by E. N. Lorenz in 1963 in a paper entitled "Deterministic Non-Periodic Flows."

5. Does the solution contain maxima or minima besides the obvious ones at the end points of the range being considered? There are two important ways in which this can be established: by physical reasoning and by inspection of the analytical expression of the quantity to be maximized or minimized. In Example 4.6, we determine the minimum escape velocity of a projectile or rocket fired from the surface of the earth. Existence of this minimum comes from the straightforward logical argument that the missile must possess a certain minimum kinetic energy to overcome gravity and air resistance (drag). Any value below this threshold will cause its return to earth, whereas values above it will project it into outer space. Note that this is not a conventional minimum of a functional relation but rather represents a minimum requirement to achieve a desired goal.

Somewhat less straightforward logic can be used to argue that an artillery piece must have an optimum angle of elevation which will maximize the range of the gun, that is, the horizontal distance it can fire a shell (Example 4.5).

The argument here runs as follows. If the gun is fired vertically, the shell will fall back to the gun, which apart from being undesirable, yields a range $r = 0$. If it is fired horizontally, the shell will hit the ground shortly after leaving the barrel, since the vertical distance is very small (equal to the elevation of the barrel). It can then be argued, somewhat tenuously, that there must be an intermediate value of the angle of elevation $90° > \alpha > 0$, which will maximize the range r. For those who find this argument too weak, there is the second route, that of scrutinizing the analytical expression for the quantity to be maximized. That expression is in this case given by an equation we shall encounter again in Chapter 4

$$r = \frac{v_0^2}{g} \sin 2\alpha$$

where v_0 is the muzzle velocity. From this it is quickly established that $\alpha_{opt} = 45°$.

The text contains several other interesting optimization problems. We draw the attention of the reader to Example 3.12 and Practice Problems 3.12 and 3.13, all of which use the technique of linear programming to minimize the costs of a particular operation or to maximize profit. In Example 5.2 we balance the increased yield obtained from large stirred tank reactors against the cost of such larger vessels to arrive at an optimum reactor size. Example 3.13 addresses the problem of minimizing adsorbent inventory in a staged purification process, while Practice Problem 6.3 deals with optimizing the thickness of pipe insulation. Two other interesting cases arise in the oxygen profile of a polluted river, which shows a minimum (Example 6.10), and in the operation of a bioreactor, which shows a maximum, in the yield at a particular flow rate (Example 5.3). In both these cases, physical reasoning as well as analytical techniques are used to seek out the extrema in question.

6. Finally, it is of some interest to scrutinize the variables themselves that appear in the solution. The importance of their functional form has already been pointed out. A second point, which is often overlooked, is the question whether all the variables contained in the original model do in fact reappear in the solution or whether certain parameters are, in the course of the solution process, eliminated from the result. Example 4.5, Path of a Projectile, presents an interesting example of this situation. Here the mass m of the projectile appears in a natural way in the two model equations (4.60) and (4.61), which are based on Newton's law. Yet if fluid drag is neglected mass does not reappear in either the expression for vertical distance (4.69) nor that for horizontal distance (4.72) covered by the projectile. Thus neither the height attained nor the range of the shell is dependent on its mass. This is contrary to conventional wisdom, which would argue that both these variables are affected by mass. We have here yet another example of the power of modeling to reveal the unexpected, which is something the analyst should always look for. The converse of this is the mere confirmation of the obvious, which as we have noted, reduces modeling to a sterile and unproductive role.

To provide an underpinning to the discourse just given, we present a number of simple illustrative examples drawn from a variety of disciplines.

◆ EXAMPLE 1.3 Discharge of Plant Effluent into a River

In this environmentally important problem we consider the case of an aqueous plant effluent containing 10 mg/L of copper being discharged into a nearby river. Environmental regulations set a maximum permissible concentration limit of 1.0 μg/L. The question to be addressed is whether the diluting effect of the river flow is sufficient to reduce the effluent concentration to a level within permissible limits.

We start, as recommended, with a sketch of the system, which is shown in Fig. 1.5. Since the determination of a concentration is involved, a material balance will likely be called for. To aid in setting up this balance, we have drawn an envelope in what appears to be an appropriate location. The rate of effluent discharge is 150 L/min and the river flow varies seasonably from 2300 to 5000 L/s.

Evidently the effluent concentration, once introduced into the river, will vary locally in a complex manner that defies precise modeling. To circumvent this difficulty, we invoke the concept of a "stirred tank" or compartment; that is, we assume the contents to be well mixed. The result will therefore be an *average* of the concentration in the river without taking into account of momentary local aberrations. This is nevertheless a useful quantity to have, since it will provide guidelines for any countermeasures that may have to be taken. We note in addition that for shallow, rapid rivers in particular, mixing will be nearly instantaneous, whereas in slower, deeper rivers, the well-mixed condition will ultimately be attained further downstream.

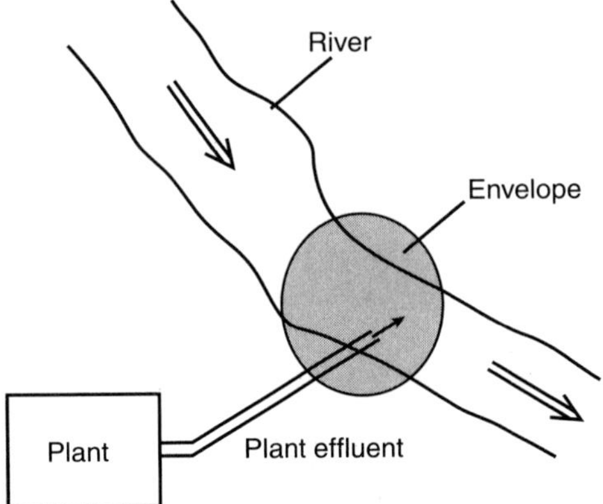

Figure 1.5 Discharge of a plant effluent into a river.

The model to be applied to the envelope shown in Fig. 1.5 consists of a copper mass balance, which can be set up in accordance with the equation presented under item 1 of the Table 1.2. In applying it, we will make the assumption that a steady state has been reached after a brief initial adjustment period (i.e., the right side of the balance will be zero). Concentrations during that period will evidently be below the ultimate steady-state level (see the similar case of lake pollution, Fig. 1.2) and are thus of no direct interest. Note that since the regulatory limit must be met at all times during the year, the lower summer flow rate of 2300 L/s will be used. We can then write

$$\text{Rate of Cu in} \quad - \quad \text{Rate of Cu out} \quad = \quad 0$$

$$\begin{bmatrix} 0\dfrac{\text{mg}}{\text{L}} \times 2300\dfrac{\text{L}}{\text{s}} \\ +10\dfrac{\text{mg}}{\text{L}} \times 150\dfrac{\text{L}}{\text{min}} \times \dfrac{1\,\text{min}}{60\,\text{s}} \end{bmatrix} - x\frac{\text{mg}}{\text{L}} \times 2302.5\frac{\text{L}}{\text{s}} = 0 \tag{1.29}$$

where x = resulting copper concentration, and the exiting river flow rate is given by 2300 + 150/60 = 2302.5 L/s. Solving for x, we obtain

$$\boxed{x = \frac{10 \times 150}{60 \times 2302.5} = 1.086 \times 10^{-2}\frac{\text{mg}}{\text{L}} = 10.86\frac{\mu\text{g}}{\text{L}}\text{Cu}}$$

Thus the regulatory limit of 1 μg Cu/L is clearly not met and the waste stream cannot be directly discharged into the river. Therefore, we must determine how much copper needs to be removed to meet these requirements. The answer is found by calculating both the discharge rate of copper from the plant and the allowable rate in the river, then subtracting the two. We obtain

Effluent

$$10\frac{\text{mg}}{\text{L}} \times 150\frac{\text{L}}{\text{min}} \times 1\frac{\text{min}}{60\,\text{s}} \times 1000\frac{\mu\text{g}}{\text{mg}} = 25{,}000\frac{\mu\text{g}}{\text{s}}$$

River (allowable)

$$1.0\frac{\mu g}{L} \times 2302.5\frac{L}{s} = 2302.5\frac{\mu g}{s}$$

Difference

$$25{,}000 - 2302.5 = 22{,}698\frac{\mu g}{s}$$

To be removed per day

$$\boxed{22{,}698\frac{\mu g}{s} \times \frac{3600\,s}{h} \times \frac{24\,h}{day} = 1.96\,kg/day}$$

Note that the copper recovered can be sold or reused, thus partially offsetting the cost of the recovery process.

◆ EXAMPLE 1.4 Electrical Field Due to a Dipole

With this example, we wish to introduce the reader to a simple application of Coulomb's law, equation (1.9). As a preliminary, we briefly discuss two simple concepts of electricity that will likely be known from other courses.

Electrical Field E

This basic electrical quantity is a vector that is defined as the force per unit positive charge placed in an electrical field. Thus

$$\mathbf{E} = \frac{\mathbf{F}}{q_0} \tag{1.30}$$

where q_0 denotes the charge in coulombs of the positive test charge. Introducing Coulomb's law into equation (1.30), we obtain, for the magnitude of **E**

$$E = \frac{F}{q_0} = \frac{1}{4\pi\varepsilon_0}\frac{q}{r^2} \tag{1.31}$$

where q is now the charge responsible for the creation of an electrical force field. We shall encounter the concept again in Chapter 4.

Vectors and Vector Sums

Since the electrical field is a vector, this quantity will have to be dealt with briefly. Much more about vectors and vector operations will appear in Chapter 2. For our present purposes we merely recall that the sum of two vectors

$$\mathbf{E} = \mathbf{E}_1 + \mathbf{E}_2 \tag{1.32}$$

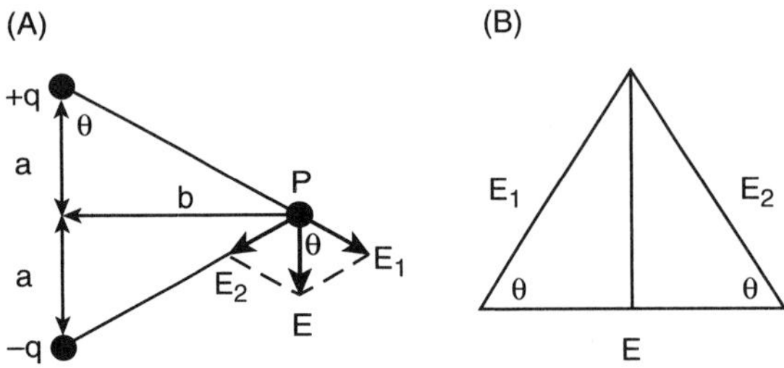

Figure 1.6 Electrical field due to a dipole: (A) schematic diagram and (B) relation between electrical field components and their resultant.

is obtained by drawing the diagonal of a parallelogram whose sides are the component vectors $\mathbf{E}_1$ and $\mathbf{E}_2$. This is shown in Fig. 1.6. We now turn to the formulation of our problem.

We consider a positive and a negative charge of equal magnitude q placed a distance 2a apart. This configuration is referred to as an electric dipole. The task we wish to address is the computation of the electrical field induced by the dipole at the position P. The pertinent quantities have been sketched in Fig. 1.6. No envelope is required because no mass and energy balances are involved.

We start by noting that a positive test charge at P will be repelled by the charge +q along the vector $\mathbf{E}_2$ and attracted by the charge −q along the vector $\mathbf{E}_1$. Since the dipole is composed of charges of equal magnitude, the magnitudes of $\mathbf{E}_1$ and $\mathbf{E}_2$ will likewise be equal. The resultant vector **E** must then point vertically downward. Its magnitude can be deduced from the isosceles triangle shown in Fig. 1.6B and is given by

$$E = 2E_1 \cos\theta = 2E_2 \cos\theta \tag{1.33}$$

To obtain an explicit expression for E, we now need to derive E_1 and $\cos\theta$. For the latter we have from the diagram

$$\cos\theta = \frac{a}{(a^2 + b^2)^{1/2}} \tag{1.34}$$

For E_1 (or E_2), we draw on Coulomb's law, equation (1.31), and the geometry of the configuration, to obtain

$$E_1 = E_2 = \frac{1}{4\pi\varepsilon_0} \frac{q}{a^2 + b^2} \tag{1.35}$$

Substituting equations (1.34) and (1.35) into equation (1.33), we obtain the final result

$$\boxed{E = \frac{1}{4\pi\varepsilon_0} \frac{2aq}{(a^2 + b^2)^{3/2}}} \tag{1.36}$$

This is the magnitude of the electrical field **E** at the position P. In other words, the force exerted by the dipole on a unit charge placed at P is given by equation (1.36).

COMMENTS

- We note that the dipole separation 2a and its charge q appear as a product in equation (1.36), so that halving the distance and doubling the charge will not change the electrical field E. This product, 2aq, is called the electric dipole moment p.

- At large distances, b ≫ a, equation (1.36) reduces to the expression

$$E = \frac{1}{4\pi\varepsilon_0}\frac{2aq}{b^3} \tag{1.37}$$

  Note that the field drops off more sharply, with distance *cubed*, than the field due to a single charge, which diminishes with distance *squared* [see equation (1.31)]. This is because the separate fields due to the individual charges of the dipole almost, but not quite, cancel each other at large distances, thereby significantly weakening the electric field. Note that we have heeded here the advice given under item 2 of "Closure" (Section 1.3) and examined the solution behavior at large parameter values.

- Inspection of equation (1.36) shows that we must restrict its validity to values of b ≫ a, for when this is not the case, the construction of Fig. 1.6 no longer holds. Thus, when b = 0, the electrical field will be zero, which is not born out by equation (1.36). This is yet another application of "closure."

◆ EXAMPLE 1.5 Design of a Thermocouple

Thermocouples are temperature-measuring devices consisting of two dissimilar thin wires joined together at the ends. They respond to changes in ambient temperature by registering a change in electrical potential, which can be measured by means of a voltmeter (Fig. 1.7). We consider here a thermocouple suspended in flowing air that undergoes a step rise in temperature from an initial temperature T_i to a new value T_a. We wish to determine the diameter d of a thermocouple whose response time is such that a 15 °C change in the air temperature will be registered within 0.5 °C of the final value in no more than 4 s. Since a dimension is being sought, the example may be viewed as a "design" problem.

Since the conductivity of the metal wires is several orders of magnitude higher than that of the air, and the wires are very thin, resistance to heat transfer within the wires can be neglected. Thus, the thermocouple becomes in a sense a "well-stirred tank" of uniform temperature T, with the heat transfer resistance residing entirely in a thin film of air surrounding the thermocouple.

The model will call for the use of an energy balance, item 2 of Table 1.2, for which we have indicated the envelope in Fig. 1.7. The rate of heat transfer through the air film, q, is given by equation (1.18) of Table 1.2, while the energy change within the thermocouple is represented by the enthalpy change, equation (1.21a). The following data are

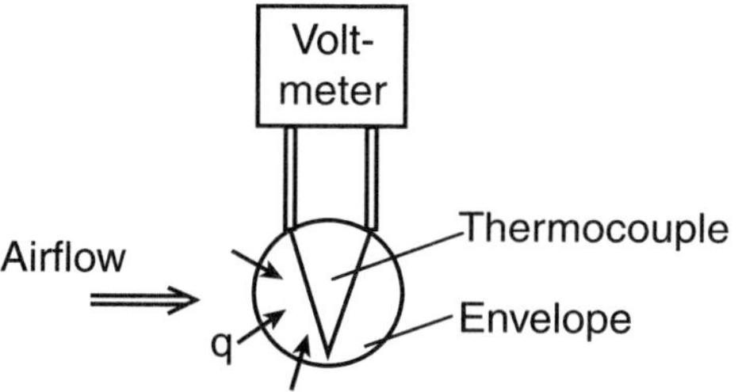

Figure 1.7 Thermocouple measuring a change in the temperature of flowing air.

provided:

Density of wires (average):	$\rho = 8800\ \text{kg/m}^3$
Specific heat of wires (average):	$C_p = 0.419\ \text{kJ/kg K}$
Heat transfer coefficient:	$h = 0.455\ \text{kJ/m}^2\ \text{s K}$

We can then write the energy balance as

$$\begin{array}{ccccc} \text{Rate of energy in} & - & \text{Rate of energy out} & = & \begin{array}{c}\text{Rate of change}\\ \text{of energy content}\end{array} \\ q & - & 0 & = & \dfrac{d}{dt}mH \end{array} \tag{1.38}$$

Drawing on the auxiliary relations (1.18) and (1.21), this becomes

$$hA(T_a - T) = \rho V C_p \frac{dT}{dt} \tag{1.39}$$

where ρ = density, V = volume, and C_p = specific heat of the thermocouple.
Integration by separation of variables yields, in the first instance

$$\int_0^t dt = \frac{\rho V C_p}{hA} \int_{T_i}^{T_f} \frac{dT}{T_a - T} \tag{1.40}$$

from which we obtain

$$t = \frac{\rho V C_p}{hA} \ln \frac{T_a - T_i}{T_a - T_f} = \frac{\rho C_p d}{4h} \ln \frac{T_a - T_i}{T_a - T_f} \tag{1.41}$$

where the subscripts i and f denote the initial and final states, respectively.
Solving for diameter d and substituting numerical values yields

$$d = \frac{4ht}{\rho C_p \ln \dfrac{T_a - T_i}{T_a - T_f}} = \frac{4 \times 0.455 \times 4}{8800 \times 0.419 \times \ln \dfrac{15}{0.5}}$$

$$\boxed{d = 5.8 \times 10^{-4}\ \text{m} = 0.58\ \text{mm}}$$

COMMENTS

- It is at first sight not evident how to go about determining the diameter of the thermocouple. One has to start by identifying the basic process involved, which is one of heat transfer. Further thought will reveal that the diameter lurks in the heat transfer area A and the mass m of the thermocouple, which was cleverly converted into the product ρV. The ratio of the two then yielded the desired quantity d/4.

- The particular simplicity of the model was due to the fact that we were able to assume uniform temperature within the thermocouple at any given moment. Had this not been the case, we would have been forced to solve a time-variant PDE in at least one dimension. Similarly, the thinness of the wires enabled us to neglect heat losses along them, which would have complicated the model further and would at any rate have been difficult to estimate.

◆ EXAMPLE 1.6 Newton's Law for Systems of Variable Mass: A False Start and the Remedy

In the classical rocket problem, one wishes to calculate, given certain data, the ultimate speed the rocket attains at burnout (i.e., the point at which the fuel is exhausted). We consider here only the basic equations that will lead to the result. An actual rocket problem is taken up in Practice Problem 1.6.

In an initial diagram of the system (Fig. 1.8A), we have indicated the changes of mass and velocity that occur during flight. A convenient starting point would be the momentum version of Newton's law, that is,

$$\sum F_{ext} = \frac{d}{dt}\mathbf{P} = \frac{\mathbf{d}}{dt}m\mathbf{v} \tag{1.1}$$

since this will allow us to account for variations in both mass and velocity. Expansion of the derivative leads to the result

$$\sum F_{ext} = m\frac{d\mathbf{v}}{dt} + \mathbf{v}\frac{dm}{dt} \tag{1.42}$$

This equation in the two variables m and $\mathbf{v}$ could be supplemented by a mass balance to complete the model. Closer inspection will show, however, that we are off to a false start. We have failed to account for the momentum of the ejected exhaust gases. To have validity, Newton's law, which here becomes a momentum balance, must be applied to the entire system (Fig. 1.8B).

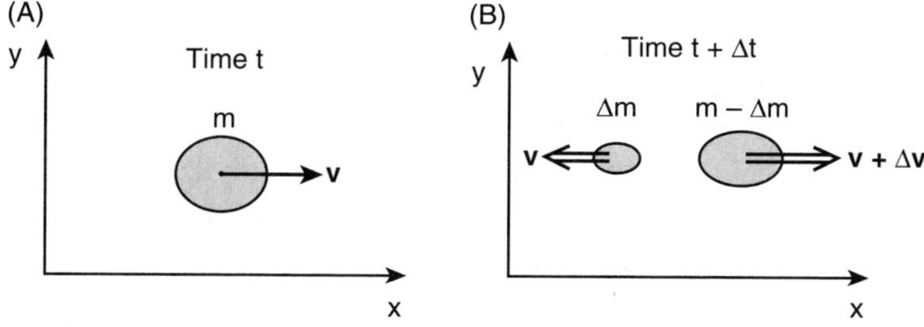

Figure 1.8 Changes in mass and velocity of a rocket: (A) initial system diagram and (B) application of momentum balance.

Let us see how that works out in practice. We start with a difference formulation of equation (1.1) and write

$$\sum F_{ext} = \frac{\Delta \mathbf{P}}{\Delta t} = \frac{\mathbf{P}_f - \mathbf{P}_i}{\Delta t} \tag{1.43}$$

where the subscripts i and f denote initial and final states. When applied to the entire system, this becomes,

$$\sum F_{ext} = \frac{[(m - \Delta m)(\mathbf{v} + \Delta \mathbf{v}) + \mathbf{u}\Delta m] - m\mathbf{v}}{\Delta \mathbf{t}} \tag{1.44}$$

or alternatively

$$\sum F_{ext} = \mathbf{m}\frac{\Delta \mathbf{v}}{\Delta t} + [\mathbf{u} - (\mathbf{v} + \Delta \mathbf{v})]\frac{\Delta m}{\Delta t} \tag{1.45}$$

In the limit of $\Delta t \to 0$, $\Delta \mathbf{v} \to 0$, we obtain

$$\sum F_{ext} = \mathbf{m}\frac{dv}{dt} + \mathbf{v}\frac{dm}{dt} - \mathbf{u}\frac{dm}{dt} \tag{1.46}$$

That is, we are now dealing with *three* derivatives, instead of the two shown in equation (1.42).

In Practice Problem 1.6 the reader is asked to show how a reduced form of this expression can be used to calculate the maximum rocket velocity at burnout.

COMMENTS

- The principal lesson to be learned from this example is that when applying conservation laws (of momentum here), care must be taken to include all pertinent entities of the system. The error made in our false start is a common one, particularly in mechanics, and such situations bear watching. Comparison of equations (1.46) and (1.42) shows that the latter would apply only in the trivial and unrealistic case of $\mathbf{u} = 0$. The results of equation (1.42) would thus be in considerable error. The closure we have referred to consists here of a careful scrutiny of the result obtained to ensure that all variables of the system are properly accounted for.

◆ EXAMPLE 1.7 Release of a Substance into a Flowing Fluid: Determination of a Mass Transfer Coefficient

In the early stages of the development of mass transfer theory, empirical values of the transfer coefficient k_C [see equation (1.17)] were often determined by casting soluble materials into various shapes, such as pipes, spheres, and disks, and passing a fluid around or through it. Values of k_C could then be obtained from an appropriate model that relates release rate to changes in concentration in the fluid. Let us see how this works for flow of a fluid through a pipe. We assume the fluid to be in turbulent flow so that radial concentration changes are confined to a thin film near the wall. We apply the linear driving force

(A)

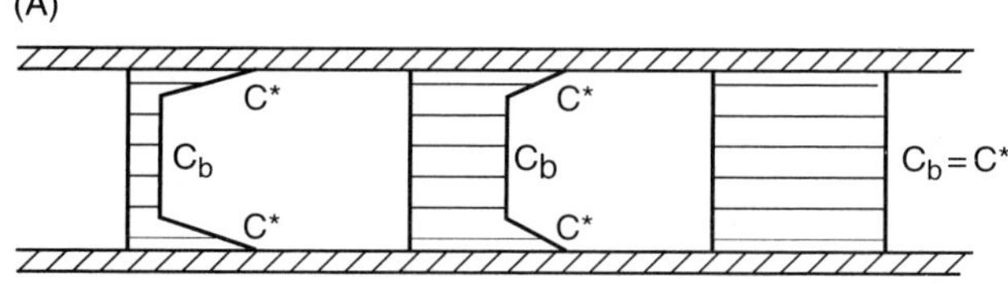

(B)

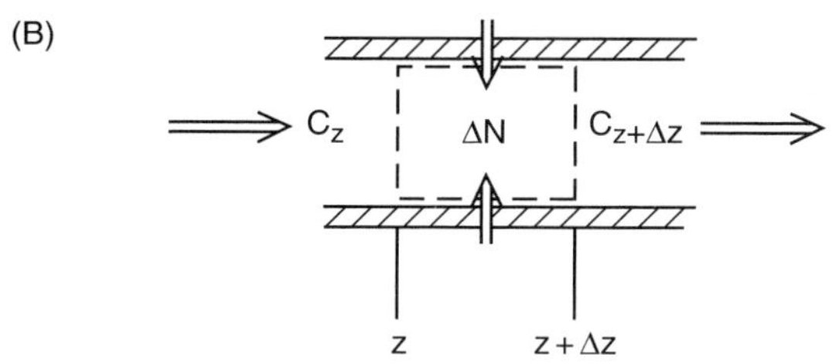

Figure 1.9 Release of a solute into a flowing fluid: (A) concentration profiles and (B) difference element for the mass balance.

concept: that is, we assume the rate of mass transfer to be given by equation (1.17). The concentration difference in that expression is given by

$$\Delta C = \text{Concentration at the wall (saturation)} - \text{Concentration in the bulk fluid}$$
$$\Delta C = C^* - C_b \tag{1.47}$$

Note that we have neatly resolved the question of the value of C^* by assuming that the concentration immediately adjacent to the soluble wall equals its saturation concentration, also known as solubility. Such solubilities of various substances in different liquids are tabulated in the literature.

Figure 1.9A presents a sketch of the concentration profile development in the flowing fluid. One notes that the bulk value is initially low, assuming that pure solvent is introduced to the pipe and that further downstream it has risen to a value that is beginning to approach the solubility prevailing at the wall. Ultimately (i.e., at sufficiently long distances from the inlet), the two values coincide, — that is, $C_b = C^*$ — and the entire liquid is fully saturated with solute (the dissolving substance).

Figure 1.9B shows the envelope over which the mass balance is to be applied. The rate at which dissolving solute enters the envelope is denoted by ΔN and the ensuing concentration changes by C_{1z} and $C_{1z+\Delta z}$.

Applying a mass balance over that envelope, we obtain

$$\begin{array}{ccccc} \text{Rate of solute in} & - & \text{Rate of solute out} & = & 0 \\ \begin{bmatrix} QC_{1z} \\ +k_c(C^* - C)_{avg} P\Delta z \end{bmatrix} & - & QC_{1z+\Delta z} & = & 0 \end{array} \tag{1.48}$$

where Q = volumetric flow rate (m^3/s) and P = perimeter (m).

Dividing by Δz and going to the limit, we obtain

$$Q\frac{dC}{dz} = k_c P(C^* - C) \tag{1.49}$$

Separating variables and integrating results in

$$\int_0^C \frac{dC}{C^* - C} = \frac{k_C P}{Q} \int_0^z dz \tag{1.50}$$

and

$$\ln \frac{C^*}{C^* - C} = \frac{k_C P z}{Q} \tag{1.51}$$

A semilog plot of $\ln[C^*/(C^* - C)]$ vs z will then yield a slope of $k_C P/Q$, from which the mass transfer coefficient k_C can be extracted.

COMMENTS

- The success of this simple model is clearly dependent on the assumption of turbulent flow in the pipe, which causes complete mixing within the bulk fluid and allows the application of the simple linear driving force concept. Had the fluid been in laminar flow, radial concentration variations would have arisen throughout the bulk fluid, leading to a PDE as the model (see Fig. 1.4A). Mass transfer coefficients can still be extracted from its solution, but this is a much more cumbersome procedure than what was presented in this example.
- The driving force $(C^* - C)$ was given the subscript "average" to indicate that this quantity varies over the increment Δz. Upon going to the limit $\Delta z \to 0$, it becomes a *point* quantity; that is, it is *not* differentiated.

With these examples, we have demonstrated to the reader the application of some of the concepts introduced earlier. Examples 1.3 and 1.5 used one of these concepts, that of a stirred tank or compartment, to arrive at simple models of processes of some complexity. The "one-dimensional pipe" was used in the last example 1.7, while in Examples 1.4 and 1.6 a first glimpse was provided of the use of vectors in describing processes involving electrical and mechanical forces. We hope with this introduction to have prepared the ground for the consideration of more complex processes, to be taken up in ensuing chapters.

PRACTICE PROBLEMS

1.1 **Fermi's estimate of the yield of the first atomic bomb** Shortly after the explosion of the first atomic bomb in New Mexico in 1945, Nobel laureate Enrico Fermi was seen outside the protective dugouts holding a clutch of paper strips. When the shock wave from the explosion arrived, he released the strips and allowed them to flutter to the ground. He then paced off the distance from the point of release to where the strips had come to rest. After some quick calculations he announced the yield of the bomb, which turned out to be remarkably close to the value ultimately established (10,000 tons of TNT).

Was any modeling involved? If so, what were the underlying principles used? Was modeling necessary? Was Fermi's result a fluke?

1.2 **Pollution of a lake** Consider item 10 of Table 1.1. Analyze the concentrations that will arise in the lake as a result of different contamination scenarios (steady, unsteady, continuous, and discontinuous). Is it appropriate to model? If so, what information would a manageable model provide?

1.3 **Laminar flow in a duct** When a fluid is in laminar or viscous flow, adjacent fluid layers do not mix but instead exhibit a continuous sliding motion with respect to each other. This sliding movement sets up a shear stress between adjacent layers that resists motion and must be overcome by applying pressure, generated, for example, by a pump. The action of these forces creates a radial velocity variation that ranges from zero at the wall to a maximum at the centerline. Assume the fluid to be an incompressible liquid and that the duct is cylindrical. The ultimate aim is to derive the radial velocity profile. A preliminary analysis leads to the following questions:

(a) What fundamental laws need to be invoked to model the process?

(b) How many equations does the model comprise, and what form are they likely to have?

(c) What is the envelope, if any, over which the laws are applied?

(d) How do velocity and pressure vary in the axial direction?

(e) What are the pertinent model equations and their solution? (*Hint*: Use Newton's viscosity law, equation (1.13), recalling that shear stress τ is a force per unit area (N/m^2).

(f) How would you proceed to obtain the more useful volumetric flow rate A (m^3/s) from the primary velocity profile? Explain in words only.

(g) What would be the model equations if we were dealing with a rectangular rather than a cylindrical duct?

ANSWER (e): $v(r) = \dfrac{\Delta p R^2}{4\mu L}\left[1 - \left(\dfrac{r}{R}\right)^2\right]$

1.4 **A financial problem: application of a dollar balance** Jane Doe, a respected financial analyst, earns $200,000/year. She pays 40% of her earnings for taxes, spends $5000 per year on food, and has other expenses (car, vacations, clothes, etc.) amounting to $50,000 per year.

(a) If she has no income now, and her pay remains steady at the same level, how long will it take her to buy a $500,000 home if she wants to pay for it all at once?

(b) At the age of 45, Jane feels that she has enough accumulated wealth to be able to partially retire and cut back on her workload. If her expenses are still the same, how much does Jane have to earn to maintain a steady-state cash position? Assume that the tax rate remains the same.

ANSWERS (a) 7.7 years
(b) $91,666

1.5 **Heating of a fluid in flow** Suppose it is desired to heat a continuous flow of fluid from a temperature T_1 to a temperature T_2. This is a frequent requirement in various industries and is usually carried out in devices called heat exchangers. A simple version, known as a shell-and-tube heat exchanger, is shown in Fig. P1.5. In it a tube, or a tube bundle, centered within a surrounding shell, receives a cold fluid that flows through and in the process is heated by a second medium introduced into the annular space between shell and tubes. The flow of the two streams is usually in opposite directions, and the device is then called a countercurrent shell-and-tube heat exchanger.

Suppose that we use steam as a heating medium and assume that it condenses with no change in temperature. The heat of condensation is transferred to the tubular fluid, which undergoes a continuous change in temperature in the direction of flow. The following are to be addressed:

(a) Derive an expression for the rate of steam flow F_h (kg/s) required to heat the cold fluid from T_1 to T_2.

(b) Derive an expression for the length of exchanger that will bring about the desired result.

To prepare the ground for the solution of these two problems, a stepwise procedure is to be followed by providing answers for the following questions:

(i) What fundamental law is to be invoked?

(ii) Over what domain should the envelopes be drawn, if any?

(iii) What are the auxiliary relations to be used?

(iv) What type of equations will result in the two cases to be considered?

ANSWERS (a) $F_h = \dfrac{F_C C_{pC}(T_2 - T_1)}{\Delta H_{cond}}$

(b) $L = \dfrac{F_C C_{pC}}{h\pi d} \ln \dfrac{T_h - T_1}{T_h - T_2}$

1.6 **Maximum speed of a rocket** A rocket weighs 15,000 kg when fueled up on the launching pad. It is fired vertically upward and, at burnout, weighs 5000 kg. Gases are

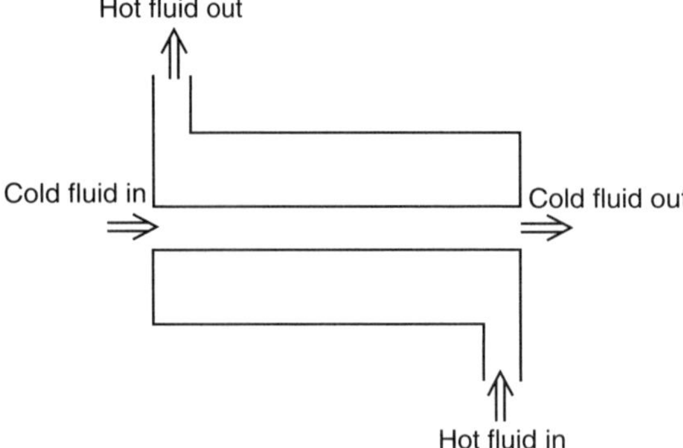

Figure P1.5 The countercurrent heat exchanger.

exhausted with a relative velocity $\mathbf{u} - \mathbf{v}$ of 2000 m/s. Calculate the maximum possible speed the rocket can attain. (*Hint*: Neglect the effect of gravity and apply equation [1.46].)

1.7 **Derivation of electrical potentials (voltage)** Electrical potential is defined as the work done per charge in moving a positive test charge from infinity to a position r through an electric field against the force represented by that field.

(a) Derive a general expression for the potential V in terms of the distance r. (*Hint*: Use a work integral, $W = \int F\,dx$.)

(b) Using the fact that for a group of charges, potential is additive, calculate the voltage at the center of the square shown in Fig. P1.7. The following data apply:

$$q_1 = +1.0 \times 10^{-8}\mathrm{C}$$

$$q_2 = -2.0 \times 10^{-8}\mathrm{C}$$

$$q_3 = +3.0 \times 10^{-8}\mathrm{C}$$

$$q_4 = +2.0 \times 10^{-8}\mathrm{C}$$

$$a = 1\ \mathrm{m}$$

ANSWER V = 506 V

1.8 **Administration of a drug** A single dose of a newly developed and approved drug is administered to a patient, and its concentration C in the blood is monitored. Since distribution in the blood is very fast, the drug can be assumed to be "well mixed" at any instant. Extrapolation of the data to time t = 0 yields a value of C = 10 mg/L. The total rate of elimination of the drug from the blood is given by $k_e CV$, where V = volume of the blood (L) and k_e = elimination rate constant (h^{-1}).

(a) If after 10 h the drug concentration is found to have dropped to 2 mg/L, what is the value of k_e?

(b) What is the ultimate steady-state concentration of the drug in the blood?

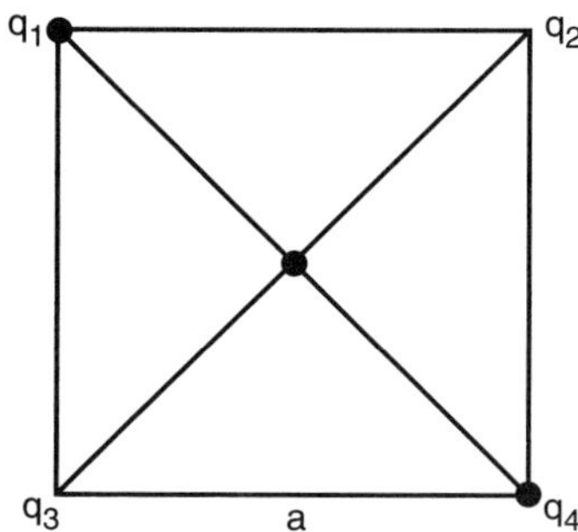

Figure P1.7 Electrical potential in a square.

(c) In some situations, it is desirable to keep the drug concentration at a single stable value. If the blood volume of a particular individual is 5 L and we wish to keep the drug concentration in the blood at 5.0 mg/L, what steady rate (mg/h) must be delivered by infusion?

ANSWERS (a) $0.161\,h^{-1}$
(b) 40.25 mg/h

CHAPTER 2

Some Mathematical Tools

This chapter briefly presents some mathematical tools that are used in modeling at an elementary level. Three topics are addressed. In the first, we examine the basic relations of vector algebra. The second topic deals with the matrix operations that arise particularly in the treatment of simultaneous equations. As the third topic, we present a compilation and analysis of important solution methods for ordinary differential equations. The exposition is accompanied by illustrative examples, and as usual there is a rich choice of practice problems at the end of the chapter.

This chapter may be omitted if its contents are already well understood and in place. It may, on the other hand, serve as a useful refresher and reference or even as a first-time exposure to the topics in question. The treatment is simple enough, and sufficiently comprehensive, to make this possible.

2.1 Vector Algebra

In what follows, we present the principal definitions and operations pertaining to vector algebra. They are also summarized in Table 2.1.

2.1.1 Definition of a Vector

A vector **A** is a mathematical entity that is defined by its length, denoted by $|\mathbf{A}| = \mathrm{A}$, and its direction. It can be represented algebraically in two different ways. In the first, the vector (**A**) is expressed as the product of a *unit* vector **a** with the same direction as **A** and its magnitude A, so that

$$\mathbf{A} = \mathrm{A}\mathbf{a} \tag{2.1}$$

where $|\mathbf{a}| = 1$.

Table 2.1 Relations of Vector Algebra

System/Operation	Defining Relation	Other Properties
Free Vector ***A***	$\mathbf{A} = A_x\mathbf{i} + A_y\mathbf{j} + A_z\mathbf{k}$ $\mathbf{A} = \lvert\mathbf{A}\rvert[\cos(\mathbf{A}, x)\mathbf{i} + \cos(\mathbf{A}, y)\mathbf{j} + \cos(\mathbf{A}, z)\mathbf{k}]$ $\mathbf{A} = \lvert\mathbf{A}\rvert\mathbf{a}$ *Magnitude* $\lvert\mathbf{A}\rvert = (A_x^2 + A_y^2 + A_z^2)^{1/2}$ *Unit Vectors* **i**, **j**, **k**: Along x, y, z axes **a**: Along vector **A**	Two vectors **A**, **B** are equal (1) If they have identical magnitude and direction ($A_x = B_x$, $A_y = B_y$, $A_z = B_z$) (2) Or if they have equal magnitude and opposite direction ($\mathbf{A} = -\mathbf{B}$, where parallel vectors are implied)
Addition and Subtraction	$(\mathbf{A} \pm \mathbf{B}) = (A_x \pm B_x)\mathbf{i} + (A_y \pm B_y)\mathbf{j} + (A_z \pm B_z)\mathbf{k}$	For graphical construction see Fig. 2.1C, 2.1D
Multiplication by Scalar m	$m\mathbf{A} = mA_x i + mA_y j + mA_z k$ $m\mathbf{A} = \lvert\mathbf{A}\rvert[\cos(\mathbf{A}, x)\mathbf{i} + \cos(\mathbf{A}, y)\mathbf{j} + \cos(\mathbf{A}, z)\mathbf{k}]$ $m\mathbf{A} = m\lvert\mathbf{A}\rvert\mathbf{a}$	
Dot Product $\boldsymbol{A} \cdot \boldsymbol{B}$ *= a Scalar*	$\mathbf{A} \cdot \mathbf{B} = \lvert\mathbf{A}\rvert\lvert\mathbf{B}\rvert \cos(\mathbf{A}, \mathbf{B})$ $\mathbf{A} \cdot \mathbf{B} = A_xB_x + A_yB_y + A_zB_z$ $\mathbf{A} \cdot \mathbf{B} = \lvert\mathbf{A}\rvert$ Projection of **B** on **A** $\mathbf{A} \cdot \mathbf{B} = \lvert\mathbf{B}\rvert$ Projection of **A** on **B** $\mathbf{A} \cdot \mathbf{B} = 0$ if **A**, **B** orthogonal $\mathbf{A} \cdot \mathbf{B} = \lvert\mathbf{A}\rvert\lvert\mathbf{B}\rvert$ if **A**, **B** parallel $\mathbf{i} \cdot \mathbf{j} = \mathbf{j} \cdot \mathbf{k} = \mathbf{i} \cdot \mathbf{k} = 0$ $\mathbf{i} \cdot \mathbf{i} = \mathbf{j} \cdot \mathbf{j} = \mathbf{k} \cdot \mathbf{k} = 1$	Dot product is *Distributive* $\mathbf{A} \cdot (\mathbf{B} + \mathbf{C}) = \mathbf{A} \cdot \mathbf{B} + \mathbf{A} \cdot \mathbf{C}$ *Commutative* $\mathbf{A} \cdot \mathbf{B} = \mathbf{B} \cdot \mathbf{A}$ *Associative* $(t\mathbf{A}) \cdot \mathbf{B} = t(\mathbf{A} \cdot \mathbf{B})$
Cross Product $A \times B$ *= a Vector*	$\mathbf{A} \times \mathbf{B} = \mathbf{C}$ = Vector normal to **A**, **B** $\mathbf{AB} = (A_yB_z - A_zB_y)\mathbf{i} + (A_zB_x - A_xB_z)\mathbf{j} + (A_xB_y - A_yB_x)\mathbf{k}$ $\lvert\mathbf{A} \times \mathbf{B}\rvert = \lvert\mathbf{A}\rvert\lvert\mathbf{B}\rvert \sin(\mathbf{A}, \mathbf{B})$ = Area of parallelogram $\mathbf{A} \times \mathbf{B} = 0$ if **A**, **B**, parallel	Cross product is *Distributive* *Associative* *It is NOT commutative* $\mathbf{A} \times \mathbf{B} = -\mathbf{B} \times \mathbf{A}$

In the second, the vector is decomposed into three directional unit vectors **i**, **j**, and **k**, which point in the positive x, y, and z directions. We obtain

$$\mathbf{A} = A_x\mathbf{i} + A_y\mathbf{j} + A_z\mathbf{k} \tag{2.2}$$

where A_x, A_y and A_z are the components of **A** along the three Cartesian coordinate axes. This situation is depicted in Fig. 2.1A. It follows that the magnitude A can be written in terms of its components as follows

$$A = (A_x^2 + A_y^2 + A_z^2)^{1/2} \tag{2.3}$$

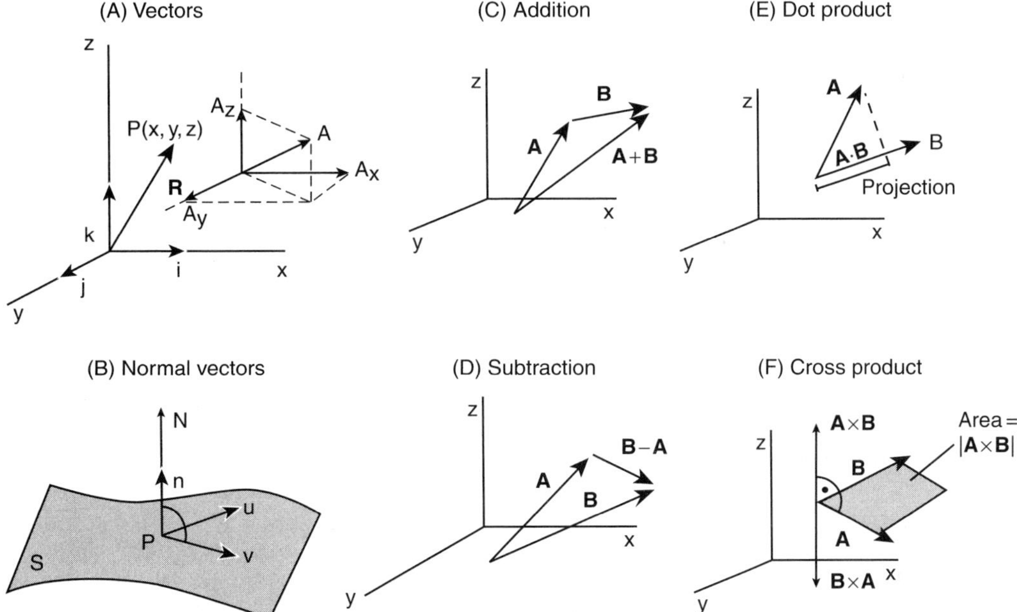

Figure 2.1 Vectors and simple vector algebra operations.

The position vector **R**, also shown in Fig. 2.1A, is a special vector, which has its starting point at the origin of the coordinate system, so that

$$\mathbf{R} = x\mathbf{i} + y\mathbf{j} + z\mathbf{k} \tag{2.4}$$

The vectors just discussed are so-called *free vectors*, which can be located anywhere in space. They are to be distinguished from so-called *line vectors*, which are defined by magnitude, direction *and* line of action; that is, two line vectors can be equal only if, in addition to having the same magnitude and direction, they also lie on the same line. Such vectors, used mainly in solid mechanics, are not addressed here.

2.1.2 Vector Equality

Two vectors **A** and **B** are equal if their components are equal. Thus

$$\mathbf{A} = \mathbf{B} \tag{2.5}$$

implies

$$A_x = B_x, \qquad A_y = B_y, \qquad A_z = B_z \tag{2.6}$$

2.1.3 Vector Addition and Subtraction

Vector addition and subtraction occasionally give rise to some confusion, and it is helpful to remember the following rules.

In *addition*, the two vectors meet head to tail, in *subtraction*, tail to tail. Thus the direction of the difference vector **B** − **A** must be such that **A** + (**B** − **A**) = **B**. The two operations are shown in Fig. 2.1C and 2.1D. The sum or difference of two vectors is obtained by adding or subtracting the components. Thus

$$\mathbf{A} \pm \mathbf{B} = (A_x \pm B_x)\mathbf{i} + (A_y \pm B_y)\mathbf{j} + (A_z \pm B_z)\mathbf{k} \tag{2.7}$$

Several illustrations in our text make use of vector addition, notably Example 3.1, which deals with a river crossing, and Examples 3.7 and 3.8, which are related to flight problems. The latter also introduce the reader to the concept of a vector derivative.

We next consider the multiplication of vectors. Because vectors have direction as well as magnitude, vector multiplication cannot follow exactly the algebraic rules of scalar multiplication. We distinguish three kinds of vector multiplication operation: multiplication of a vector by a scalar, multiplication of two vectors in a way that yields a scalar, and multiplication of two vectors in a way that yields a new vector. These three operations are taken up in turn.

2.1.4 Multiplication by a Scalar m

The simplest operation of the three, multiplication by a scalar m, defines a new vector m**A**, whose magnitude is m times that of **A**. Thus

$$m\mathbf{A} = mA_x\mathbf{i} + mA_y\mathbf{j} + mA_z\mathbf{k} \tag{2.8}$$

2.1.5 The Scalar or Dot Product

The scalar product of two vectors **A** and **B** is defined to be

$$\boxed{\mathbf{A} \cdot \mathbf{B} = |\mathbf{A}\,||\,\mathbf{B}| \cos(A, B) = AB\cos(A, B)} \tag{2.9}$$

That is, it equals the product of the magnitude of the two vectors times the cosine of the angle between them. It is therefore a scalar, and furthermore the operation is commutative: one can change the order of multiplication without affecting the result.

Now, we could have defined the dot product in any number of other ways, for example, $A^{1/3}B^{1/2}\tan(A, B)$, but those definitions would not have been of much use in solving practical problems. By defining it as shown in equation (2.9), we create a convenient shorthand to represent the following.

- The component of a vector in a particular direction; that is, its projection onto another vector that points in that direction (see Fig. 2.1E). This is a useful property, which is used in problems of mechanics.
- Two lines that intersect at right angles. We then have

$$\mathbf{A} \cdot \mathbf{B} = 0 \qquad \mathbf{A} \perp \mathbf{B} \tag{2.10}$$

  This particular property is also useful in mechanics, as well as in problems of geometry and analytical geometry.

- Various forms of work or energy, including mechanical work, gravitational potential energy, and electrical potential.

Some applications along these lines will appear in the examples that follow, as well as in later cases, Examples 3.6 and 4.9.

Equation (2.9) serves as a definition as well as a working equation. When the angle (A, B) is not known, however, an alternative expression, based only on the components of the two vectors, is required. This can be arrived at as follows. We start by applying the dot product to the unit vectors i, j, and k and obtain

$$\mathbf{i} \cdot \mathbf{i} = \mathbf{j} \cdot \mathbf{j} = \mathbf{k} \cdot \mathbf{k} = 1 \tag{2.11}$$

$$\mathbf{i} \cdot \mathbf{j} = \mathbf{j} \cdot \mathbf{k} = \mathbf{k} \cdot \mathbf{i} = 0 \tag{2.12}$$

The first of these relations follows from equation (2.9) and from the fact that for a zero angle between vectors, $\cos(\mathrm{A}, \mathrm{B}) = 1$. Conversely, for orthogonal vectors, the cosine vanishes [see equation (2.9)] and the expression (2.12) results. We next write a general dot product in terms of the vector components. Thus

$$\mathbf{A} \cdot \mathbf{B} = (\mathrm{A_x}\mathbf{i} + \mathrm{A_y}\mathbf{j} + \mathrm{A_z}\mathbf{k}) \cdot (\mathrm{B_x}\mathbf{i} + \mathrm{B_y}\mathbf{j} + \mathrm{B_z}\mathbf{k}) \tag{2.13}$$

Upon multiplying out and introducing the relations (2.10) and (2.11), we have

$$\boxed{\mathbf{A} \cdot \mathbf{B} = \mathrm{A_x B_x} + \mathrm{A_y B_y} + \mathrm{A_z B_z}} \tag{2.14}$$

This provides us with a second working equation, which together with equation (2.9) comprises the tools needed in calculations involving the scalar product.

2.1.6 The Vector or Cross Product

The cross product of two vectors **A** and **B** is no longer a scalar but results in a new vector **C** instead. Thus

$$\mathbf{A} \times \mathbf{B} = \mathbf{C} \tag{2.15}$$

Since the cross product represents a vector, both its magnitude and direction need to be defined. For the *magnitude*, we use the definition

$$|\mathbf{A} \times \mathbf{B}| = \mathrm{C} = \mathrm{AB}\sin(\mathrm{A}, \mathrm{B}) \tag{2.16}$$

That is, the magnitude of a cross product equals the product of the magnitudes of the two vectors times the sine of the angle between them. Another way of putting this is to say that the magnitude of a cross product of two vectors equals the area of the parallelogram formed by them (see Fig. 2.1E).

The *direction* of **C** is defined to be perpendicular to the plane formed by **A** and **B**. To specify the *sense* of the vector, imagine rotating a right-handed screw perpendicular to **A** and **B**, so as to turn it *from **A** to **B*** through the angle (A, B). The direction of advance of the screw gives the direction of the vector product $\mathbf{A} \times \mathbf{B}$. Thus the product $\mathbf{A} \times \mathbf{B}$ in Fig. 2.1E points upward, while the product $\mathbf{B} \times \mathbf{A}$ points downward. It follows that

$$\mathbf{A} \times \mathbf{B} = -\mathbf{B} \times \mathbf{A} \tag{2.17}$$

which implies that the cross-product operation is *not* commutative; that is, the order of multiplication cannot be changed. This is in contrast to the scalar product, which is commutative, as we have seen.

The reason for defining the vector product in this and only this fashion again arises from the fact that certain important physical quantities can be represented as vector products. These include torque, angular momentum, the force on a moving charge in an electrical field, and the flow of electromagnetic energy. Example 4.3, Thomson's Determination of e/m, provides an example of its application.

When the angle (A, B) in equation (2.16) is not known, we again require an alternative expression in terms of the vector components. We proceed in a manner similar to what was done for the dot product and obtain

$$\mathbf{i} \times \mathbf{j} = \mathbf{k}, \qquad \mathbf{j} \times \mathbf{k} = \mathbf{i}, \qquad \mathbf{k} \times \mathbf{i} = \mathbf{j} \tag{2.18}$$

as well as

$$\mathbf{i} \times \mathbf{i} = \mathbf{j} \times \mathbf{j} = \mathbf{k} \times \mathbf{k} = 0 \tag{2.19}$$

These two relations are introduced into the following expanded version of the cross product of two vectors **A** and **B**

$$\mathbf{A} \times \mathbf{B} = (A_x\mathbf{i} + A_y\mathbf{j} + A_z\mathbf{k}) \times (B_x\mathbf{i} + B_y\mathbf{j} + B_z\mathbf{k}) \tag{2.20}$$

Upon cross-multiplying out and introducing the relations (2.17) and (2.18), we obtain

$$\boxed{\mathbf{A} \times \mathbf{B} = (A_yB_z - A_zB_y)\mathbf{i} + (A_xB_z - A_zB_x)\mathbf{j} + (A_xB_y - A_yB_x)\mathbf{k}} \tag{2.21}$$

Both equations (2.16) and (2.21) are used as working expressions in computations involving cross products.

Example 2.1 provides an illustration of the use of both scalar and vector products.

◆ EXAMPLE 2.1 Distance of a Point from a Plane

We are given a point P(1, −2, 1) and a plane defined by the three points A(2, 4, 1), B(−1, 0, 1) and C(−1, 4, 2); see Fig. 2.2. The task is to find the shortest distance from the point to the plane.

We start by noting that the shortest distance d must be measured along a perpendicular line from P to the plane. To establish that perpendicular, we make clever use of the properties of the vector product by applying it to two vectors lying in the plane, for example, $\overrightarrow{\mathbf{AB}}$ and $\overrightarrow{\mathbf{AC}}$. The perpendicular we seek is then given by $\overrightarrow{\mathbf{AB}} \times \overrightarrow{\mathbf{AC}}$, and the shortest distance d will lie on that perpendicular. We next draw an arbitrary vector from the point P to the plane, for example, to the point A, yielding $\overrightarrow{\mathbf{AP}}$. When that vector is projected onto the

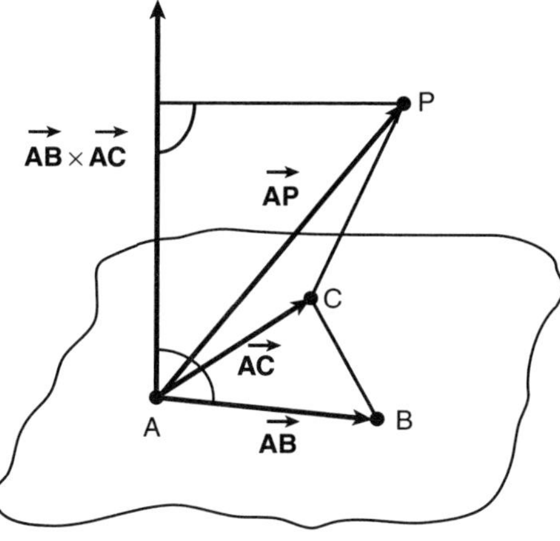

Figure 2.2 Distance of a point from a plane.

perpendicular we have already obtained, it establishes the shortest distance d. The question that now arises is how to find that projection. Here we make the second clever use of vector products, but this time we reach for the scalar product. The reason is that this product, as expressed by equation (2.9), can be viewed as the projection of a vector, say **A** onto another vector **B**, with the projection given by A cos(A,B). For the case at hand, this means that we have to form the dot product of the vector $\overrightarrow{\mathbf{AP}}$ and the perpendicular to the plane, $\overrightarrow{\mathbf{AB}} \times \overrightarrow{\mathbf{AC}}$. That is, we write

$$(\overrightarrow{\mathbf{AB}} \times \overrightarrow{\mathbf{AC}}) \cdot \overrightarrow{\mathbf{AP}} = |\overrightarrow{\mathbf{AB}} \times \overrightarrow{\mathbf{AC}}\,||\,\overrightarrow{\mathbf{AP}}|\cos(\overrightarrow{\mathbf{AB}} \times \overrightarrow{\mathbf{AC}}, \overrightarrow{\mathbf{AP}}) \tag{2.22}$$

where we recognize the last two terms on the right to represent d, the shortest distance being sought, that is, $d = |\overrightarrow{\mathbf{AP}}|\cos(\overrightarrow{\mathbf{AB}} \times \overrightarrow{\mathbf{AC}}, \overrightarrow{\mathbf{PA}})$. Solving for d, we then obtain

$$d = \frac{(\overrightarrow{\mathbf{AB}} \times \overrightarrow{\mathbf{AC}}) \cdot \overrightarrow{\mathbf{AP}}}{|\overrightarrow{\mathbf{AB}} \times \overrightarrow{\mathbf{AC}}|} \tag{2.23}$$

This is the expression that now must be evaluated. Before doing this we note that vectors going from one known point to another, such as $\overrightarrow{\mathbf{AP}}$, are evaluated by forming the *differences* of their coordinates. This follows because they can be viewed as differences of two *position vectors*. For example, for $\overrightarrow{\mathbf{AB}}$ we can write

$$\overrightarrow{\mathbf{AB}} = \overrightarrow{\mathbf{OB}} - \overrightarrow{\mathbf{OA}} \tag{2.24}$$

where $\overrightarrow{\mathbf{OB}}$ and $\overrightarrow{\mathbf{OA}}$ extend from the origin to the points B and A respectively. Thus for $\overrightarrow{\mathbf{AB}}$ in our problem we obtain

$$\overrightarrow{\mathbf{AB}} = [(-1\ -2),\ (0-4),\ (2-1)] = [-3, -4, 0] \tag{2.25}$$

and similarly

$$\overrightarrow{\mathbf{AC}} = [(-1-2),\ (4-4),\ (2-1)] = [-3, 0, 1] \tag{2.26}$$

We can now turn to the evaluation of the various terms in equations (2.23). Thus for the first term of the numerator we have

$$\overrightarrow{\mathbf{AB}} \times \overrightarrow{\mathbf{AC}} = [-3, -4, 0] \times [-3, 0, 1] \tag{2.27}$$

which upon application of equation (2.21) yields

$$\overrightarrow{\mathbf{AB}} \times \overrightarrow{\mathbf{AC}} = [-4, 3, -12] \tag{2.28}$$

This vector has the magnitude

$$|\overrightarrow{\mathbf{AB}} \times \overrightarrow{\mathbf{AC}}| = [(-4)^2 + (3)^2 + (-12)^2]^{1/2} = 13 \tag{2.29}$$

Equation (2.23) has now been reduced to the expression

$$d = \frac{[-4, 3, -12] \cdot [1, 6, 0]}{13} \tag{2.30}$$

and there remains the final step of evaluating the dot product. This is quickly done by using equation (2.14). We obtain

$$[-4, 3, -12] \cdot [1, 6, 0] = (-4)(1) + (3)(6) + (-12)(0) = 14$$

so that the final result for the distance is given by

$$\boxed{d = \frac{14}{13} = 1.077} \tag{2.31}$$

COMMENT

- One notes here the ingenious use of both scalar and vector products. A vector product was used to establish the perpendicular to the plane that coincides with the shortest distance being sought. That distance was then obtained by projecting a line from the point P to the plane onto the perpendicular, making use of the properties of scalar products. Thus a combination of the two vector products yielded the desired result. A similar approach will be used in the example that follows.

◆ EXAMPLE 2.2 Shortest Distance Between Two Lines

Two lines $\overrightarrow{\mathbf{AB}}$ and $\overrightarrow{\mathbf{CD}}$ are defined by the four points

$$A(1, -2, -1), B(4, 0, -3), C(1, 2, -1), D(2, -4, -5)$$

We must find the shortest distance between the two lines.

The procedure here is identical to that used in Example 2.1. We erect the normal to the two lines by forming the cross product $\overrightarrow{\mathbf{AB}} \times \overrightarrow{\mathbf{CD}}$. Then we project an arbitrary vector

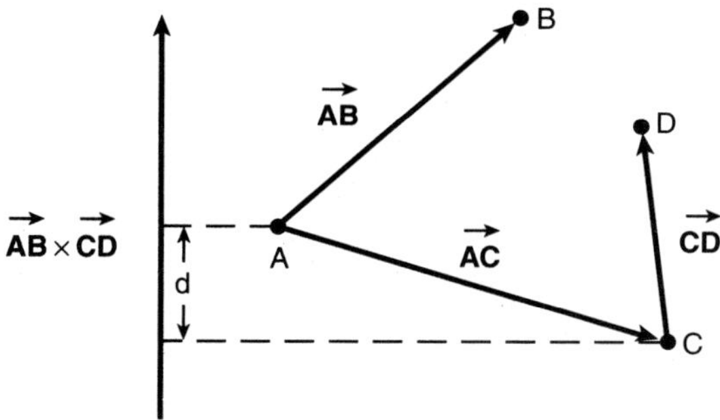

Figure 2.3 Shortest distance between two lines.

connecting the two lines, say $\overrightarrow{\mathbf{AC}}$, onto the perpendicular $\overrightarrow{\mathbf{AB}} \times \overrightarrow{\mathbf{CD}}$, as shown in Fig. 2.3. The resulting expression for d is analogous to that of equation (2.23) in Example 2.1. We have

$$d = \frac{(\overrightarrow{\mathbf{AB}} \times \overrightarrow{\mathbf{CD}}) \cdot (\overrightarrow{\mathbf{AC}})}{|\overrightarrow{\mathbf{AB}} \times \overrightarrow{\mathbf{CD}}|} \tag{2.32}$$

Numerical evaluation leads to the following results, obtained by using equation (2.21):

$$\overrightarrow{\mathbf{AB}} \times \overrightarrow{\mathbf{AC}} = [3, 2, -2] \times [1, -6, -4] = [-20, 10, -20]$$

from which we obtain the magnitude

$$|\overrightarrow{\mathbf{AB}} \times \overrightarrow{\mathbf{CD}}| = [(-20)^2 + (10)^2 + (-20)^2]^{1/2} = 30$$

The final step consists of evaluating the dot product in equation (2.32), which yields

$$d = \frac{[-20, 10, -20] \cdot [0, 4, 0]}{30} = \frac{(-20)0 + (10)4 + (-20)0}{30} = \frac{40}{30}$$

$$\boxed{d = 1.33} \tag{2.33}$$

◆ EXAMPLE 2.3 Work as an Application of the Scalar Product

A block of mass 10.0 kg is to be raised from the bottom to the top of an incline 5.00 m long and 3.00 m off the ground at the top (Fig. 2.4). Assuming frictionless surfaces, how much work must be done by a force F parallel to the incline if the block is pushed at a constant velocity?

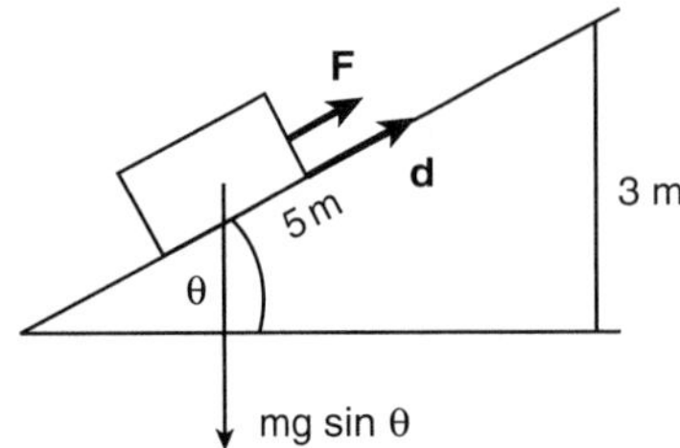

Figure 2.4 Raising a load up an incline.

Since work contains force as a component, we must start by finding F. Because the motion is not accelerated, we have from Newton's law

$$\sum \text{Forces} = 0$$

or

$$F = mg \sin \theta \tag{2.34}$$

It follows that

$$F = mg \sin \theta = 10 \times 9.81 \times \tfrac{3}{5} = 58.8\,\text{N} \tag{2.35}$$

Now work is defined as the scalar product of force and distance covered, or

$$W = \mathbf{F} \cdot \mathbf{d} \tag{2.36}$$

Applied to the present case, we obtain

$$\boxed{W = \mathbf{F} \cdot \mathbf{d} = Fd \cos(F, d) = 58.8 \times 5 \times 1 = 294\,\text{J}} \tag{2.37}$$

If the block were raised vertically, the work done would be given by

$$\boxed{W = \mathbf{F_g} \cdot \mathbf{h} = mg\,h \cos(\mathbf{F_g}, \mathbf{h}) = 10 \times 9.81 \times 3 \times 1 = 294\,\text{J}} \tag{2.38}$$

which is the same as before, as indeed it must be. The only difference is that with the incline, one can apply a smaller force ($F = 58.8$ N) than is required without the incline ($F_g = 98.1$ N).

COMMENT

- Friction, neglected in this case, but can be taken into account using the relation

$$F_f = \mu F_g \tag{2.39}$$

where F_f and F_g are the friction and gravitational forces, and μ is an empirical friction coefficient. For smooth surfaces, μ is typically 0.1. The inclusion of friction in equation (2.35) would then yield

$$F = 10 \times 9.81 \times \left(0.1 + \tfrac{3}{5}\right) = 68.7\,\text{N}$$

which is higher by some 20% than the frictionless force of 58.8 N, but still well below the force required to raise the load vertically. Note that a smooth surface can be created by covering the work area with oilcloth or similar materials.

◆ EXAMPLE 2.4 Extension of the Scalar Product to n Dimensions: A Sale of Stocks

We have seen that one of the working expressions for the scalar product, equation (2.14), consists of the sum of the products of the three components. It frequently happens in practical applications that more than three component products make their appearance. This has prompted an extension of the scalar product to n dimensions, which we formulate as follows:

$$\mathbf{A} \cdot \mathbf{B} = A_1B_1 + A_2B_2 + \cdots + A_nB_n = \sum_{k=1}^{n} A_kB_k \tag{2.40}$$

We illustrate the use of this equation with the following simple example of a financial transaction. An investment company decides to sell four of its stocks. In one transaction, 200 shares of stock A, 300 shares of stock B, 100 shares of stock C, and 200 shares of stock D were sold. The selling prices per share were \$20, \$30, \$50, and \$10, respectively. We represent the quantity of stocks by the vector $\mathbf{A} = (200, 300, 100, 200)$ and the selling prices by the vector $\mathbf{B} = (20, 30, 50, 10)$. The total receipts from the transaction are then given by

$$\mathbf{A} \cdot \mathbf{B} = (200, 300, 100, 200) \cdot (20, 30, 50, 10) \tag{2.41}$$

$$= 200 \times 20 + 300 \times 30 + 100 \times 50 + 200 \times 10$$

$$\mathbf{A} \cdot \mathbf{B} = \$20{,}000 \tag{2.42}$$

COMMENT

- Evidently, such uses of the scalar product are not confined to financial transactions. Extensions to a multitude of other commercial activities are obvious, as shown in Example 2.5, which for simplicity, involves only three-dimensional vectors.

◆ EXAMPLE 2.5 A Simple Model Economy

Consider an economic system that comprises three types of industry and three types of consumer. The industries are the crude oil industry, which produces crude oil, the refining industry, which produces gasoline, and the utility industry, which supplies electricity. The consumers are the general public, the government, and the export firms. *Both* the consumers and the industries use some or all of the products. For example, the crude oil industry uses no oil, but needs 4 units of gasoline to run its pumps and 2 units of electricity. Its demand vector is then

$\mathbf{A}_1 = (0, 4, 2)$ crude oil industry

We specify in similar fashion the remaining five demand vectors:

$\mathbf{A}_2 = (8, 0, 6)$ for the refining industry
$\mathbf{A}_3 = (1, 6, 0)$ for the utility industry
$\mathbf{A}_4 = (1, 9, 5)$ for the public
$\mathbf{A}_5 = (5, 8, 8)$ for the government
$\mathbf{A}_6 = (7, 2, 0)$ for the export firms

The total demand from both the consumers and the industries is then

$$\mathbf{A}_{\text{tot}} = \mathbf{A}_1 + \mathbf{A}_2 + \mathbf{A}_3 + \mathbf{A}_4 + \mathbf{A}_5 + \mathbf{A}_6$$

which turns out to be

$$\mathbf{A}_{\text{tot}} = (\text{oil, gasoline, utility}) = (25, 29, 21) \tag{2.43}$$

Now suppose the price of crude oil is \$4 per unit, the price of gasoline \$3 per unit, and the price of electricity \$2 per unit. This can be expressed as an additional cost vector

$$\mathbf{B} = (4, 3, 2)$$

Let us compute the operating costs of the crude oil industry. Since they are composed of the sum of the products of demand and cost, they will equal the dot product of the demand and cost vectors. Thus

$$\mathbf{A} \cdot \mathbf{B} = (0, 4, 2) \cdot (4, 3, 2) = (0 \times 4 + 4 \times 3 + 2 \times 2) = \$16 \tag{2.44}$$

Now the same industry has sold 25 units of crude oil [first entry of $\mathbf{A}_{\text{tot}}$, equation (2.43)] at a price of \$4, giving an income of $4 \times 25 = \$100$. Hence the gross profit of the industry is

$$\text{P} = \$100 - \$16 = \$84$$

Profits for the other industries can be computed in similar fashion.

2.2 Matrices

We start this section with the definition of a matrix and a description of some important special matrices. These are summarized for convenience in Table 2.2.

Table 2.2 Types of Matrix

1. m × n Matrix **A**

$$\mathbf{A} = \begin{bmatrix} a_{11} & a_{12} & \cdots & a_{1n} \\ a_{21} & a_{22} & \cdots & a_{2n} \\ \vdots & \vdots & & \\ a_{m1} & a_{m2} & \cdots & a_{mn} \end{bmatrix}$$

2. Square Matrix

$$\mathbf{A} = \begin{bmatrix} 24 & 1 & 3 \\ 6 & 0 & 5 \\ -13 & 7 & -6 \end{bmatrix}$$

3. Diagonal Matrix

$$\mathbf{A} = \begin{bmatrix} 5 & 0 & 0 \\ 0 & -3 & 0 \\ 0 & 0 & 7 \end{bmatrix}$$

4. Upper Triangular Matrix

$$\mathbf{U} = \begin{bmatrix} 5 & 57 & -6 \\ 0 & 15 & 3 \\ 0 & 0 & 100 \end{bmatrix}$$

5. Lower Triangular Matrix

$$\mathbf{U} = \begin{bmatrix} 6 & 0 & 0 \\ 3 & 4 & 0 \\ -2 & 67 & 1 \end{bmatrix}$$

6. Identity Matrix

$$\mathbf{I} = \begin{bmatrix} 1 & 0 & 0 \\ 0 & 1 & 0 \\ 0 & 0 & 1 \end{bmatrix}$$

7. Null Matrix

$$\mathbf{O} = \begin{bmatrix} 0 & 0 & 0 \\ 0 & 0 & 0 \\ 0 & 0 & 0 \end{bmatrix}$$

8. Column Vector

$$\mathbf{A} = \begin{bmatrix} 5 \\ 63 \\ -6 \end{bmatrix}$$

9. Row Vector

$$\mathbf{A} = \begin{bmatrix} 5 & 0 & 17 \end{bmatrix}$$

2.2.1 Types of Matrix

The matrix itself is *an array of entries, usually numerical, which are arranged in m rows and n columns* (see item 1 of Table 2.2). Among the special matrices we note the square matrix (item 2) with m = n, and the diagonal matrix (item 3), which is a square matrix with nonzero entries confined to the diagonal. When these entries are all unity, the array becomes the identity matrix (item 6). Upper and lower triangular matrices (items 4 and 5) are square matrices that have zero entries respectively below and above the diagonal. Finally we note two one-dimensional versions of the general matrix that take the form of column and row vectors (items 8 and 9). Although the examples given in the Table 2.2 are all 3 × 3 matrices for reasons of convenience, the order of the matrix can in principle comprise any number of rows and columns. Further, the entries are not confined to integers, as was done here for brevity; an entry can be any positive or negative fraction or integer, including of course zero. Matrices with imaginary or complex entries also exist but are outside the scope of our tale.

2.2.2 The Echelon Form, Rank r

Table 2.3 introduces the reader to a special matrix array termed the echelon form. Here zero entries are mandatory, and each one increases in number in a prescribed fashion with each successive row.

After a specific row r, termed the rank of a matrix, all entries become zero. Rank is an important property of matrices and equals the number of linearly independent rows. Another way

Table 2.3 Matrices in Echelon Form

Properties

- First r rows must have nonzero entries.
- All remaining m − r rows must have *only* zero entries.
- r is the *rank* of the matrix and equals the number of linearly independent rows.
- The first nonzero entries in a row are termed **pivots**. The number of zeros in front of the pivots must increase in successive rows.
- All upper triangular and diagonal matrices are already in echelon form.

Examples

I

$$\mathbf{E} = \begin{bmatrix} 2 & 5 & 1 & 0 & 6 \\ 0 & 7 & 3 & 2 & 0 \\ 0 & 0 & 0 & 0 & 5 \\ 0 & 0 & 0 & 0 & 0 \end{bmatrix}$$

$r = 3$

II

$$\begin{bmatrix} 3 & 2 & -6 \\ 6 & 4 & -12 \\ 21 & 17 & 1 \end{bmatrix}$$

$r = 2$

III

$$\mathbf{U} = \begin{bmatrix} 3 & 2 & -5 & 6 \\ 0 & 5 & 7 & 3 \\ 0 & 0 & 16 & 8 \\ 0 & 0 & 0 & -2 \end{bmatrix}$$

$r = 4$

IV

$$\mathbf{A} = \begin{bmatrix} 5 & 1 & 0 & 6 \\ 7 & 8 & 5 & 2 \\ 0 & 0 & 6 & 2 \\ 0 & 0 & 0 & 1 \end{bmatrix}$$

Not in echelon form

of defining it is to state that the rank of a matrix equals the order of the highest nonvanishing determinant made up of the entries of the matrix. In the examples provided in Table 2.3, items I and III are in proper echelon form because the number of zeros increases with each successive row, starting with the second row, as required. In contrast, item IV lacks a leading zero entry in the second row and is therefore not in echelon form. In item II, one notes that the first and second rows are linearly dependent, the latter being a multiple of the first. Hence there are only two independent rows, and the matrix has rank 2.

2.2.3 Matrix Equality

Two matrices **A** and **B** are equal if all corresponding entries are equal. That is, we must have

$$a_{ij} = b_{ij} \tag{2.45}$$

2.2.4 Matrix Addition

Matrices are added by forming the sum of corresponding entries. Thus

$$\mathbf{A} + \mathbf{B} = \mathbf{C} \tag{2.46}$$

implies

$$a_{ij} + b_{ij} = c_{ij} \tag{2.47}$$

Example 2.6 illustrates the addition of matrices.

◆ EXAMPLE 2.6 Acquisition Costs

A manufacturer produces a certain product for which he requires three types of raw material (R, S, and T), which are drawn from two different suppliers. The costs of purchasing and transporting specific amounts of these materials are given by the following matrices

$$\text{Supplier 1} \qquad \mathbf{A} = \begin{array}{c} \begin{matrix} \text{PC} & \text{TC} \end{matrix} \\ \begin{bmatrix} 16 & 20 \\ 10 & 16 \\ 9 & 4 \end{bmatrix} \end{array} \begin{matrix} \\ \text{R} \\ \text{S} \\ \text{T} \end{matrix}$$

$$\text{Supplier 2} \qquad \mathbf{B} = \begin{bmatrix} 12 & 10 \\ 14 & 14 \\ 12 & 10 \end{bmatrix} \begin{matrix} \text{R} \\ \text{S} \\ \text{T} \end{matrix}$$

where PC is purchase cost and TC is transportation cost.

The matrix representing total purchase and transportation costs from the two suppliers is then given by the sum of the two matrices:

$$\mathbf{A} + \mathbf{B} = \begin{bmatrix} (16+12) & (20+10) \\ (10+14) & (16+14) \\ (9+12) & (4+10) \end{bmatrix} = \begin{bmatrix} 28 & 30 \\ 24 & 30 \\ 21 & 14 \end{bmatrix}$$

2.2.5 *Multiplication by a Scalar*

A matrix **A** is multiplied by a scalar k by multiplying each entry by k. Thus

$$\mathbf{kA} = [\mathrm{ka_{ij}}]_{(m,n)} \tag{2.48}$$

2.2.6 *Matrix Multiplication*

Multiplication of two matrices is a more complex operation than the multiplication of a scalar and a matrix. We have set out some of the pertinent rules in Table 2.4, from which it is seen that the product of two matrices equals the sum of prescribed products of row entries of the first matrix by column entries of the second matrix. We note in addition:

- There is only one type of multiplication of matrices. This is in contrast to vector multiplication, which gave rise to two modes, the scalar product and the vector product.
- Like the vector product, the operation is not commutative; that is, $\mathbf{AB} \neq \mathbf{BA}$.

Examples 2.7 and 2.8 illustrate these features.

Table 2.4 Matrix Multiplication

AB = **C** implies that the entry C_{ij} of **C** = $\sum$ [(entries of ith row in **A**) × (entries of jth column in **B**).

Number of columns in **A** must equal number of rows in **B**. In particular,

[Row Vector][Column Vector] → Scalar Product

[Matrix][Column Vector] → Column Vector

[Column Vector][Row Vector] → Matrix

[Column Vector][Matrix] → Does not exist

[Matrix][Row Vector] → Does not exist

Multiplication of matrices is noncommutative: **AB** ≠ **BA**. Products of the identify matrix (**AI** = **IA**) are an important exception.

Other relations

$$(\mathbf{A}+\mathbf{B})\mathbf{C} = \mathbf{AC}+\mathbf{BC}$$

$$\mathbf{A}(\mathbf{BC}) = (\mathbf{AB})(\mathbf{C})$$

◆ EXAMPLE 2.7 The Product of Two Matrices

Given the two matrices

$$\mathbf{A} = \begin{bmatrix} 2 & 1 \\ 3 & 2 \end{bmatrix} \quad \mathbf{B} = \begin{bmatrix} 0 & 4 \\ 1 & 3 \end{bmatrix}$$

Then

$$\mathbf{AB} = \begin{bmatrix} 2 & 1 \\ 3 & 2 \end{bmatrix}\begin{bmatrix} 0 & 4 \\ 1 & 3 \end{bmatrix} = \begin{bmatrix} (2\times 0+1\times 1) & (2\times 4+1\times 3) \\ (3\times 0+2\times 1) & (3\times 4+2\times 3) \end{bmatrix} = \begin{bmatrix} 1 & 11 \\ 2 & 18 \end{bmatrix}$$

By contrast, if we reverse the order of multiplication, we obtain

$$\mathbf{BA} = \begin{bmatrix} 0 & 4 \\ 1 & 3 \end{bmatrix}\begin{bmatrix} 1 & 2 \\ 3 & 2 \end{bmatrix} = \begin{bmatrix} (0\times 2+4\times 3) & (0\times 1+4\times 2) \\ (1\times 2+3\times 3) & (1\times 1+3\times 2) \end{bmatrix} = \begin{bmatrix} 12 & 8 \\ 11 & 7 \end{bmatrix}$$

The two results are seen to be totally different, proving that matrix multiplication is not commutative. Neither is there a simple change in sign involved, as in the case of vector cross products. Examples 7.8 and 7.9, dealing with manufacturing costs and electrical networks, provide further illustrations of matrix multiplication.

Matrix notation was introduced in 1858 by the English mathematician Arthur Cayley, who used it as a shorthand for systems of simultaneous linear algebraic equations. Example 2.8 illustrates the methodology developed by Cayley.

◆ EXAMPLE 2.8 Matrix–Vector Representation of Linear Algebraic Equations

Consider the following set of linear algebraic equations

$$\begin{aligned} 2x + 3y &= 3 \\ x + 4y &= 1 \end{aligned} \tag{2.49}$$

We assert that these equations can be represented by a matrix–vector product as follows:

$$\begin{bmatrix} 2 & 3 \\ 1 & 4 \end{bmatrix} \begin{bmatrix} x \\ y \end{bmatrix} = \begin{bmatrix} 3 \\ 1 \end{bmatrix} \tag{2.50}$$

The proof follows by carrying out the indicated multiplication. We obtain

$$\begin{bmatrix} (2x) + (3y) \\ (x) + (4y) \end{bmatrix} = \begin{bmatrix} 3 \\ 1 \end{bmatrix} \tag{2.51}$$

which is identical to the set of equation (2.49).

This result can be generalized by expressing the set of linear algebraic equations

$$\begin{aligned} a_{11}x_1 \; + a_{12}x_2 \; + \cdots + a_{1n}x_n \; &= \; b_1 \\ a_{21}x_1 \; + a_{22}x_2 \; + \cdots + a_{2n}x_n \; &= \; b_2 \\ \vdots \qquad \quad \vdots \qquad \qquad \quad & \vdots \\ a_{m1}x_1 + a_{m2}x_2 + \cdots + a_{mn}x_n &= b_{mn} \end{aligned} \tag{2.52}$$

as

$$\mathbf{AX} = \mathbf{B} \tag{2.53}$$

where

$$\mathbf{A} = \begin{bmatrix} a_{11} & a_{12} & \dots & a_{1n} \\ a_{21} & a_{22} & \dots & a_{2n} \\ \vdots & \vdots & & \vdots \\ a_{m1} & a_{m2} & \dots & a_{mn} \end{bmatrix} \qquad \mathbf{X} = \begin{bmatrix} X_1 \\ X_2 \\ \vdots \\ X_n \end{bmatrix} \qquad \mathbf{B} = \begin{bmatrix} b_1 \\ b_2 \\ \vdots \\ b_n \end{bmatrix}$$

Here **A** is referred to as the *coefficient matrix* of the system, while **X** and **B** are column vectors of the unknowns x_i and the nonhomogeneous terms b_i, respectively. The combination $\mathbf{A:B}$, which will also be encountered, is called an *augmented matrix*. It consists of an extension of the entries of **A** with those of **B**.

Thus, if $\mathbf{A} = \begin{bmatrix} 0 & 2 \\ 1 & 3 \end{bmatrix}$ and $\mathbf{B} = \begin{bmatrix} 5 \\ 6 \end{bmatrix}$, we have $\mathbf{A:B} = \begin{bmatrix} 0 & 2 & 5 \\ 1 & 3 & 6 \end{bmatrix}$.

2.2.7 Elementary Row Operations

Elementary row operations are central to the solution of linear algebraic equations by matrix methods. They have the property of leaving unchanged both the rank of a matrix and its determinant. This implies that when such operations are applied to the matrix equation (2.53), which represents a set of simultaneous linear algebraic equations, the solution to that set remains unchanged. We shall shortly see that the repeated application of such operations can reduce equation (2.53), hence the underlying set of equations, to a form in which the solution becomes transparent. The operations which will accomplish this are three in number and are termed elementary row operations. They consist of the following:

An interchange of any two rows

Multiplication of any row by a nonzero scalar

Addition to any row of a scalar multiple of another row

Example 2.9 will serve to illustrate these operations.

◆ EXAMPLE 2.9 Application of Elementary Row Operations: Algebraic Equivalence

Consider the matrix

$$\mathbf{A} = \begin{bmatrix} 1 & 3 & 4 \\ -2 & 3 & 1 \end{bmatrix} \tag{2.54}$$

We wish to reduce it to echelon form with unit pivots. This can be accomplished by performing the following two elementary row operations.

Multiply the first row by 2 and add to the second row:

$$\begin{bmatrix} 1 & 3 & 4 \\ -2 & 3 & 1 \end{bmatrix} \underset{2R_1+R_2}{\Rightarrow} \begin{bmatrix} 1 & 3 & 4 \\ 0 & 9 & 9 \end{bmatrix}$$

Multiply the second row by 1/9:

$$\begin{bmatrix} 1 & 3 & 4 \\ 0 & 9 & 9 \end{bmatrix} \underset{\frac{1}{9}R_2}{\Rightarrow} \underset{\text{Echelon form}}{\begin{bmatrix} 1 & 3 & 4 \\ 0 & 1 & 1 \end{bmatrix}}$$

The last matrix is then in echelon form with unit pivots. Let us now consider the initial matrix given by equation (2.54) to be the augmented matrix $\mathbf{A}:\mathbf{B}$ of the following set of equations:

$$\begin{aligned} x_1 + 3x_2 &= 4 \\ -2x_1 + 3x_2 &= 1 \end{aligned} \tag{2.55}$$

It will be seen that the elementary row operations we performed on the foregoing matrix correspond to the operations that are used to obtain a solution of a system of linear equations by the so-called addition–subtraction method. Thus by multiplying the first equation by 2 and adding it to the second equation, we obtain:

$$\begin{aligned} x_1 + 3x_2 &= 4 \\ 9x_2 &= 9 \end{aligned} \tag{2.56}$$

Multiplication of the second equation by 1/9 leads to the result

$$\begin{aligned} x_1 + 3x_2 &= 4 \\ x_2 &= 1 \end{aligned} \tag{2.57}$$

The solution is then essentially complete, since x_1 is easily obtained by back-substitution of the second into the first equation. Note that the operations performed on the matrix and on the equations are identical, but the matrix approach is the terser of the two and has a systematic formalism that lends itself well to computer solutions. We have here a first glimpse of the use of matrix operations in the solution of linear algebraic equations which will ultimately be enshrined in a procedure called Gaussian elimination. That method is discussed in the next section.

2.2.8 Solution of Sets of Linear Algebraic Equations: Gaussian Elimination

Systems of linear algebraic equations are generally classified according to the number of equations m and the number of unknowns n they contain. Three cases are distinguished:

1. Number of equations m > number of unknowns n (i.e., the system is overspecified).
2. Number of equations m = number of unknowns n. This is the case with which the reader will be most familiar and is also the one that is most frequently encountered.
3. Number of equations m < number of unknowns n (i.e., the system is underspecified).

The number of solutions exhibited by these systems also falls into three categories. We can have

1. No solution
2. One solution (unique solution)
3. An infinite number of solutions

The applicable case is determined by the *rank of the coefficient matrix*. In the case of square coefficient matrices, case 2, the number of solutions obtained can also be related to the *determinant* of the coefficient matrix.

The three systems in question, and the criteria that apply for the number of solutions obtained, are listed in Table 2.5. One notes from these tabulations that a unique solution, as well as no solution and an infinite number of solutions, can arise in overspecified systems. This comes as something of a surprise, since one would have expected only the latter case to apply. An explanation is provided in Fig. 2.5, in which we give a geometrical interpretation of the behavior

Table 2.5 Solutions of Systems of Linear Algebraic Equations

System	Number of Solutions	Criteria
1. Normal number of equations m = Number of unknowns n	0	det $\mathbf{A} = 0$ Rank $\mathbf{A}:\mathbf{B} \neq$ Rank $\mathbf{A}$
	1	det $\mathbf{A} \neq 0$ Rank $\mathbf{A}:\mathbf{B} =$ Rank $\mathbf{A} = n$
	∞	det $\mathbf{A} = 0$ Rank $\mathbf{A}:\mathbf{B} =$ Rank $\mathbf{A} < n$
2. Overspecified number of equations m > Number of unknowns n	0	Rank $\mathbf{A}:\mathbf{B} \neq$ Rank $\mathbf{A}$
	1	Rank $\mathbf{A}:\mathbf{B} =$ Rank $\mathbf{A} = n$
	∞	Rank $\mathbf{A}:\mathbf{B} =$ Rank $\mathbf{A} < n$
3. Underspecified number of equations m < Number of unknowns n	0	Rank $\mathbf{A}:\mathbf{B} =$ Rank $\mathbf{A}$
	∞	Ranks $\mathbf{A}:\mathbf{B} =$ Rank $\mathbf{A} < n$

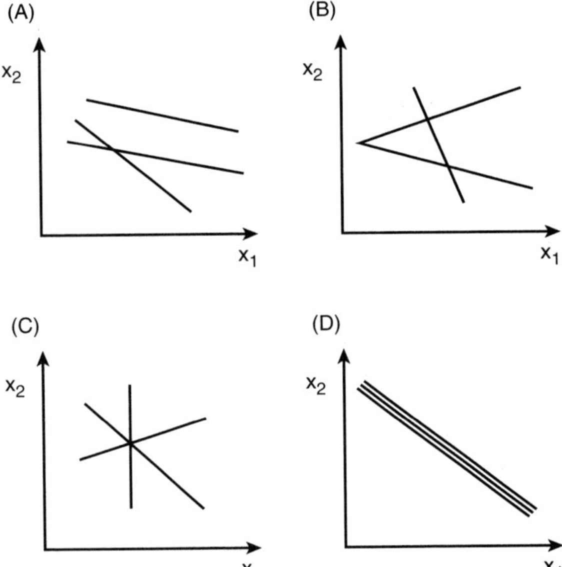

Figure 2.5 Solutions to three equations in two unknowns.

of a system consisting of three equations and two unknowns. In cases A and B, the three lines representing the three equations do not intersect at a single point, hence have no solution. A unique solution exists in case C, where intersection at a single point occurs. Case D, consisting of three parallel lines, again has no solution, but if the three lines coincide, an infinite number of solutions result. This latter case arises when the three equations are multiples of one another.

Unique solutions evidently also arise when $m = n$, but none exist for an under-specified system $m < n$. We can summarize these results of Table 2.5 as follows:

1. When the number of equations m equals the number of unknowns n, a unique solution exists only if the determinant of the coefficient matrix is nonzero (i.e., det $\mathbf{A} \neq 0$).

2. When the number of equations m is *greater* than the number of unknowns n (i.e., the system is overdefined), a unique solution exists only under the restrictive condition rank $\mathbf{A} = \text{rank}\ \mathbf{A:B} = n$.
3. When the number of equations m is *less* than the number of unknowns n (i.e., the system is underdefined), *no unique* solution can exist. Example 7.6 illustrate an underdefined system that arises in stoichiometric calculations, while Examples 3.5 and 7.7 deal with the case $m = n$.

We now demonstrate these features with two examples, using Gaussian elimination. This procedure, as indicated earlier, consists of reducing the augmented matrix $\mathbf{A:B}$ to echelon form with unit pivots and obtaining the full solution by back-substitution.

◆ EXAMPLE 2.10 An Overspecified System of Equations with a Unique Solution

It is required to solve the overspecified system

$$\begin{aligned} 2x_1 + x_2 &= 1 \\ -4x_1 + 6x_2 &= 6 \\ -2x_1 + 7x_2 &= 7 \end{aligned} \tag{2.58}$$

We proceed to reduce the augmented matrix to echelon form.

$$\mathbf{A:B} = \begin{bmatrix} 2 & 1 & 1 \\ -4 & 6 & 6 \\ -2 & 7 & 7 \end{bmatrix} \underset{\substack{2R_1+R_2 \\ R_1+R_3}}{\Rightarrow} \begin{bmatrix} 2 & 1 & 1 \\ 0 & 8 & 8 \\ 0 & 8 & 8 \end{bmatrix} \underset{-R_2+R_3}{\Rightarrow} \begin{bmatrix} 2 & 1 & 1 \\ 0 & 8 & 8 \\ 0 & 0 & 0 \end{bmatrix} \underset{\substack{\frac{1}{2}R_1 \\ \frac{1}{8}R_2}}{\Rightarrow} \begin{bmatrix} 1 & \frac{1}{2} & \frac{1}{2} \\ 0 & 1 & 1 \\ 0 & 0 & 0 \end{bmatrix} \tag{2.59}$$

It can be seen most easily from the final matrix that $\mathbf{A:B}$ has a rank $r = 2$ equal to the number of unknowns. This is because it contains two independent rows or, alternatively, because the highest nonvanishing determinant is of order 2, which follows from the fact that the third row has only zero entries. We further note that the same rank applies to both the coefficient and the augmented matrices. Hence the two criteria for a unique solution, listed in Table 2.5 (item 2) are satisfied. A closer inspection of the set in equation (2.58) also reveals that the third equation is composed of the sum of the first two equations and is therefore redundant. Consequently the system effectively consists of two independent equations with a unique solution.

◆ EXAMPLE 2.11 A Normal System of Equations with No Solution

We consider the following "normal" set in which the number of equations equals the number of unknowns.

$$\begin{aligned} x_1 + 3x_2 - x_3 &= 4 \\ x_1 + 2x_2 + x_3 &= 2 \\ 3x_1 + 7x_2 + x_3 &= 9 \end{aligned} \tag{2.60}$$

As before, we apply elementary row operations to reduce the augmented matrix to echelon form with unit pivots, obtaining

$$\mathbf{A:B} = \begin{bmatrix} 1 & 3 & -1 & 4 \\ 1 & 2 & 1 & 2 \\ 3 & 7 & 1 & 9 \end{bmatrix} \underset{\substack{-R_1+R_2 \\ -3R_1+R_3}}{\Rightarrow} \begin{bmatrix} 1 & 3 & -1 & 4 \\ 0 & -1 & 2 & -2 \\ 0 & -2 & 4 & -3 \end{bmatrix}$$

$$\underset{-R_2}{\Rightarrow} \begin{bmatrix} 1 & 3 & -1 & 4 \\ 0 & 1 & -2 & 2 \\ 0 & -2 & 4 & -3 \end{bmatrix} \underset{2R_2+R_3}{\Rightarrow} \begin{bmatrix} 1 & 3 & -1 & 4 \\ 0 & 1 & -2 & 2 \\ 0 & 0 & 0 & 1 \end{bmatrix}$$

Here we see that the coefficient matrix has only zeros in the last row; hence its rank is 2 and the determinant = 0. The augmented matrix, on the other hand, has a rank of 3 because of the nonzero entry in the last row. Hence

$$\text{Rank } \mathbf{A:B} \neq \text{Rank } \mathbf{A} \tag{2.61}$$

which implies, according to Table 2.5 (item 1) that the system has *no solution*. Such systems are also referred to as *inconsistent*.

We recall here for the benefit of readers with only a hazy memory of the properties of determinants that a determinant is zero, det $\mathbf{A} = 0$, if one of the following conditions holds:

- All entries in a row or column are zero.
- Two rows or columns are proportional.
- A diagonal entry of a triangular or diagonal matrix is zero.

It was the first of these properties that led us to conclude that the determinant of the coefficient matrix vanishes. A further property that is often useful is the following:

- The value of a diagonal determinant equals the product of its entries. Thus

$$\begin{vmatrix} a_{11} & & 0 & \\ & a_{22} & & \ddots \\ & & \ddots & 0 \\ 0 & & a_{nn} & \end{vmatrix} = a_{11} \times a_{22} \times \cdots \times a_{nn} \tag{2.62}$$

This concludes our brief synopsis of matrix properties and operations. We shall have occasion to use these concepts in Chapter 7, which deals with simple networks and sets of simultaneous equations, and in Chapter 8, which deals with more complex systems of this type.

2.3 Ordinary Differential Equations (ODEs)

Differential equations, both of the ordinary and partial type, arise with extraordinary frequency in all sciences and in engineering, as well as in a host of other disciplines. The present text confines itself to ordinary differential equations (ODEs) but will attempt to cover a broad range of disciplines illustrated with representative examples. To give the reader a flavor of what is to come, we start with a pair of illustrative examples drawn from two different disciplines. These are simple problems leading to ODEs that can be integrated by the elementary method of separation of variables. We will subsequently enter into a more detailed discussion both of the classification of ODEs as well as their solution methods. This will put the topic on a firmer footing.

◆ EXAMPLE 2.12 A Population Model

Malthus' law states that the time rate of change of a population p(t) is proportional to p(t). This holds for many populations as long as they are not too large. A more refined model is the logistic law, given by

$$\frac{dp}{dt} = ap - bp^2 \tag{2.63}$$

For the United States, the following parameter values were established in 1845: $a = 0.03$ and $b = 1.6 \times 10^{-4}$, with p being measured in millions of people and t in years. Given that the U.S. population in 1850 was 23 million, we are asked to predict the population in 1930, and compare it with the actual value of 123 million.

Integration of equation (2.63) by separation of variables yields

$$\int_0^{80} dt = \int_{23}^{x} \frac{dp}{p(a - bp)} \tag{2.64a}$$

where the integral on the right is obtained from tables and is given by

$$\int \frac{dp}{p(a - bp)} = -\frac{1}{a} \ln \frac{a - bp}{p} \tag{2.64b}$$

Evaluation of the integrals then yields

$$80 = -\frac{1}{a} \ln \frac{a - bx}{x} \frac{23}{a - 23b} \tag{2.65}$$

Substitution of numerical values for a and b and solving for x leads to the result

$$x = \frac{23 \times 0.03}{(0.03 - 23 \times 1.6 \times 10^{-4}) \exp[-80 \times 0.03] + 23 \times 1.6 \times 10^{-4}} \tag{2.66}$$

from which we finally obtain

$$\boxed{x = 113.6}$$

Thus the model predicts a 1930 U.S. population of 113.6 million, whereas the actual value was 123 million.

COMMENTS

- Differential equations of the type proposed by Malthus arise in a number of disciplines and are referred to as *equations of exponential growth or decay*. This designation follows from the solution of the ODE that is of an exponential form. Thus

$$\frac{dy}{dx} = \pm ky \tag{2.67}$$

yields, upon separating variables and integrating between the limits (1) and (2),

$$\int_{y_1}^{y_2} \frac{dy}{y} = \pm k \int_{x_1}^{x_2} dx \tag{2.68}$$

$$\ln \frac{y_2}{y_1} = \pm k(x_2 - x_1) \tag{2.69}$$

or alternatively

$$\frac{y_2}{y_1} = \exp[\pm k(x_2 - x_1)] \tag{2.70}$$

Hence the term exponential growth (+) or decay (−).

In all the following examples involving exponential growth, the rate of change of a variable is postulated to be proportional to the variable itself.

Population Growth (Malthus)

$$\frac{dp}{dt} = k_1 p \tag{2.71}$$

Bacterial Growth

$$\frac{dN}{dt} = k_2 N \tag{2.72}$$

First-Order Chemical Reaction

$$\frac{dC}{dt} = -k_3 C \tag{2.73}$$

Radioactive Decay

$$\frac{dN'}{dt} = -k_4 N' \tag{2.74}$$

Cooling of a Body (Newton)

$$\frac{d}{dt}\Delta T = -k_5 \Delta T \tag{2.75}$$

A further application of these laws is given in Example 2.13, which is a use of equation (2.75).

◆ EXAMPLE 2.13 Newton's Law of Cooling

We consider here a copper ball 1 cm in diameter that has been heated to a temperature of 100 °C and at time t = 0 is placed in a water bath large enough to ensure that its temperature will remain constant at 30 °C. At the end of 3 min, the temperature of the ball is reduced to 70 °C. We are required to find the time at which the temperature of the ball has dropped to 31 °C (i.e., within 1 °C of the water bath temperature).

Analysis

We start by noting that the model calls for the use of an energy balance around the ball, which we will assume to have a uniform temperature because of its high thermal conductivity and small size. This is the same assumption we used to derive the response and dimension of a thermocouple in Example 1.5. In other words, in both these cases we make use of the concept of a compartmental model, or "stirred tank." This enables us to formulate the following energy balance around the copper ball:

Rate of energy in − Rate of energy out = Rate of change of energy contents

or, using the auxiliary relations from chapter 1, equations (1.18) and (1.21), in Table 1.3,

$$0 - hA\Delta T = \frac{d}{dt} m\Delta H = mC_p \frac{d}{dt}\Delta T \tag{2.76}$$

The expression can be recast in the form

$$-k\Delta T = \frac{d}{dt}\Delta T \tag{2.77}$$

and then becomes identical to Newton's law of cooling, given by equations (1.12) and (2.75).

A further point of note is that the rate constant k is unknown and must be extracted from the given information. Hence the solution will consist of the following two steps:

1. Integrate equation (2.77) between the limits T = 100 to 70 °C, and t = 0 to 3 min. The result enables us to evaluate the unknown cooling constant k.
2. Integrate the same equation again, this time between the limits T = 100 to 31 °C and t = 0 to t, using the cooling constant derived earlier. Solve the result for the required time t.

Model Solution

Step 1. We rewrite equation (2.75) in the form

$$-k(T-30) = \frac{d}{dt}T \tag{2.78}$$

and obtain by separation of variables

$$-\int_{100}^{70} \frac{dT}{T-30} = k\int_0^3 dt \tag{2.79}$$

$$\ln\frac{100-30}{70-30} = 3k \tag{2.80}$$

$$\boxed{k = 0.187\ h^{-1}} \tag{2.81}$$

Step 2. The integration limits are now changed and we obtain

$$-\int_{100}^{31} \frac{dT}{T-30} = 0.187\int_0^t dt \tag{2.82}$$

$$\ln\frac{100-30}{31-30} = 0.187\,t \tag{2.83}$$

from which

$$\boxed{t = \frac{\ln 70}{0.187} = 22.78\,\text{min}} \tag{2.84}$$

Thus, it takes approximately 23 min to cool the sphere within 1° of the bath temperature.

COMMENTS

- The reader will have noted that except for a sign change, the model equation (2.77) is identical in form to that used to describe the thermocouple response, equation (1.39). This is as it should be, since the problems describe similar physical processes that differ only in the direction of energy transfer and the geometry of the system. There is, however, another aspect in which the two differ. In the case of the thermocouple, the quantity required was the diameter of the device that would give a prescribed response. We were, in a sense, dealing with what is often referred to as a *design problem*. In the present example, two different pieces of information were extracted from the solution. In the first step we extracted the value of the cooling constant, for which we drew on *experimental observations*. This process of evaluating unknown coefficients in the model from experimental data is referred to as *parameter estimation*. The second step consisted of determining the time necessary to attain a certain temperature. This is a more common exercise, which may be referred to as *time estimation*. The lesson to be drawn here is that a model can serve a multitude of purposes: any of the quantities appearing in it or in its formal solution, including integration limits, may represent the desired unknown. We may require values for the independent variable or for the dependent variables, for example. Or we may wish to estimate empirical constants for which experimental data will be required. In yet another example, that of systems in which flow occurs, the flow rate itself may be the desired quantity. Flow rates are usually neither a dependent nor an independent variable; as a rules they merely lurk in the model as adjustable parameters. They reappear in the solution and can then be extracted from it.

What we have said for differential equations applies equally well to algebraic equations, of which we have seen a number of examples in the preceding sections on vector algebra and matrices. Again, any of the quantities that appear in these equations may serve as the unknown being sought, be it the magnitude of a force, the distance between two objects, or an angle. The art of modeling, of course, consists of choosing the appropriate equations that will contain the desired unknowns and allow their correct evaluation. This is something we will practice in this book.

We now turn to the task of examining the properties and the principal solution methods of ordinary differential equations. Wherever possible, we want to establish a relation between the mathematical concepts presented and the physical problems to which they may be applied. This will enable the reader who confronts a problem involving ODEs to *anticipate* the mathematics to be dealt with, the degree of difficulty to be expected, and the solution methods to be applied.

2.3.1 Order of an ODE

The order of a differential equation is equal to the order of its highest derivative. This applies to ordinary as well as partial differential equations. Thus the exponential growth and decay equations presented earlier, equations (2.71) to (2.75), are all first-order ODEs. As another example, consider Newton's law in one dimension, which can be written in algebraic form as

$$\sum F = ma \tag{2.85}$$

Upon replacing acceleration by its equivalent velocity derivative, however, this equation becomes

$$\sum F = m\frac{dv}{dt} \tag{2.86}$$

which is a first-order ODE. If we further replace velocity by the distance derivative, it becomes a second-order ODE, that is,

$$\sum F = m\frac{d^2x}{dt^2} \tag{2.87}$$

Note that distance x is a dependent variable here. This is relatively rare. A more common occurrence is for distance to be the independent variable. This happens in the "one-dimensional pipe" model set out in Chapter 1, or whenever a state variable changes with distance. We saw an example of this in Example 1.75, which dealt with the release of a substance from a tubular wall into a flowing fluid. The model which resulted was given by

$$Q\frac{dC}{dz} = kP(C^* - C)$$

which is again a first-order ODE.

Imagine now that we superpose a diffusional flux onto the convective flow that is taking place in that configuration. Diffusion does in fact occur, and it takes place in the opposite direction to that of the flow, that is, in the direction of diminishing concentration. Then by using equation (1.10) and making a diffusional mass balance over the increment Δz, we obtain

$$\text{Rate in by diffusion} - \text{Rate out by diffusion}$$

$$\left(-DA\left.\frac{dC}{dz}\right|_{z+\Delta z}\right) - \left(-DA\left.\frac{dC}{dz}\right|_{z}\right)$$

where A is the cross-sectional area of the tube.

Upon dividing this expression by Δz and letting $\Delta z \to 0$, we obtain

$$-DA\frac{d^2C}{dz^2}$$

This term has to be added to equation (1.49), which now becomes

$$-DA\frac{d^2C}{dz^2} + Q\frac{dC}{dz} = kP(C^* - C) \tag{2.88}$$

That is, the first-order equation (1.49) has been transformed into a second-order ODE.

We are now in a position to generalize these results into the following statements:

- All unsteady integral balances lead to first-order ODEs in time, as we saw in Examples 1.5 (Design of a Thermocouple) and 2.13 (Newton's Law of Cooling).
- All direct applications of the equation of exponential growth or decay, equations (2.71) to (2.75), likewise lead to first-order equations in time. An extended example of this category is the logistic equation (2.63).
- Steady-state differential balances in one direction and *involving no diffusive transport* always yield first-order ODEs in distance. See Example 1.7 (Release of a Substance into a Flowing Fluid).
- When diffusive transport involving Fick's law, Fourier's law, or Newton's viscosity law does occur, and the balances are taken over a distance increment, the result is a second-order equation in distance; see equation (2.88). *In general, whenever the auxiliary relations used in the model already contain a nth-order derivative, incorporation into a differential balance will increase the order of the equation by n.*
- Newton's law is a first-order ODE when velocity is the dependent variable or a second-order ODE when distance is the dependent variable.

- Higher order ODEs result whenever several lower order equations can be combined into a single one. Consider as an example the set

$$\frac{dy}{dx} + K_1(y - z) = 0 \tag{2.89a}$$

$$\frac{dz}{dx} + K_2(y - z) = 0 \tag{2.89b}$$

Here we can solve either equation for one of the dependent variables and substitute the result into the second equation. Thus if we solve equation (2.89a) for z, we obtain

$$z = \frac{1}{K_1}\frac{dy}{dx} + y \tag{2.90}$$

and substitution into equation (2.89b) then yields

$$\frac{d^2y}{dx^2} + K_1(1 - K_2)\frac{dy}{dx} = 0 \tag{2.91}$$

We have thus managed to simplify the system represented by equations (2.89a and b) to a single equation, but the price to be paid was an increase in the order of the ODEs. We note in this connection that whenever analytical solutions are being sought, combining lower order equations in this fashion is a fruitful approach. In *numerical* work, on the other hand, the reverse procedure is often preferred; that is one decomposes higher order ODEs into equivalent sets of first-order ODEs. This is done to take advantage of standard ODE solver packages, which are specifically designed to solve sets of first-order ODEs. We shall have occasion to use these concepts in Example 5.13, dealing with the dynamics of HIV infections, and in Example 8.9, dealing with the effect of drag on a projectile.

- ODEs of order higher than 2, although less common, also arise in certain applications of solid mechanics, examples of which are taken up in Chapter 4. In the vast majority of cases, however—upward of 90% of all applications that result in ODEs—the equations are either first or second order. This eases our task considerably.

We now turn to an examination of additional properties of ODEs.

2.3.2 Linear and Nonlinear ODEs

The classification of ODEs as linear and nonlinear is of great importance in determining the method of solution, and indeed the ease of solution of these equations. As a general rule, linear ODEs are easier to solve by analytical means than nonlinear systems. The latter usually requires numerical methods. This is no different from what one sees in algebra, where matrix methods can be applied routinely to solve linear systems (see Section 2.2), but no convenient analytical methods exist for solving sets of nonlinear equations.

An *informal* definition of linear ODEs is that all *dependent variables and their derivatives* must appear in linear form; that is, they are not multiplied or divided by each other, nor are they raised to a power other than one. Note that this definition does not require the *independent* variable

Table 2.6 Example of Linear and Nonlinear ODE classification

Linear ODEs	
$\frac{d^3y}{dx^3} + K_1\frac{d^2y}{dx^2} + K_2\frac{dy}{dx} + K_3y = 0$	Homogeneous ODE with constant coefficients
$\frac{d^2y}{dx_2} + e^x\frac{dy}{dx} + y = 0$	Homogeneous ODE with variable coefficients
$\frac{dy}{dx} + K_1y = f(x)$	Nonhomogeneous ODE with constant coefficients
$\frac{d^2y}{dr^2} + \frac{1}{r}\frac{dy}{dr} + Ky = f(r)$	Nonhomogeneous ODE with variable coefficients
$\frac{dy}{dx} = y + z$ $\frac{dz}{dx} = y - z$	Set of homogeneous ODEs with constant coefficients
$\frac{dy}{dx} = y + z + f(x)$ $\frac{dz}{dx} = y - z + g(x)$	Set of nonhomogeneous ODEs with constant coefficients

Nonlinear ODEs	Source of Nonlinearity
$y\frac{dy}{dz} + x = 0$	$y\frac{dy}{dx}$
$\frac{d^2y}{dx^2} + \frac{dy}{dx} + y^2 = 0$	y^2
$\frac{d^2y}{dx^2} + \left(\frac{dy}{dx}\right)^2 + y = 0$	$\left(\frac{dy}{dx}\right)^2$
$\frac{dy}{dx} + y/z = 0$ $\frac{dz}{dx} - yz = 0$	y/z, yz

to be linear. In fact, the latter can be arbitrarily complex without violating linearity. When the independent variable or some functional form of it appears as a coefficient of the derivatives of a *linear* ODE, one speaks of a *linear differential equation with variable coefficients*. When it appears as an *isolated term* (i.e., not in conjunction with the dependent variable or its derivative), the ODE is referred to as being *nonhomogeneous*. Otherwise it is termed homogeneous.

The examples in Table 2.6 serve to introduce the reader to these classifications.

For linear ODEs, or sets of linear ODEs, we have two powerful methods available which will be taken up in Section 2.3.5: the D-operator or matrix eigenvalue methods, and the Laplace transformation. Both these analytical procedures handle linear systems with ease and can often be applied painlessly and in nearly mechanical fashion. Nonlinear systems pose greater difficulties, just as they do in algebraic systems. If one is fortunate and the equation is separable, the method of separation of variables can be applied with fruitful results. We had seen an inst of this in Example 2.12, which involved the logistic equation. A number of ad hoc procedures also exist that can be applied to specialized cases. These are briefly taken up in Section 2.3.5 and tabulated there, along with other analytical solution methods.

2.3.3 Boundary and Initial Conditions

Boundary and initial conditions (BCs and ICs), usually expressed as equations, are needed to evaluate integration constants, or integration limits, and to provide starting values for numerical integration procedures. The number of such conditions required equals the order of the ODE.

For *first-order ODEs* in time or distance, these conditions are usually (but not always) identified from the prevailing values of the state variable at the start, that is, at time $t = 0$, or at the position $z = 0$ that coincides with the origin of the coordinate system one chooses. The case $t = 0$ results in an *initial condition*. The case $z = 0$ is usually identified as a *boundary condition* but is occasionally referred to as an initial condition as well. Some typical examples of the information used to obtain these conditions are the following.

- Initial velocity and position of a rising or falling object. This information can be used in the integration of Newton's law.
- The initial level, concentration, or temperature in a "stirred tank" or compartment. In Example 2.13, Newton's Law of Cooling, for example, the IC was given by the initial temperature of 100 °C of the copper sphere, which was our compartment there, and used as an integration limit in equations (2.79) and (2.82).
- The initial dollar values of a financial system. Here we refer the reader to Practice Problem 1.4, which requires the evaluation of the time required for a person with a certain income and expenses to accumulate a prescribed amount of cash. Since the process starts with no cash on hand, the initial condition is clearly given by zero dollars at time zero and gives us a lower integration limit.
- Concentration, temperature or pressure at the inlet of a "one-dimensional pipe" (i.e., at $z = 0$). An example of this condition was seen in Example 1.7, Release of a Substance into a Flowing Fluid. The tacit assumption there was that the fluid enters the pipe with no prior content in the soluble materials, so that at the inlet $z = 0$, the concentration $C = 0$.
- Initial number of entities in systems where these entities vary with time. This was the case in Example 2.12, where an initial value of the U.S. population in 1850 was given at 23 million. This information provided us again with a lower integration limit, which was used in equation (2.64).

For second-order ODEs, one requires two boundary conditions. For these and higher order equations, one distinguishes between two situations. In the first case, *all* boundary or initial conditions are given at the same point in time or space. One then speaks of an initial value problem, or IVP. Applications of Newton's law in its second-order form [i.e., with distance as the dependent variable (equation (2.87)] lead to IVPs, since one is usually able to specify the system's initial velocity and position. Yet another category of initial value problems arises in "one-dimensional pipe" problems in which all the dependent variables are specified at the inlet.

The second category of ODEs involves systems in which at least two of the boundary conditions are given at *different* locations in space or time. One then speaks of boundary value problems, or BVPs. Such problems typically arise in *cylindrical* or *spherical* geometries, with one boundary condition specified at the surface and the second one at some other location,

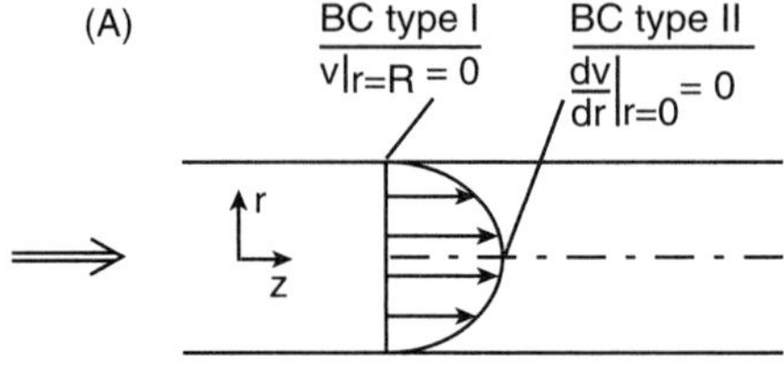

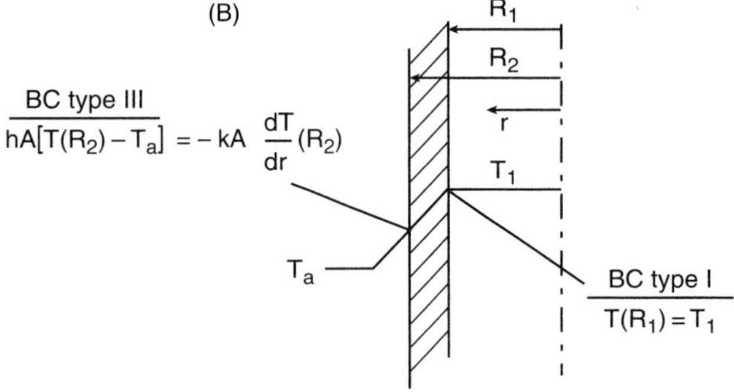

Figure 2.6 Boundary conditions in two physical systems: (A) laminar flow in a pipe and (B) conduction through a pipe wall.

usually the central axis or the center of the sphere. Two such physical systems, which lead to boundary value problems, are shown in Fig. 2.6.

Figure 2.6A displays the parabolic velocity profile that arises in laminar flow through a cylindrical pipe. The boundary conditions here are given at the axis, where the velocity gradient is zero due to symmetry, and at the wall, where the velocity itself vanishes.

Figure 2.6B shows the conditions associated with heat conduction through a cylindrical pipe. Here the condition at the inner wall is one of constant temperature $T(R_1) = T_1$, a situation that arises, for example, when steam is condensing in the pipe. Heat is conducted through the wall to the outer surface, where it is dissipated to the atmosphere, which is assumed to be at a constant temperature T_a. The boundary condition here expresses the requirement that the heat reaching the surface by conduction must equal the heat lost to the atmosphere; that is, we must have

$$\underbrace{hA[T(R_2) - T_a]}_{\text{Rate of heat loss to the atmosphere}} = \underbrace{-kA\frac{dT}{dr}(R_2)}_{\text{Rate of heat arrival by conduction}} \tag{2.92}$$

The reader will note that the boundary conditions differ in form depending on the physical situation they describe. Some contain the dependent variable only, others the derivative only, and a third category, represented by equation (2.90), contains both the derivative and the dependent variables. These three forms are referred to, respectively, as type I, II, and III boundary conditions. We have summarized their properties in Table 2.7. Applications of different boundary conditions to the same problem appear later in Example 4.2 and in Practice Problem 4.4, dealing with the buckling of a strut.

Let us now consider some illustrative examples of the ODE properties just described.

Table 2.7 Types of Boundary Conditions

Type	Property	Occurrence
I	Contains y only	Specification of y at a particular location
II	Contains dy/dz only	Symmetrical profile
		Constant or zero diffusional flux
III	Contains both dy/dx and y	Diffusional flux with film resistance

◆ EXAMPLE 2.14 Classification of ODEs and Boundary Conditions

We examine here a number of ODEs that arise in different physical situations and classify them with respect to order, linearity, homogeneity, and type of coefficient.

Type I

$$\underset{\text{Conduction}}{k\left[\frac{d^2T}{dr^2}+\frac{1}{r}\frac{dT}{dr}\right]} - \underset{\text{Convection}}{\frac{h}{L}(T-T_a)} - \underset{\text{Radiation}}{\frac{\varepsilon\sigma}{L}\left(T^4-T_a^4\right)} = 0 \tag{2.93}$$

This equation describes heat conduction in the radial direction, with an attendant heat loss to the atmosphere by convection and radiation. It arises, for example, in the description of heat losses from a furnace wall with metallic inserts.

- The equation is second order by virtue of its highest derivative, d^2T/dr^2.
- It is nonlinear by virtue of the term T^4.
- It is nonhomogeneous because of the terms in T_a.
- A classification with respect to variable coefficients does not arise because the equation is nonlinear.

Type II

$$\frac{d^2x}{dt^2}+\frac{k}{m}x = g \tag{2.94}$$

This equation describes a mass suspended from a spring vibrating under its own weight. It is a second-order linear ODE with constant coefficients. It is also nonhomogeneous because of the term g.

Two boundary conditions are required, which are typically given by

Initial velocity: $dx/dt = v_0$

Initial position: $x(0) = x_0$

Both these conditions are of type I, and the problem is an initial value problem.

Type III

$$k^2x^2\left(\frac{d^2y}{dx^2}\right)^2 = 1 + \left(\frac{dy}{dx}\right) \tag{2.95}$$

This equation is known as the path of pursuit, or simply the pursuit equation, and it is taken up in Chapter 3 (Example 9). It describes the optimum path for a hunter to follow to intercept a prey moving at a constant velocity. The term $(d^2y/dx^2)^2$ in equation (2.95) makes this a second-order nonlinear ODE that is also nonhomogeneous by virtue of the constant term 1.

Although the coefficient k^2x^2 is a variable one, no classification in this regard is called for because the equation is nonlinear.

Type IV

$$D\left[\frac{d^2C}{dr^2} + \frac{2}{r}\frac{dC}{dr}\right] + k_rC = 0 \tag{2.96}$$

Diffusion Reaction

BC1

$$\frac{dC}{dr}(0) = 0$$

BC2

$$kA[C_b - C(R)] = -DA\frac{dC}{dr}(R) \tag{2.97}$$

We have here a linear second-order ODE that is also homogeneous and contains a variable coefficient 2/r. Among the physical situations it describes is that of a spherical catalyst pellet in which a first-order reaction is taking place. The first boundary condition is of type II and accounts for the symmetricality of the reactant concentration profile about the center of the sphere. The second boundary condition equates the external rate of reactant supply to the surface with the diffusive transport into the interior of the sphere. In this it resembles the corresponding heat transfer boundary condition, equation (2.92), and, like it, is of type III.

Type V

$$\begin{aligned} -k_1C_AC_B &= \frac{dC_A}{dt} \\ k_1C_AC_B - k_2C_RC_B &= \frac{dC_B}{dt} \end{aligned} \tag{2.98}$$

The situation described by this set is that of two parallel reactions

$$A + B \xrightarrow{k_1} R$$

$$R + B \xrightarrow{k_2} S$$

which are both second order in the reactants. The set is one of two homogeneous first-order ODEs, both of which are nonlinear by virtue of the product terms C_AC_B and C_RC_B. Note that since the system contains three dependent variables C_A, C_B, and C_R it is underspecified. A third equation, which can be algebraic or an ODE, is therefore required.

2.3.4 *Equivalent Systems*

In this section we introduce the reader to the notion that physical systems that are entirely different can nevertheless result in models of identical form. This may be because the underlying physical laws are identical *in form*, as is the case with Fick's law and Fourier's law, but can also arise when no such correspondence exists. We term such systems *equivalent systems*. A particular case is given in Example 2.15.

◆ EXAMPLE 2.15 Equivalence of Vibrating Mechanical Systems and an Electrical RLC Circuit

The intent here is to demonstrate the equivalence of the equations that describe mechanical vibrations and the oscillations that arise in so-called RLC electrical circuits composed of a resistor, a capacitor, and an inductor. The two systems are displayed in Fig. 2.7.

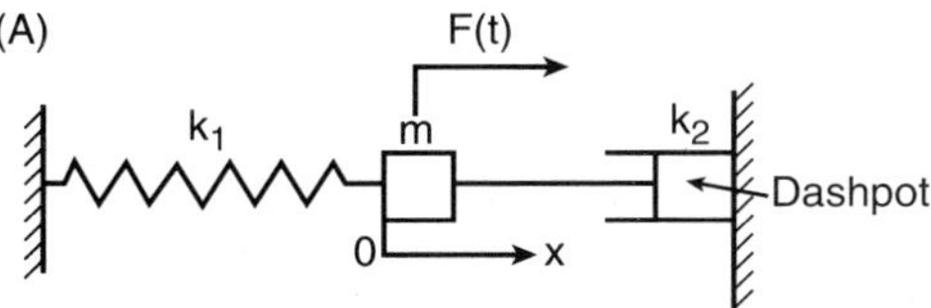

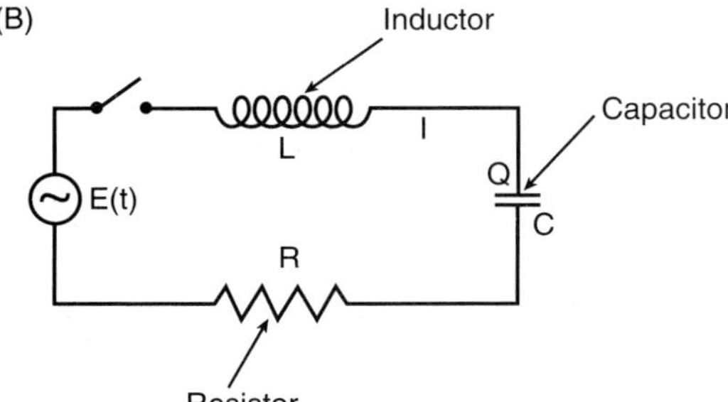

Figure 2.7 Two different physical systems that give rise to identical forms of second-order ODE: (A) a vibrative mass and (B) an oscillating electrical circuit.

The mechanical system (Fig. 2.7A) is composed of a mass m attached to a spring (constant k_1) and connected to a viscous damper termed a dashpot, which consists of a cylinder filled with viscous oil. The mass is acted on by an arbitrary time-dependent force F(t), which is oriented in the positive x direction. Acting in the negative direction are the restoring force of the spring $F_s = -k_1 x$ and the damping effect of the dashpot, which we assume to be proportional to the velocity of the mass. Thus $F_d = -k_2(dx/dt)$.

Applying Newton's law to the system, we obtain

$$F(t) - F_s - F_d = ma \tag{2.99a}$$

$$F(t) - k_1 x - k_2 \frac{dx}{dt} = m \frac{dx^2}{dt^2} \tag{2.99b}$$

or

$$\underset{\text{Acceleration}}{m\frac{d^2x}{dt^2}} + \underset{\text{Damper}}{k_2\frac{dx}{dt}} + \underset{\text{Spring}}{k_2 x} = \underset{\text{Forcing function}}{F(t)} \tag{2.99c}$$

We note that this is a linear, nonhomogeneous second-order ODE with constant coefficients.

The electrical circuit is made up of a coil with inductance L, a resistor with resistance R, a capacitor with capacitance C, and a time-varying voltage supply E(t). From elementary electricity theory it is known that for the first three items, the voltage drop is given by

Inductor

$$V_I = \frac{Ldi}{dt} \tag{2.100a}$$

Resistor

$$V_R = iR \tag{2.100b}$$

Capacitor

$$V_C = \frac{q}{C} \tag{2.100c}$$

According to Kirchhoff's second rule, equation (1.8), the three voltage drops must equal the imposed voltage, so that

$$L\frac{di}{dt} + \frac{q}{C} + iR = E(t) \tag{2.101}$$

Since current equals the rate of flow of charge at any time, we have $i = dq/dt$. Introducing this relation into equation (2.101), we obtain

$$\underset{\text{Inductor}}{L\frac{d^2q}{dt^2}} + \underset{\text{Resistor}}{R\frac{dq}{dt}} + \underset{\text{Capacitor}}{\frac{q}{C}} = \underset{\text{Imposed voltage}}{E(t)} \tag{2.102}$$

This equation is seen to be identical in form to that describing the mechanical system, equation (2.99c), and is again a linear, nonhomogeneous, second-order ODE with constant coefficients. We can expect the corresponding solutions to be likewise identical in form, differing only in the nature of the dependent variables and coefficients.

This ends our discussion of properties of ODEs. We next turn to an examination of analytical solution methods.

2.3.5 Analytical Solution Methods

We start by providing the reader with a summary, shown in Table 2.8, of the more important analytical solution methods. A distinction is made between "major methods", which are encountered with the greatest frequency in various disciplines, and "other methods" of less frequent occurrence. Among the major methods, the method of separation of variables (item 1), the D-operator or matrix eigenvalue method (item 2), and the Laplace transformation (item 5) stand out, and these will be examined in some detail. Among "other methods", we shall have occasion to use item 6, the general solution to linear first-order nonhomogeneous ODEs with variable coefficients, in which both the nonhomogeneous term and the variable coefficient can assume an arbitrary functional form (see Example 6.10). Mention should also be made of the so-called p-substitution method (items 9 and 10), which serves to solve second-order ODEs with missing terms. This method must be used to solve Practice Problem 3.8. With the exception of the Laplace transformation, these methods are limited to the solution of *single* ODEs. This still makes them highly useful tools, as we shall see from the numerous problems we are able to solve by these methods. For *sets* of ODEs, we are able to use the Laplace transformation, which applies to linear ODEs only, and the occasional ad hoc methods, one of which is given by equations (2.89) to (2.91). With one or two exceptions, sets of nonlinear ODEs tend to require the application of numerical methods, which are taken up in Chapter 8.

Solution by Separation of Variables Separation of variables is a powerful and sometimes underrated method that is capable of solving both linear and nonlinear ODEs. It is the only method used so far in this text to solve a range of different problems, which attests to its power and versatility. Further applications of the method appear in Examples 5.9, 5.10, and 6.4. The *repeated* use of separation of variables is seen in Examples 5.5 and 6.5.

The following points should be noted in connection with this method.

- Separation of variables is the preferred method of solving single ODEs and should always be tried first.
- Single second-order ODEs may be amenable to solution by separation of variables, provided one of the boundary conditions is given in terms of the first derivative (see item 1 of Table 2.7).
- ODEs that at first glance appear to defy separation of the variables may on closer inspection and by proper manipulation be reduced to separable form.
- We illustrate these points with Example 2.16.

Table 2.8 Analytical Solutions of ODEs

System	Solution
Major Methods	
1. Separable ODEs $y' = \frac{f(x)}{g(y)}$ $y'' = \frac{f(x)}{g(y')}$	By separation of variables $\int g(y)\,dy = \int f(x)\,dx + C$ $\int g(y')\,dy' = \int f(x)\,dx + C$
2. Linear homogeneous second-order ODEs with constant coefficients $y'' + ay' + by = 0$	By D-operator method or matrix eigenvalues $y = C_1 \exp(D_1 x) + C_2 \exp(D_2 x)$
3. Linear nonhomogeneous second-order ODEs with constant coefficients $ay'' + by' + cy = f(x)$	Solution of homogeneous ODE + particular integral y_P: see Table 2.9 for listing of y_P
4. Linear homogeneous second-order ODEs with variable coefficients $a(x)y'' + b(x)y' + c(x)y = 0$	By power series $y = C_1 \sum_1^{\infty} a_n x^{n+k_1} + C_2 \sum_1^{\infty} b_n x^{n+k_2}$
5. Nonhomogeneous initial value ODEs with constant coefficients	By Laplace transformation
Other Methods	
6. Linear first-order nonhomogeneous ODEs with variable coefficients $y' + f(x)y = g(x)$	Directly given by formula $y = \exp \int -f(x)\,dx \left[\int g(x) \exp \int f(x)\,dx\,dx + C\right]$
7. General first-order ODEs. $\frac{y' + f(x, y)}{g(x, y)} = 0$	Various solutions arise depending on the method used:
(a) By substitution $y = v\,x$	(a) If substitution yields a separable ODE in v, x only, the solution is $\ln x = C - \int \frac{g(1, v)\,dv}{f(1, v) + vg(1, v)}$
(b) ODE is exact, that is, $\frac{\partial f}{\partial y} = \frac{\partial g}{\partial x}$	(b) Directly given by formula $\int f(x, c)\,dx + \int g(c, y) \frac{\partial}{\partial y} \int f(x, c)\,dx\,dy = C$
8. Nonlinear second-order ODE with first derivative and terms in x missing $y'' = f(y)$	Multiply equation by $2(dy/dx)\,dx = 2\,dy$ to obtain $y' = \left[2 \int f(y)\,dy + C\right]^{1/2}$ and then apply separation of variables
9. Nonlinear second-order ODE with missing dependent variable $f(y'', y', x) = 0$	Reduce to first-order ODE by "p-substitution" $p = dy/dx$ and attempt integration by one of preceding methods, followed by a second integration
10. Nonlinear second-order ODE with missing independent variable $f(y'', y', y) = 0$	As for 9; note that second-order derivative becomes $\frac{d^2y}{dx^2} = \frac{dp}{dx} = \frac{dp}{dy}\frac{dy}{dx} = p\frac{dp}{dy}$

◆ EXAMPLE 2.16 Solution of Nonlinear ODEs by Separation of Variables

We attempt here to solve the following nonlinear first-order ODEs by separation of variables:

$$x \sin y \frac{dy}{dx} + \frac{ye^x}{1 - 0.5y} = e^x(y^3 + y) \tag{2.103}$$

$$\frac{dy}{dx} = ay^2 + by + c \tag{2.104}$$

Equation (2.103) is highly nonlinear, not because of the exponential terms in the independent variable x, but by virtue of the y^3 and $y(1 - 0.5\,y)$ terms, as well as the expression $\sin y(dy/dx)$. The equation does not at first sight appear to be separable. The situation is improved, however, by factoring out e^x and collecting terms. We obtain

$$e^x\left(y^3 + y - \frac{y}{1 - 0.5y}\right) = x \sin y \frac{dy}{dx} \tag{2.105}$$

which is clearly separable. Formal integration leads to the result

$$\int \frac{e^x}{x}\,dx = \int \frac{\sin y\,dy}{y^3 + y - y/(1 - 0.5y)} \tag{2.106}$$

Analytical evaluation of the left side is possible, while the right side requires numerical integration. The question then arises, whether numerical methods are not better applied at the source (i.e., at the ODE level). We prefer to answer this question in the negative. The integrated form [equation (2.106)], although not fully evaluated, may nevertheless provide a better picture of solution behavior. In particular, it reveals that the integral on the right diverges (i.e., becomes infinity) when $y = 1$. Although numerical integration of the ODE would yield a similar result, it would not be clear whether this was due to numerical instability, nor at precisely what value of y the divergence occurred. For this and other reasons, we prefer to push an analytical solution to its limits rather than make immediate use of numerical methods.

Equation (2.104) is a reduced version of the Riccati equation, which has the form

$$y^1 + f(x)y + g(x)y^2 = h(x) \tag{2.107}$$

and is nonlinear by virtue of the term y^2. Solutions to various special cases of this expression exist but involve fairly lengthy and cumbersome procedures. In most instances numerical solutions are resorted to. We can avoid both these avenues by recognizing that the equation is separable and that it leads in the first instance to the expression

$$\int_{y_0}^{y} \frac{dy}{ay^2 + by + c} = \int_{x_0}^{x} dx \tag{2.108}$$

To aid in the evaluation of the integral on the left, we write it in the factored form

$$\int \frac{dy}{(y - y_1)(y - y_2)} = \int dx + c \tag{2.109}$$

where y_1 and y_2 are the roots of the denominator in equation (2.108), both of which are real for $b^2 > 4ac$. We can then make use of the following integration formula available in mathematical tables

$$\int \frac{dx}{(a+bx)(a'+b'x)} = \frac{1}{ab'-a'b}\ln\frac{a'+b'x}{a+bx} \tag{2.110}$$

where for our case, $b = b' = 1$, $a = -y_1$, and $a' = -y_2$. This leads to the integrated expression

$$x - x_0 = \frac{1}{y_2 - y_1}\ln\frac{y_2 - y}{y_1 - y}\frac{y_1 - y_0}{y_2 - y_0} \tag{2.111a}$$

or alternatively

$$y = \frac{y_2 - y_1\dfrac{y_2 - y_0}{y_1 - y_0}\exp[-(y_1 - y_2)(x - x_0)]}{1 - \dfrac{y_2 - y_0}{y_1 - y_0}\exp[-(y_1 - y_2)(x - x_0)]} \tag{2.111b}$$

This simple application of the method of separation of variables stands in stark contrast to the complexities of a formal integration of the Riccati equation. We note that for $b^2 < 4ac$, the solution ceases to be real. We are thus able not only to establish a simple exponential relation between the variables but to delineate also conditions for the existence of the solution. One cannot ask much more of a solution method.

The D-Operator and Eigenvalue Methods, Particular Integrals The D-operator and eigenvalue methods are used to solve linear ODEs with constant coefficients. When that equation is nonhomogeneous as well, a so-called particular integral must be added to the solution of the homogeneous problem.

Let us start with a simple example as an introduction to the concepts of characteristic roots or eigenvalues of a linear ODE, and to the important superposition principle. Consider the equation

$$\frac{d^2y}{dx^2} - y = 0 \tag{2.112}$$

A relative novice to the field might attempt a solution to the problem by substituting trial functions into the ODE and seeing whether the ODE is satisfied. It might further be argued that since the equation is a second-order one, two integration constants will have to be evaluated from two given boundary conditions. These integration constants must be associated with two *independent* functions, for if they were not, the two constants would coalesce into a single one and we would be unable to satisfy the two boundary conditions.

Let us attempt a solution with some simple trial functions. Neither sin x nor cos x satisfies equation (2.112). However, both e^x and e^{-x} do. Furthermore, these are two independent functions. One might therefore formulate a general solution of the form

$$y = C_1e^x + C_2e^{-x} \tag{2.113}$$

This sum also satisfies the ODE. Thus we have, somewhat inadvertently, discovered the *superposition principle*, at least as it applies to this example. That principle in essence states that the general solution to an nth-order *linear* ODE is composed of the *sum* of n independent functions. Uniqueness of the solution is guaranteed by uniqueness theorems described in most texts dealing with ODEs.

If one were to conduct extensive trials of this type, one would discover that the solution of any linear homogeneous ODE with constant coefficients, of whatever order, is always composed of the sum of exponential functions with either real or imaginary arguments. Early workers in the field were well aware of this fact. They had further noted that a precise connection existed between the *constant coefficients* of the ODE and the *arguments* of the exponential functions. This led to a formalism known as the D-operator method. In it, the operational part of the derivative, d^n/dx^n, is replaced by the operator symbol D^n, and that symbol is treated as an *algebraic* entity, subject to the rules of algebra. Equation (2.113) can then be written in the form

$$(D^2 - 1)y = 0 \tag{2.114}$$

Since $y \neq 0$, we must, by the rules of algebra, have

$$D^2 - 1 = 0 \tag{2.115}$$

with the solutions

$$D_1 = 1, \qquad D_2 = -1 \tag{2.116}$$

Equation (2.115) is termed the *characteristic equation* of the ODE equation (2.114), and its solution its *characteristic roots*. These roots, which are functions of the constant coefficients of the ODE, are also *identical* to the coefficients of the arguments of the exponential functions that appear in equation (2.113). When the coefficients are altered, a corresponding change occurs in the values of the characteristic roots and, by extension in the values of the exponential arguments. Thus, a direct relation between the two is established.

Table 2.9 lists a compilation of characteristic roots and the corresponding solutions for the most frequently encountered case of a second-order ODE. For real and distinct roots, the solution

Table 2.9 Solutions of the Second-Order ODE $ay'' + by' + cy = 0$

Characteristic Roots or Eigenvalues		Solution[a,b]
1. Distinct and Real: $D_{1,2}$		$y = C_1e^{D_1x} + C_2e^{C_2x}$
	or	$y = C_1' \sinh D_1x + C_2' \cosh D_2x$
2. Identical and Real: $D_1 = D_2 = D$		$y = C_1e^{Dx} + C_2xe^{Dx}$
3. Imaginary: $C_{1,2} = \pm bi$		$y = C_1e^{bix} + C_2e^{-bix}$
	or	$y = C_1' \cos bx + C_2' \sin bx$
4. Complex conjugate: $D_{1,2} = a \pm bi$		$y = C_1e^{(a+bi)x} + C_2e^{(a-bi)x}$
	or	$y = (C_1' \cos bx + C_2' \sin bx)e^{ax}$

[a] $\sinh x = \frac{1}{2}(e^x - e^{-x})$, $\cosh x = \frac{1}{2}(e^x + e^{-x})$. [b] *Euler's formula:* $e^{ix} = \cos x + i \sin x$.

is the sum of the corresponding exponential functions, which can also be expressed in terms of equivalent hyperbolic functions. When the roots are identical, one of the exponential functions is premultiplied by the independent variable. Exponential functions with imaginary arguments that result from complex conjugate characteristic roots are converted to trigonometric functions with real arguments using the Euler formula given in Table 2.9. Applications of the D-operator method are seen in Examples 4.2, 5.13, 6.6 to 6.8, and 6.11. In all these cases, the underlying model consists of a second-order linear ODE with constant coefficients.

These characteristic roots can also be obtained by matrix methods. To accomplish this, we decompose the nth-order equation into an equivalent set of first-order ODEs and evaluate the so-called eigenvalues λ of the matrix formed by the coefficients of the algebraic terms. To apply this procedure to equation (2.112), we set $dy/dx = z$, a new dependent variable that furnishes one of the two ODEs resulting from the decomposition. We obtain the equivalent set

$$\begin{aligned} \frac{dy}{dx} &= 0 + z \\ \frac{dz}{dx} &= y + 0 \end{aligned} \tag{2.117}$$

for which the coefficient matrix reads

$$\mathbf{A} = \begin{bmatrix} 0 & 1 \\ 1 & 0 \end{bmatrix} \tag{2.118}$$

The eigenvalues, denoted by λ, are obtained from the relation

$$\det[\mathbf{A} - \lambda\mathbf{I}] = 0 \tag{2.119}$$

where $\mathbf{I}$ is the identity matrix we displayed earlier (Table 2.2, item 6). Using the rules given in Section 2.2 for the addition of matrices and multiplication of a matrix by a scalar, we can expand the determinant of equation (2.119) into the form

$$\begin{vmatrix} 0 - \lambda & 1 \\ 1 & 0 - \lambda \end{vmatrix} = 0 \tag{2.120}$$

which yields the quadratic equation

$$\lambda^2 - 1 = 0 \tag{2.121}$$

with roots $\lambda_{1,2} = \pm 1$. The eigenvalues of the coefficient matrix of the set (2.117) are thus seen to be identical to the characteristic rots obtained by the D-operator method, equation (2.116).

We next examine the case of *nonhomogeneous* ODEs with constant coefficients. Consider as an example a second-order equation of the form

$$a\frac{d^2y}{dx^2} + b\frac{dy}{dx} + cy = f(x) \tag{2.122}$$

Table 2.10 Particular Integrals y_p of the Second-Order ODE $ay'' + by' + cy = f(x)$

f(x)	Form of y_p	Coefficients of y_p
1. a_0	K	$K = \frac{a_0}{c}$
2. $a_0 + a_1x + \cdots + a_nx^n$	$A_0 + A_1x + \cdots + A_nx_n$	Determined by substituting into ODE and equating coefficients
3. a_0e^{rx}	A_0e^{rx}	$A_0 = \frac{a_0}{ar^2 + br + c}$
4. $a_0 \sin nx$	$A_0 \sin nx$	$A_0 = \frac{(c - n^2a)a_0}{(c - n^2a)^2 + n^2b^2}$
5. $a_0 \cos nx$	$A_0 \cos nx$	$A_0 = \frac{(c - n^2a)a_0}{(c - n^2a)^2 + n^2b^2}$
6. $a_0 \sin nx + b_0 \cos nx$	$A_0 \sin nx + B_0 \cos nx$	$A_0 = \frac{(c - n^2a)a_0 + nbb_0}{(c - n^2a)^2 + n^2b^2}$ $B_0 = \frac{(c - n^2a)b_0 - nba_0}{(c - n^2a)^2 + n^2b^2}$

where a, b, and c are constants, and f(x) represents the nonhomogeneous term, often called a *forcing function*. It can be shown by means of the superposition principle that the solution will in this case be made up of the sum of the solution of the *homogeneous* form of equation (2.122), termed the *complementary function*, plus a *particular integral* y_p, which has the same functional form as the nonhomogeneous term f(x).

Thus

$$\text{General solution} = \text{Complementary function} + \text{Particular integral}$$

$$y = y_c + y_p \tag{2.123}$$

The particular integral can be evaluated by the so-called method of undetermined coefficients. This method consists of substituting the known form of y_p, which is identical to that of f(x) but contains some unknown coefficients, into the ODE (2.122) and setting the coefficients of a particular function equal to zero. This results in several algebraic equations in the unknown coefficients, which are subsequently solved for. To avoid this somewhat cumbersome procedure, we have provided in Table 2.10 a listing of the most frequently required particular integrals applicable to second-order ODEs with nonhomogeneous terms (equation [2.122]). We demonstrate their use with the following example.

◆ EXAMPLE 2.17 Mass on a Spring Subjected to a Sinusoidal Forcing Function

We return for this example to the mass–spring system we examined to show its equivalence to an electrical RLC circuit (Fig. 2.7). The operating equation describing its motion, x = f(t), was given earlier by

$$m\frac{d^2x}{dt^2} + k_1\frac{dx}{dt} + k_2x = F(t) \tag{2.99c}$$

where the second and third term represent the damping action of the dashpot and the restoring force of the spring, respectively. We consider here a reduced version of this expression, with the first-order term omitted (no damping), and the general nonhomogeneous term replaced by a specific sinusoidal forcing function. We obtain

$$m\frac{d^2x}{dt^2} + kx = F_0 \sin \omega t \tag{2.124}$$

or alternatively

$$\frac{d^2x}{dt^2} + \omega_0^2 x = \frac{F_0}{m} \sin \omega t \tag{2.125}$$

where $\omega_0 = (k/m)^{1/2}$ is the so-called natural frequency of the system, that is, the frequency with which the system would oscillate if no external force were applied.

We first derive a solution to the homogeneous form of equation (2.125) by writing it in D-operator form:

$$(D^2 + \omega_0^2)x = 0 \tag{2.126}$$

which has the two imaginary roots $D_{1,2} = \pm i\omega_0$. The corresponding solution will consequently be periodic, and is given by (see Table 2.9)

$$x_c = A \sin \omega_0 + B \cos \omega_0 t \tag{2.127}$$

This is what we had termed the complementary function x_c of the ODE (2.126). The particular integral x_p, which has to be added to it, is obtained from Table 2.10. We have from item 5 of that table

$$A_0 = \frac{(c - n^2 a)a_0}{(c - n^2 a)^2 + n^2 b^2} \tag{2.128}$$

where for our case $a = 1$, $b = 0$, $c = \omega_0^2$, $a_0 = F_0/m$, and $n = \omega$. We obtain

$$A_0 = \frac{(\omega_0^2 - \omega^2)(F_0/m)}{(\omega_0^2 - \omega^2)^2} = \frac{(F_0/m)}{\omega_0^2 - \omega^2} \tag{2.129}$$

and for the particular integral

$$x_p = \frac{(F_0/m)}{(\omega_0^2 - \omega^2)} \sin \omega t \tag{2.130}$$

The general solution is then given by

$$x(t) = x_c + x_p = A \sin \omega_0 t + B \cos \omega_0 t + \frac{(F_0/m)}{\omega_0^2 - \omega^2} \sin \omega t \tag{2.131}$$

where A and B are the integration constants, which must now be evaluated. We set, for simplicity,

$$x(0) = \frac{dx}{dt}(0) = 0 \tag{2.132}$$

That is, we assume the motion to start at the origin with an initial velocity of zero. For the first condition $x(0) = 0$, the sine terms vanish and it follows that

$$B = 0 \tag{2.133}$$

From the second condition of zero initial velocity, we have

$$\frac{dx}{dt}(0) = 0 = A\omega_0 + \frac{\omega(F_0/m)}{\omega_0^2 - \omega^2} \tag{2.134}$$

so that

$$A = -\frac{\omega(F_0/m)}{\omega_0\left(\omega_0^2 - \omega^2\right)} \tag{2.135}$$

With the integration constants in hand, the general solution now becomes

$$\boxed{x(t) = \frac{\omega(F_0 - m)}{\omega_0\left(\omega^2 - \omega_0^2\right)} \sin \omega_0 t - \frac{(F_0/m)}{\left(\omega^2 - \omega_0^2\right)} \sin \omega t} \tag{2.136}$$

This is the final expression, which describes the motion of the mass x(t) when acted on by a sinusoidal forcing function $F_0 \sin \omega t$.

COMMENTS

- The solution is seen to be composed of the *difference* of the two sine waves, one with the natural frequency of the system, ω_0, the other with the frequency ω of the forcing function. At $t = 0$, both sine terms vanish, satisfying the first initial condition $x(0) = 0$. The second initial condition, $(dx/dt)_{t=0} = 0$, is likewise satisfied, since the two cosine terms that result exactly cancel each other.
- In Section 1.3, entitled Closure, we urged the reader to carefully examine a solution for unusual behavior. Among the candidates for special scrutiny, we mentioned in particular solutions containing fractions whose denominator was composed of a *difference* [cf. equation (1.27)]. This case is seen to arise here in equation (2.137), with the term $\omega^2 - \omega_0^2$ representing the difference in question. It follows from this expression that when the frequency ω of the imposed oscillation equals the "natural frequency" of the system [i.e., when $\omega_0 = (k/m)^{1/2}$], the solution becomes unbounded, or it ceases to exist. Intuitively one senses that this cannot be true for all values of time t and that there will

be at least an initial period during which the extension x(t) remains bounded. Evidently, this case, a singularity as it is often termed, requires special treatment. It turns out that the Laplace transformation, to be taken up next, provides us with a better tool for analyzing this special case as well as the case $\omega \neq \omega_0$. We now turn to a more detailed examination of this solution method.

The Laplace Transformation The Laplace transformation (or Laplace transform, for short) belongs to a broader class of integral transform operations in which a function F(t) is multiplied by a "kernel" K(s, t) and integrated between the limits t = a and t = b. Thus

$$T\{F(t)\} = \int_a^b F(t)K(s,t)\,dt = f(s) \tag{2.137}$$

where T is the operational symbol for the transformation and f(s) denotes the transform of F(t).

The function operated on, F(t), is quite arbitrary in form and can be an ordinary function in t, such as sin at, a derivative such as d^2F/dt^2, or even an integral. The kernel and the integration limits (a,b) define the type of transform. For the Laplace transform, the kernel K(s, t) is the exponential e^{-st} and the integration limits are from zero to infinity. Thus the Laplace transform takes the form

$$L\{F(t)\} = \int_0^\infty F(t)e^{-st}\,dt = f(s) \tag{2.138}$$

where L is the operational symbol for the transformation and L^{-1} denotes the inverse process, that is,

$$L^{-1}\{f(s)\} = F(t) \tag{2.139}$$

The transformation (2.138), when applied to each term of an ordinary differential equation, has the effect of eliminating the independent variable t, thus reducing the ODE to an algebraic equation (AE) in the transformed state variable y(s). That variable can also be denoted by $\bar{y}$. After solving for y(s), the result, which is still algebraic, is translated back into solution space by means of appropriate "dictionaries," the Laplace transform tables. The reader will note that this procedure bears a resemblance to the D-operator method, which likewise transforms an ODE to an AE and, after solving the latter, translates the result back into the solution domain by means of an appropriate dictionary, Table 2.8.

The Laplace transform is generally used to solve linear initial value ODEs, both singly and in sets. Its advantages are the following:

- It can directly solve a nonhomogeneous problem without having to establish the solution of the corresponding homogeneous problem.
- It can handle a wide array of nonhomogeneous terms without recourse to particular integrals. This includes the important class of discontinuous forcing functions, which can be handled by the D-operator method only with difficulty, if at all.
- The initial conditions are automatically incorporated in the solution, thus avoiding the need to evaluate integration constants. The application of the method will be

Table 2.11 General Properties of the Laplace Transform

F(t)	f(s)
1. *Laplace Transform*	
$F(t)$	$\int_0^\infty F(t)e^{-st}\,dt = f(s)$
2. *Inverse Transform*	
$\frac{1}{2\pi i}\lim_{\beta\to\infty}\int_{\gamma-i\beta}^{\gamma+i\beta} e^{tz}f(z)\,dz$	$f(s)$
3. *Transform of a Constant*	
C	C/s
4. *Transform of a Sum*	
$aF(t) + bG(t)$	$af(s) + bg(s)$
5. *Transform of Derivatives*	
$F'(t)$	$sf(s) - F(0)$
$F''(t)$	$s^2f(s) - sF(0) - F'(0)$
$F^{(n)}(t)$	$s^nf(s) - s^{n-1}F(0) - s^{n-2}F'(0)\cdots F^{(n-1)}(0)$
6. *Transform of an Integral*	
$\int_0^t F(\tau)\,d\tau$	$\frac{1}{s}f(s)$
7. *Inverse of a Product: The Convolution Theorem*	
$\int_0^t F(\tau)G(t-\tau)\,d\tau = F(t) * G(t)$	$f(s)\,g(s)$
8. *Inverse of a Ratio of Polynomials (Heaviside Expansion)*	
$\sum_{n=1}^{\infty}\frac{(s-a_n)p(a_n)}{q(a_n)}e^{a_n t} = \sum_{n=1}^{\infty}\frac{p(a_n)}{q'(a_n)}e^{a_n t}$ where a_n = roots of $q(s)$	$\frac{p(s)}{q(s)}$ [order of $p(s) < q(s)$]
9. *Inverse of Derivatives*	
$-t\,F(t)$	$f'(s)$
$(-1)^n t^n\,F(t)$	$f^{(n)}(s)$
10. *Inverse of an Integral*	
$\frac{1}{t}F(t)$	$\int_s^\infty f(x)\,dx$
11. *Translation or Shifting Properties*	
$e^{at}\,F(t)$	(a) $f(s-a)$
$\begin{cases} F(t-a) & t > a \\ 0 & t < a \end{cases}$	(b) $e^{-as}f(s)$
12. *Initial Value Theorem*	
$F(0)$	$\lim_{s\to\infty} sf(s)$
13. *Final Value Theorem*	
$\lim_{t\to\infty} F(t)$	$\lim_{s\to 0} sf(s)$

illustrated later in this chapter, as well as in the Examples 7.2, 7.4, and 7.5 dealing with networks.

We have tabulated in Table 2.11 some general properties of the Laplace transform and supplement these, in Table 2.12, with tabulations of the transforms of a number of common functions.

Table 2.12 Laplace Transforms of Some Functions

Function	Transform
1. 1	$\frac{1}{s}$
2. t	$\frac{1}{s^2}$
3. e^{-at}	$\frac{1}{s+a}$
4. t^n	$\frac{n!}{s^{n+1}}$
5. sin kt	$\frac{k}{s^2+k^2}$
6. $t^n e^{-at}$	$\frac{n!}{(s+a)^{n+1}}$
7. cos kt	$\frac{s}{s^2+k^2}$
8. sinh kt	$\frac{k}{s^2-k^2}$
9. e^{-at} cos kt	$\frac{s+a}{(s+a)^2+k^2}$
10. cosh kt	$\frac{s}{s^2-k^2}$
11. e^{-at} sin kt	$\frac{k}{(s+a)^2+k^2}$

We note that both tables are akin to a table of integrals in that one can move from left to right or from right to left, depending on the information sought. In this instance, one moves from left to right to obtain the transform f(s) of a function F(t) and from right to left to obtain what is termed the inverse transform, or inverse, of f(s), that is, the function F(t) itself. The latter process is commonly referred to as an inversion. The formula given for the general inversion, item 2 of Table 2.11, involves a line integral in the complex plane and requires some background knowledge of complex variable theory. It is used only sparingly, and we shall here limit ourselves to making use of more convenient Items.

Item 5 of Table 2.11, the transform of derivatives, requires special mention. Its principal feature is that it converts the derivatives into *algebraic* expressions in s. In this it resembles the D-operator method, but with the additional advantage of incorporating the initial conditions in the resulting algebraic expression. This was not the case in the D-operator method, which required the somewhat cumbersome evaluation of integration constants.

Two additional items call for special mention. One is the convolution integral, item 7, which allows the inversion of the *product* of two arbitrary functions, f(s) and g(s). It is frequently used to carry a general and unspecified function, say F(t), into the integration process and return it upon inversion as an integrand of the convolution integral. The second item of special interest is the Heaviside expansion, item 8. To use it in the inversion of a ratio of polynomials, one first has to evaluate the roots of the denominator q(s), which are then substituted into the inversion formula given on the left. The formula applies only to denominators with distinct roots.

To guide the reader in the use of the transform in solving ODEs, we summarize details of the procedure as follows.

1. Apply the Laplace transform in turn to each term of the ODE in Y(t). If that term is a *derivative*, one obtains a composite of the transform of the unknown function, y(s), and its initial values (see item 5 of Table 2.11). For example, dY/dt becomes, when transformed, y(s) – Y(0). Terms containing directly the unknown state variable become the transform of that variable. Thus $L\{kY(t)\} = ky(s)$. Finally, the nonhomogeneous terms, or forcing functions, are directly reduced to explicit functions of s. For example, the forcing function e^{-at} is transformed directly into the explicit form 1/(s – a) (see Table 2.12, item 3).
2. Solve the algebraic equation in y(s) that resulted from step 1. This yields expressions of the form

$$y(s) = G(s) \tag{2.140}$$

where y(s) is the transformed state variable Y(t), and G(s) is an explicit function of s containing, among other things, the transforms of the forcing functions and the initial conditions. An example is

$$y(s) = G(s) = \frac{sY(0)}{s-1} \tag{2.141}$$

3. Invert the expression (2.141): that is, apply the operator L^{-1} to each side. For y(s), this yields the desired state variable Y(t). Inversion of the right-side term G(s) is accomplished by means of tables, such as Table 2.12, and by one or more of the procedures listed in Table 2.11, such as the use of the convolution integral or the Heaviside expansion. Together these two inversions yield the desired solution in the final form

$$Y(t) = f(t, \text{initial conditions}) \tag{2.142}$$

For simultaneous ODEs, the procedure is similar but requires some additional steps. We proceed as follows.

1. Apply the Laplace transform to each of the n ODEs of the system. Instead of a single algebraic equation in y(s), we now obtain n such equations, containing n transforms, $y_1(s), y_2(s), \ldots, y_n(s)$.
2. Eliminate algebraically $n - 1$ of the transforms, reducing the system to a *single* algebraic equation in $y_1(s)$.
3. Solve the equation resulted in Step 2 for $y_1(s)$.
4. Invert the result from Step 3 to obtain $Y_1(t)$.
5. Repeat the procedure for the remaining $n - 1$ equations.

We note that the process essentially involves the solution of n simultaneous algebraic equations in the n unknowns $y_1(s), y_2(s), \ldots, y_n(s)$. When n is large, one may wish to use the matrix methods outlined in Section 2.2.

◆ EXAMPLE 2.18 Application of Inversion Procedures

Suppose that in the course of solving an ODE, the following expression is obtained

$$y(s) = \frac{3s+1}{(s-1)(s^2+1)} \tag{2.143}$$

which we now wish to invert. We recognize this as a ratio of polynomials p(s)/q(s), where $p(s) = 3s + 1$, $q(s) = (s - 1)(s^2 + 1)$, and the roots of q(s) are $a_1 = 1$, $a_2 = i$, $a_3 = -i$. Since the roots are not repeated and the order of p(s) is less than that of q(s), the Heaviside expansion may be applied.

We have

$$q(s) = s^3 - s^2 + s - 1 \tag{2.144}$$

and

$$q'(s) = 3s^2 - 2s + 1 \tag{2.145}$$

Expanding the sum of the Heaviside expression, we obtain

$$\begin{aligned} &\frac{p(1)}{q'(1)}e^t + \frac{p(i)}{q'(i)}e^{it} + \frac{p(-i)}{q'(-i)}e^{-it} \\ &= 2e^t + \left(-1 - \tfrac{1}{2}i\right)(\cos t + i\sin t) + \left[\left(-1 + \tfrac{1}{2}i\right)(\cos t - i\sin t)\right] \end{aligned} \tag{2.146}$$

where we have used Euler's formula given in Table 2.9 to convert the imaginary exponentials into trigonometric form.

Final evaluation of the products then leads to the expression

$$L^{-1}\left\{\frac{3s+1}{(s-1)(s^2+1)}\right\} = 2e^t - 2\cos t + \sin t \tag{2.147}$$

◆ EXAMPLE 2.19 The Mass–Spring System Revisited: Resonance

We return here to the situation we considered in Example 2.17 to examine more fully the singularity that acrose when the imposed frequency ω equaled the natural frequency of the system ω_0. We start with a more general formulation by letting the forcing term become an arbitrary function and by removing the restriction to zero initial conditions. We have

$$\frac{d^2x}{dt^2} + \omega_0^2 x = \frac{F(t)}{m} \tag{2.148}$$

and

$$\begin{aligned} x(0) &= x_0 \\ x'(0) &= v_0 \end{aligned} \tag{2.149}$$

with $\omega = (k/m)^{1/2}$ as before.

This more general formulation will enable us to obtain additional practice in the use of inversion techniques.

We carry the forcing function in unspecified form into the Laplace transformation process and obtain, using items 1, 4, and 5 of Table 2.11:

$$s^2x(s) \; - \; sx_0 - v_0 + \omega_0^2 x(s) = \frac{f(s)}{m} \tag{2.150}$$

Solving for x(s) we obtain

$$x(s) = \frac{x_0 s + v_0}{s^2 + \omega_0^2} + \frac{f(s)}{m} \frac{1}{s^2 + \omega_0^2} \tag{2.151}$$

This is the expression which now has to be inverted. Inversion of the first fraction is by items 5 and 7 of the Table 2.12, which leads to a cosine and a sine term, respectively. For the second term, we use the convolution integral, item 7 of the Table 2.11. This leads to the result

$$x(t) = x_0 \cos \omega_0 t + \frac{v_0}{\omega_0} \sin \omega_0 t + \frac{1}{m\omega_0} \int_0^t \sin \omega_0 \tau F(t - \tau)\, d\tau \tag{2.152}$$

This, then, is the most general expression describing the oscillations of a mass-spring system subjected to a forcing term of arbitrary functional form.

Let us now consider the special case $F(t) = F_0 \sin \omega t$. The transform of this function is obtained from item 5 of Table 2.11 and its introduction into equation (2.151) leads to the result

$$x(s) = \frac{x_0 s + v_0}{s^2 + \omega_0^2} + \frac{F_0}{m} \frac{\omega}{(s^2 + \omega_0^2)(s^2 + \omega^2)} \tag{2.153}$$

The first term on the right is again inverted by items 5 and 7 of Table 2.12, while the second term yields to a Heaviside expansion, item 8 of Table 2.11. Details of this last step are left to the exercises (see Practice Problem 2.13). We obtain

$$x(t) = x_0 \cos \omega_0 t + \frac{1}{\omega_0} \left[v_0 + \frac{F_0 \omega}{m(\omega^2 - \omega_0^2)} \right] \sin \omega_0 t - \frac{F_0}{m(\omega^2 - \omega_0^2)} \sin \omega t \tag{2.154}$$

This is the general solution for a mass–spring system subjected to a sinusoidal forcing function. At this point we return to equation (2.153) and investigate the previous singularity by setting $\omega = \omega_0$. This leads to the expression

$$x(s) = \frac{x_0 s + v_0}{s^2 + \omega_0^2} + \frac{F_0 \omega_0}{m(s^2 + \omega_0^2)^2} \tag{2.155}$$

One notices that no singularity arises here, and thus one can proceed with the inversion. Inversion of the first term is done as in the earlier cases, while the second term can be inverted by a clever application of the convolution integral. The details are again left to the exercises (Practice Problem 2.13). The result is given by the expression

$$x(t) = x_0 \cos \omega_0 t + \frac{1}{\omega_0^2}\left[v_0\omega_0 + \frac{F_0}{m}\right] \sin \omega_0 t - \frac{F_0}{2m\omega_0} t \cos \omega_0 t \tag{2.156}$$

One notices here immediately the appearance of a factor t in the second cosine term, which indicates that the amplitude of the oscillations increases indefinitely with time. One speaks of the forcing function F(t) as being in *resonance* with the system frequency. In particular, for $x_0 = 0$ and $v_0 = -F_0/(2m\omega_0)$ one obtains the simple expression

$$x(t) = -\frac{F_0}{2m\omega_0} t \cos \omega_0 t \tag{2.157}$$

That expression clearly shows a linear increase with t of the amplitude of the oscillation and would in time lead to a rupture of the mass–spring system. A similar but more complex problem, involving two linked mass–spring systems, is taken up in Example 7.5. It is found there that the combination gives rise to a pair of distinct resonating frequencies, which can be established by the method given in the present example.

COMMENTS

- The successful analysis of the singular case of resonance not only reveals the power of the Laplace transformation but also demonstrates that a change in tools or a change in approach may sometimes resolve the difficulties experienced with a particular procedure. The D-operator method having clearly failed to provide a general answer to the problem of forced vibration, we were able to obtain one by reaching for the alternate method for the Laplace transformation.

PRACTICE PROBLEMS

2.1 Properties of vectors

(a) Can a scalar product be negative?

(b) If $\mathbf{A} \cdot \mathbf{B} = \mathbf{A} \cdot \mathbf{C}$, does it necessarily follow that **B** equals **C**?

(c) If $\mathbf{A} \times \mathbf{B} = 0$, must **A** and **B** be parallel to each other? Is the converse true?

(d) If three vectors add up to zero, they must all be in the same plane. Make this plausible.

(e) What are the properties of two vectors **A** and **B** such that

$$\mathbf{A} + \mathbf{B} = \mathbf{C}$$

$$\mathbf{A} + \mathbf{B} = \mathbf{A} - \mathbf{B}?$$

2.2 **Distance of a point from a line** Consider a line defined by the two points A(2, 3, 0) and B(−1, 2, 4). Find the shortest distance d to this line from a point P with the coordinates (3, 1, −1). (*Hint*: Show that $d = |\overrightarrow{\mathbf{PA}} \times \overrightarrow{\mathbf{AB}}|/|\overrightarrow{\mathbf{AB}}|$).

ANSWER 2.245

2.3 **Torque: Balancing a teeter-totter** If a force **F** acts on a body at a point P, which is located by the position vector **R**, then the torque τ acting on the body with respect to the origin 0 is defined as

$$\tau = \mathbf{R} \times \mathbf{F} \tag{1}$$

with a magnitude τ of

$$\tau = RF \sin\theta \tag{2}$$

we apply the relation (1) to Newton's law

$$\mathbf{F} = m\mathbf{a} \tag{3}$$

and obtain, by cross-multiplying each side by the position vector **R** and summing over all forces,

$$\sum \tau = \sum \mathbf{R} \times \mathbf{F} = \sum \mathbf{R} \times \mathbf{a} \tag{4}$$

For a static system at equilibrium, $\mathbf{a} = 0$, and we obtain

$$\sum \tau = \sum \mathbf{R} \times \mathbf{F} = 0 \tag{5}$$

That is, the sum of the torques acting on a static body must equal zero. This is sometimes referred to as the lever rule.

Consider the teeter-totter shown in Fig. P2.3, seating two persons weighing 80 kg (A) and 100 kg (B) each at identical distances of 2 m from the fulcrum. To keep the board from touching the ground, B proposes to lean forward toward the fulcrum, thus reducing the effective lever arm length on that side. At what angle θ to the board should B incline the upper body to maintain equilibrium? Assume the length of the upper body to be 1 m and that its center of gravity is at the midpoint.

ANSWER 0°

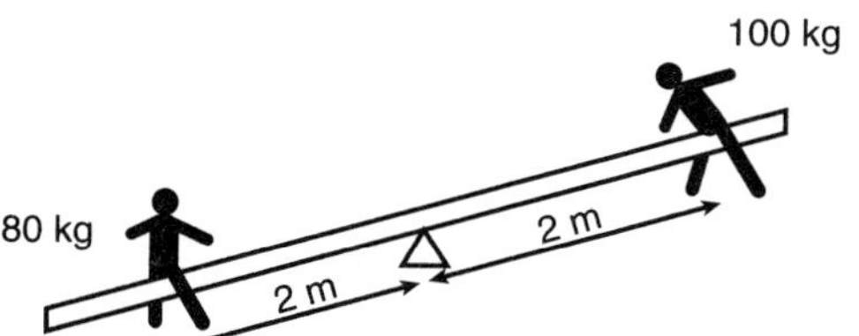

Figure P2.3 The teeter-totter.

2.4 **Matrix multiplication** We are given the following row vector and matrix

$$\mathbf{A} = \begin{bmatrix} 2 & 1 & 0 \end{bmatrix} \qquad \mathbf{B} = \begin{bmatrix} 4 & 0 \\ 0 & 2 \\ -1 & 1 \end{bmatrix}$$

Evaluate **AB** and **BA**.

2.5 **Factory output** Let the matrix

$$\mathbf{A} = \begin{bmatrix} 2 & 1 \\ 4 & 3 \end{bmatrix}$$

represent the number of gadgets R and S that factories P and Q can produce in a day, according to Table P2.5. Let

$$\mathbf{N} = \begin{bmatrix} 5 \\ 6 \end{bmatrix}$$

represent the number of days the two factories operate; that is, P operates 5 days a week and Q operates 6 days a week. Find **AN** and state what it represents.

ANSWER $\mathbf{AN} = \begin{bmatrix} 16 \\ 38 \end{bmatrix}$

2.6 **Solution of an underspecified system of linear algebraic equations** Show that the following system has an infinite number of solutions:

$$\begin{aligned} x_1 + x_2 + x_3 + x_4 &= 4 \\ x_1 + 3x_2 + 3x_3 &= 2 \\ x_1 + x_2 + 2x_3 - x_4 &= 6 \end{aligned}$$

Follow this up by expressing x_1, x_2, x_3 in terms of x_4.

ANSWER

$$\begin{aligned} x_1 &= 5 - \tfrac{3}{2}x_4 \\ x_2 &= -3 - \tfrac{3}{2}x_4 \\ x_3 &= 2 + 2x_4 \end{aligned}$$

Table 2.5

	Factory P	Factory Q
Gadget R	2 per day	1 per day
Gadget S	4 per day	3 per day

2.7 Classification of ODEs and boundary conditions. Possible solution methods The following systems are to be classified and possible solution methods suggested.

(a)

$$k\left[\frac{d^2T}{dr^2} + \frac{1}{r}\frac{dT}{dr}\right] + q = 0$$

$$T(R_1) = T_1$$

$$T(R_2) = T_2$$

where q is a constant.

This equation describes conduction in a radial or cylindrical geometry with internal heat generation q (J/s).

(b)

$$1 + \left(\frac{dy}{dx}\right)^2 = k^2x^2\left(\frac{d^2y}{dx^2}\right)^2$$

Expressions of this type arise in the analysis of "path of pursuit" problems in which the path of a pursuer aiming at the interception of a prey is to be established.

(c)

$$y' + (y - z) = 0$$

$$z' + (y - z) = 0$$

(d)

$$\frac{d^2y}{dx^2} - K_1y^2 = 0$$

$$\frac{dy}{dx}(0) = 0$$

$$-K_2\frac{dy}{dx}(\pm L) = y(\pm L) - y_0$$

This set describes the simultaneous occurrence of diffusion and a second-order chemical reaction in a slab of catalyst, with convective mass transport from the external fluid.

2.8 Equivalent systems Consider two systems in which flow takes place inside two concentric tubes. The flow is in the same direction in the inner tube and in the annular space outside it. In the first of the two processes considered, heat is transferred from the annular space to the inner tube with a driving force $(T_s - T_t)$. This type of system corresponds to a shell-and-tube heat exchanger of the type shown earlier (Fig. P1.5).

In the second process, mass is transferred through the inner tube wall, assumed to be permeable, to the annular fluid with a driving force $(C_t - C_s)$. A process of this kind occurs in blood dialysis, described in Chapter 1 (see Fig. 1.4).

Show that the two processes are described by two ODEs that are identical in form (i.e., are equivalent).

2.9 **The pendulum. Decomposition of a second-order ODE** The equation describing the motion of a pendulum is given by

$$mL\frac{d^2\theta}{dt^2} + kL\frac{d\theta}{dt} + mg\sin\theta = 0$$

or equivalently

$$a_0\ddot{\theta} + a_1\dot{\theta} + a_2\sin\theta = 0$$

Show that this equation can be reduced to the form

$$\frac{d\dot{\theta}}{d\theta} = -\frac{a_1\dot{\theta} + a_2\sin\theta}{a_0\dot{\theta}}$$

2.10 **Separation of variables** Solve the following system by separation of variables.

$$-u = \frac{dv}{dt}$$

$$-k(uw) = \frac{d(vw)}{dt}$$

where u, v, w are all functions of t. Note that the system is underspecified. A reduced solution of the form $v = f(w)$ can nevertheless be obtained.
(*Hint*: Divide the two equations.)

ANSWER $\frac{w_2}{w_1} = \left(\frac{v_2}{v_1}\right)^{k-1}$

2.11 **Mass on a spring vibrating under its own weight. Application of the D-operator method** The equation describing the motion of a mass attached to a spring and vibrating under its own weight is given by

$$m\frac{d^2x}{dt^2} + kx = mg$$

where mg is the forcing function. Solve this equation by the D-operator method and by determining the particular integral for the nonhomogeneous term. Assume zero initial velocity, with an initial position of $x = x_0$.

ANSWER $x = (x_0 - mg/k)\cos[(k/m)^{1/2}\,t]$

2.12 Comparison of the D-operator and Laplace transformation methods Use both the D-operator and Laplace transformation method to solve the following ODE:

$$\frac{d^2x}{dt^2} + 5\frac{dx}{dt} + 4x = e^{-5}$$

Assume the initial conditions to be homogeneous; that is, they are given by

$$x(0) = 0, \qquad \frac{dx}{dt}(0) = 0$$

ANSWER $y = \frac{1}{12}e^{-t} - \frac{1}{3}e^{-4t} + \frac{1}{4}e^{-5t}$

2.13 Inversions

(a) Use the Heaviside Expansion to invert the expression that arises in equation (2.153).

$$\frac{1}{(s^2 + \omega^2)\left(s^2 + \omega_0^2\right)}$$

(b) Invert

$$\frac{1}{\left(s^2 + \omega_0^2\right)^2}$$

which appears in equation (2.155). (*Hint*: Apply the convolution integral.)

ANSWER (b) $\frac{1}{2}t\cos\omega_0 t$

CHAPTER 3

Geometrical Concepts

In the two preceding chapters we provided the reader with some guidelines for initiating and implementing the modeling process. Chapter 1 dealt with certain initial steps that can be taken to overcome what is often the greatest hurdle in modeling, getting off to a good start. We also alluded to the importance of concluding the process with an analysis of the results obtained and drawing the appropriate lessons. Chapter 2 provided a compendium of mathematical tools likely to be of use in modeling at the introductory level. Although we shall have occasion to use tools in addition to those presented there, our main working kit will comprise vector and matrix algebra, plus the essentials of ordinary differential equations. Supplementary topics will be addressed as they arise (see Chapter 8).

The present chapter inaugurates the modeling process in a more comprehensive manner. The common theme is the use of geometrical concepts, which we apply to a range of problems drawn from a wide variety of disciplines. The concepts referred to are not confined to pure geometry as it is commonly understood; rather, they encompass both plane and solid geometry, trigonometry, unusual modes of graphical representation, and certain simple topological concepts.

In Example 3.1 to 3.3, we start our deliberations at a deceptively simple level, but the reader will quickly note that even with these simplest of tools, the results obtained are not only highly useful but on occasion startling in the features they reveal and in the elegance of the end result. Thus in Example 3.1, Crossing of a River, only the most elementary trigonometric and vector relations are used to determine the heading and speed required for a boat to reach a prescribed destination. Even simpler geometry is used in Example 3.2, on the structure of quasi crystals. We apply the rule that the sum of the angles of a N-sided polygon equals $N \times 180° - 360°$ to construct the structure of amorphous metallic glasses. Similarly simple concepts are used in Example 3.3, which deals with establishing financial market trends using the so-called Japanese candlestick method.

The next two examples share the common feature of the determination of locations in space but use different tools to arrive at the result. In Example 3.4 we use the concept of similar triangles and some simple trigonometric relations to establish the coordinates of a point in a plane. The methods are those used in surveying. In Example 3.5, The Global Positioning System, solid geometry is invoked to determine a position on the globe by taking a fix on satellites.

Examples 3.5 to 3.8 all make use of vector operations to establish a variety of results. In Example 3.6, The Orthocenter of a Triangle, we draw on the scalar vector product to prove that the altitudes of a triangle have a common point of intersection, the so-called orthocenter. Example 3.7 discusses absolute and relative velocities in terms of the derivative of the position vector and establishes the so-called wind triangle used in aerial navigation. The latter is applied in Example 3.8 to determine the straight-path interception of a plane.

Interception of a different type is taken up in Example 3.9, Path of Pursuit. Here the interception path is curved to minimize the time of interception and leads to a nonlinear ordinary differential equation.

Example 3.10, which considers an unusual coordinate system, introduces the use of trilinear coordinates. The task is to determine the trajectory required to obtain a fixed measure of liquid using three different sized jugs, one of which is full.

Example 3.11 draws the reader's attention to the existence of multiple solutions of nonlinear algebraic equations. The occurrence of these in a chemical and biological context is discussed. This is followed by Example 3.12, Linear Programming, in which an optimization problem is solved by means of a graphical construction in rectangular coordinates. Another unusual graphical construction, again in two-dimensional rectangular coordinates, is taken up in Example 3.13, Stagewise Adsorption Purification of Liquids. Example 3.14, finally, uses some simple geometrical and topological concepts to unravel the structure of DNA.

The chapter concludes with a rich selection of practice problems that deserve the reader's attention.

◆ EXAMPLE 3.1 A Simple Geometry Problem: Crossing of a River

A river flows due south at 40 m/min. A motorboat, moving at 150 m/min in still water, is headed due east across the river (i.e., its heading is due east). We address the following problems.

(a) Find the direction in which the boat moves and its speed.

(b) For the boat to move due east, in what direction must it be headed and at what speed?

Solution of this problem requires the construction of a diagram in which the velocities of the boat and the current are entered as vectors (Fig. 3.1). The unknowns are then obtained by applying the principle of vector addition.

(a) Refer to Fig. 3.1A. Then in the right triangle OAB,

$$\mathrm{OB} = [(150)^2 + (40)^2]^{1/2} = 155\ \mathrm{m/min} \tag{3.1}$$

$$\tan\theta = \frac{40}{150} = 0.267 \qquad \theta = 14°54' \tag{3.2}$$

Thus the boat moves at 155 m/min in the direction S 75°6′ E.

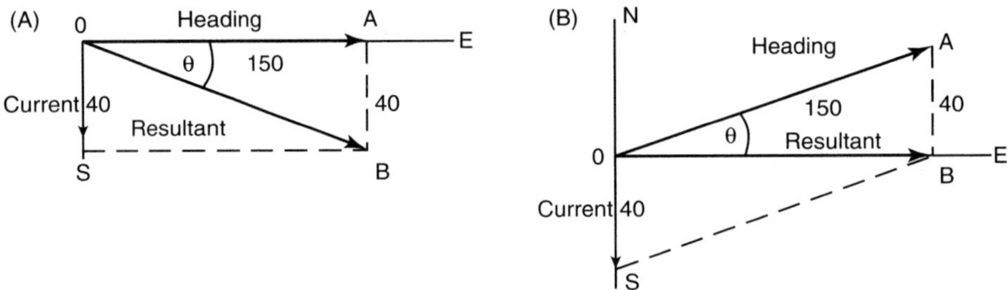

Figure 3.1 River crossing: (A) finding the direction the boat must take and (B) finding the speed.

(b) Refer to Fig. 3.1B. Then in the right triangle OAB

$$OB = [(150)^2 - (40)^2]^{1/2} = 144.6\,\text{m/min} \tag{3.3}$$

$$\sin\theta = \frac{40}{150} = 0.267 \qquad \theta = 15°30' \tag{3.4}$$

Thus to reach the other side directly east of the starting point, the boat must be headed N 74°30′ E with a speed of 144.6 m/min.

COMMENTS

- We have here the application of some simple notions taken from trigonometry and vector algebra to a problem of a certain practical interest. Only the most basic relations are involved, those of vector and addition and the definition of the trigonometric sine and tangent. Yet it still requires some thought to arrive at the correct formulation of the vector diagrams. For example, it is customary to think of the resultant, or the *diagonal* in the diagram as the unknown vector. This is in fact the case for part a, but in part b it is one of the sides of the triangle that becomes the unknown.
- We also have here our first encounter with a system containing both a fixed reference frame (the river bed) and a moving reference frame (the river itself). This leads to the notion of *absolute* velocity, taken with respect to the fixed coordinate system, and of a *relative* velocity with respect to the moving coordinates. We had briefly encountered these concepts in Example 1.6 of Chapter 1 and will return to them for a closer examination in Example 3.7. The need to distinguish between absolute and relative velocities did not arise in the present illustration since all quantities were taken with respect to fixed coordinates, i.e., we were dealing with absolute velocities throughout.

◆ EXAMPLE 3.2 The Formation of Quasi Crystals and Tilings from Two Quadrilateral Polygons

Most crystalline substances can be viewed as being composed of identical rhombic shapes. The cubic and hexagonal crystal units, which will be familiar to the reader, are of this type. Floor and wall tilings are usually and similarly composed of identical rhombic subunits (Fig. 3.2A).

There is a certain category of quasi crystals, or amorphous crystals, that lack this simple structure. Attempts have been made to model these quasi crystals and to generalize tiling

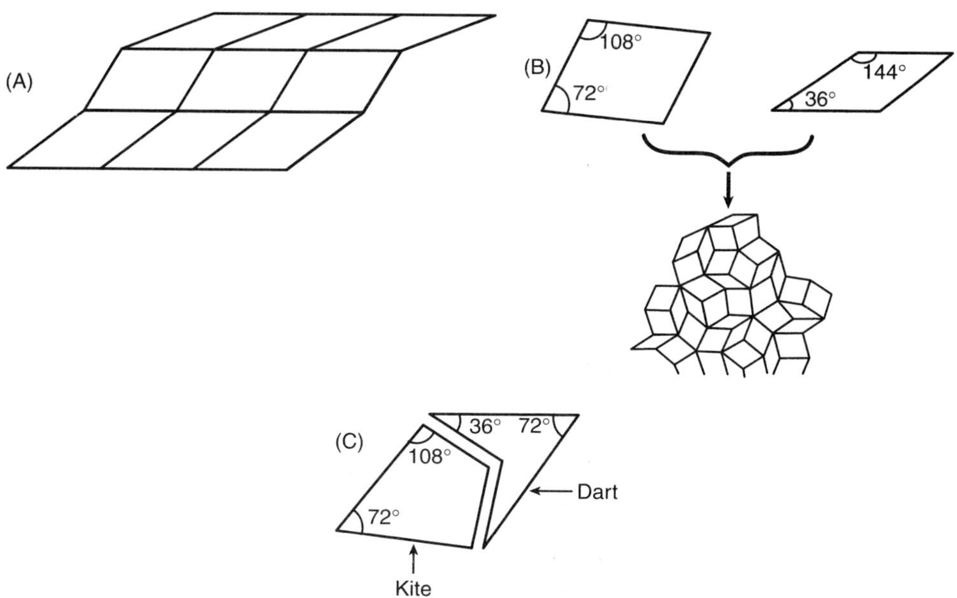

Figure 3.2 Formulation of quasi crystals: (A) regular crystals composed of identical rhombic units, and (B) and (C) combination of two quadrilateral polygons leading to quasi crystals.

patterns. The basis of these procedures is to formulate two quadrilateral substructures that have identical corresponding sides but differ in the magnitude of the internal angles. These two shapes are then joined in such a way that a new polygon, usually a rhombus or hexagon (Fig. 3.2, B and C) is produced. That polygon can subsequently be folded across two corners to construct three-dimensional structures (Fig. 3.2B). The only geometrical consideration needed in this procedure is the rule that the sum of the internal angles of a N-sided polygon equals $N \times 180° - 360°$. This is clearly satisfied in the two examples shown (Fig. 3.2, B and C). The result in Fig. 3B of joining two parallelograms, for example, is a hexagon with the summed internal angles $2(108+144+72+36) = 720$, in agreement with the required sum of $6 \times 180° - 360° = 720$.

COMMENTS

- We start by noting that the entire procedure for arriving at startlingly new geometrical structures was based on one simple geometrical rule, supplemented by a good dose of imagination. The resulting configurations, while seemingly irregular in appearance, have a certain internal regularity brought about by the procedure used in constructing them. The term "quasi crystals" was coined to reflect this property.

◆ EXAMPLE 3.3 Charting of Market Price Dynamics: The Japanese Candlestick Method

Market analysis consists largely of the charting of market prices, such as those of stocks and currencies, and the prediction, whenever possible, of trends in those prices. The so-called

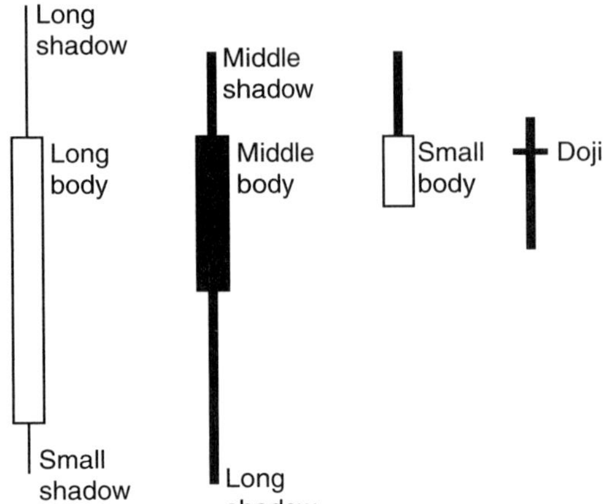

Figure 3.3 Candlesticks of varying body length, color, and shadow length.

Japanese candlestick, which is one of several existing tools for market analysis, owes its popularity to the fact that it expresses key factors of market dynamics in an attractive visual or geometric way, which can be quantified by means of a decimal code.

There are four key features associated with a Japanese candlestick, several examples of which appear in Fig. 3.3. Before proceeding to the specifics of this example, we present a brief explanation of these features.

The body length of the candle, which is proportional to the absolute difference between opening and closing prices. Thus a large rise or drop in the day's market price is associated with a long or large body, a moderate change with a body of medium size, and a modest change with a small body. Examples of each of these appear in Fig. 3.3. The fourth example shown has no body length or only minute length, signifying an open and a close at the same price, or nearly the same price. Such candles are referred to as *dojis*.

The shadows of the candle, comprising two features: the upper and the lower shadow. The upper shadow, which tops the candle body in Fig. 3.3, is proportional to the difference between the day's high and the higher of the two terminal prices, that is, high – max (open, close). Depending on the size of these differences, one speaks of a large or long shadow, a middle shadow, and a small shadow. A doji can have any size of shadow even though it has little or no body length. This follows from the definitions of candle and shadow length.

The candle body color, which can be black or white, depending on whether the market closed lower or higher during a day's activities. For dojis associated with no price change, the relative length of the shadows determines the color. Thus a doji candlestick is white if its upper shadow is longer than its lower shadow.

Color and body length determine the nature of the market, that is, whether it is bullish or bearish. Thus a long black body is a strong bearish signal, a middle black body less and white bodies convey bullish signals of various strengths, depending on the length of the candlestick.

To further illustrate these features, we consider in Fig. 3.4 a typical daily price fluctuation and construct the candlestick expressing these fluctuations. We start with color and quickly establish that it must be black, since the closing level is lower than the opening price. Body length is given by the absolute difference between open and close, abs(open – close), which is here of middle size. The upper shadow is determined by the difference [high – max(open, close)], which is also of middle length, while the lower shadow length,

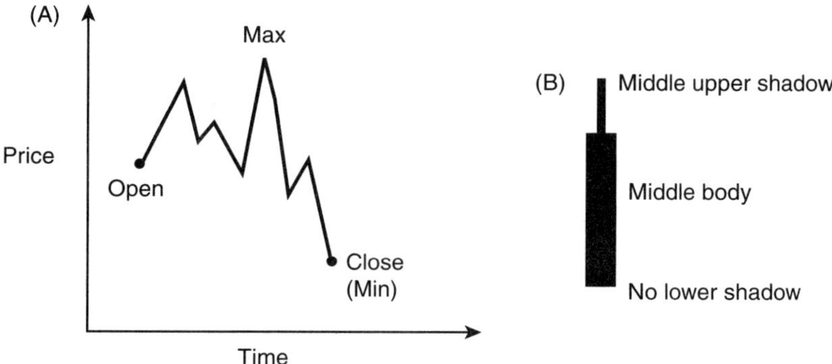

Figure 3.4 (A) A typical daily price fluctuation. (B) The associated Japanese candlestick.

defined by [low − min(open, close)] is zero, since the low and the minimum coincide. We note here that what constitutes "large," "middle," and "small" is determined by the analyst by tuning body and shadow lengths during a lengthy period of activity (say 6 months) to determine upper and lower threshold values for these quantities. Once these threshold values have been established, they become fixed features of the analysis and the determination of length. Thus any body or shadow that registers above the upper threshold is declared large, small if below the lower threshold, and middle if it falls between the two thresholds.

It has recently become possible to quantify the four characteristic features of a candlestick by means of a single binary code. This code is defined as the sequence of seven binary digits, which are given a position number as shown in the accompanying illustration.

6	5	4	3	2	1	0
X	X	X	X	X	X	X
Color	Body		Upper shadow		Lower shadow	

The first binary digit of the code represents the color of the candlestick and is set at one (1) for a white one, and zero (0) for a black one. The second and third positions, composed of a pair of binary digits, denote body size and depend on body color according to the convention shown in Table 3.1. The fourth to seventh positions, consisting of two pairs of binary digits, code the sizes of the upper and lower shadows, respectively, according to the scheme of Table 3.2. To translate the binary code into an actual decimal code, we proceed as follows.

For each of the seven positions, the number 2 is raised to the power represented by that position if there is a "1" in that position. If there is a "0," the position is omitted; that is, the term is set equal to zero. The *sum* of these terms composed of powers of 2 and zeros constitutes the decimal code. The decimal value thus obtained ranges from zero to a maximum of $2^6 + 2^5 + 2^4 + 2^3 + 2^2 + 2^1 + 2^0 = 127$.

Let us see how this works out in practice. Figure 3.5 shows the computation of a decimal code for a white candlestick with a long body and upper shadow and a small lower shadow. We start with the color, which is white, and is therefore assigned the digital code 1. Color is in position 6, that is, it is given the heaviest weight of all characteristics

Table 3.1 Digital Code for Body Size

Candlesticks	White	Black
No body size (doji)	00	11
Small body	01	10
Middle body	10	01
Large body	11	00

Table 3.2 Digital Code for Shadow Size

Shadows	Upper	Lower
None	00	11
Small	01	10
Middle	10	01
Large	11	00

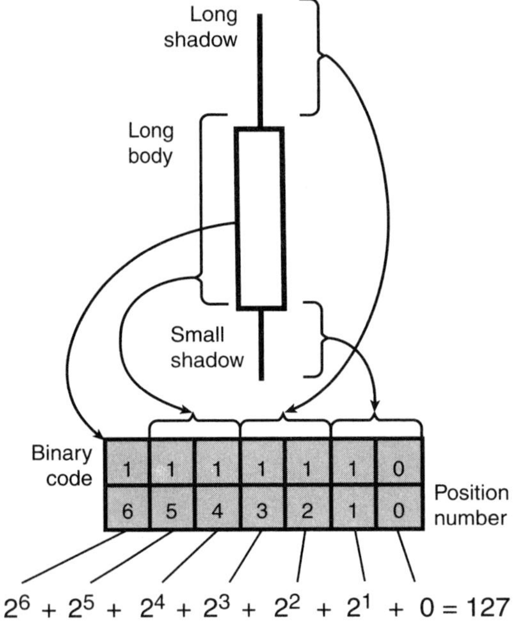

Figure 3.5 Candlestick coding: translation of the binary code of a typical candlestick into decimal code.

and yields the term 2^6. Body length follows, with the assigned digital code 11 (see Table 3.1), which leads to the terms 2^5 and 2^4. The upper shadow, which comes next and is large, takes the digital code 11 (see Table 3.2) and results in the terms 2^3 and 2^2. The lower shadow, finally, is given the lowest weight and has, because of its small size, the digital code 10 (see Table 3.2). This leads to the last two terms of the sum, 2^1 and 0. The total decimal sum is then given by $2^6 + 2^5 + 2^4 + 2^3 + 2^2 + 2^1 + 0 = 126$.

Let us next examine what the geometry of the candlestick and its decimal code can convey to us. Figure 3.6 shows three sets of candlesticks, each with a specific characteristic.

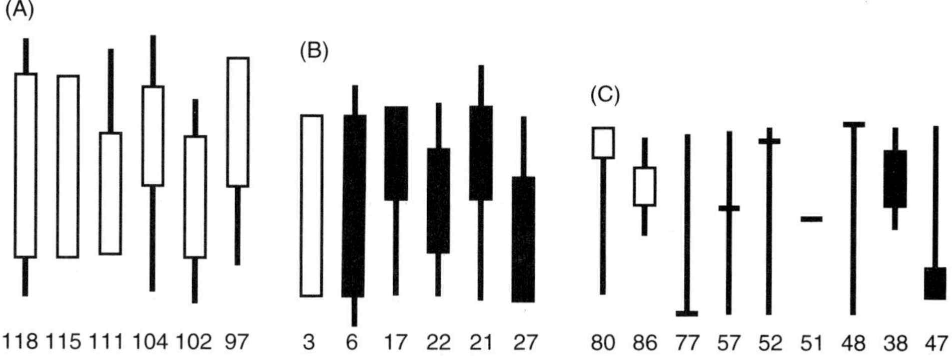

Figure 3.6 (A) Bullish candles. (B) Bearish candles. (C) Reversal.

The candles in group A are all white and have high decimal codes close to the upper limit of 127. These are all bullish candlesticks. Generally, the more bullish a candle, the higher the value of its decimal code. Typically, these values are in the range 90–127. By contrast, the candlesticks of Fig. 3.6B are all strong bearish signals, with decimal codes between 0 and 35. In general, the lower the code value, the more bearish the candlestick.

Further important information can be gleaned from the candlesticks shown in Fig. 3.6C, which fall in the intermediate code range (35–90). Candlestick 80 is referred to as a *karakasa* ("empty umbrella"). At the top of the market, it is called a *hanging man*; at the bottom, it is referred to as a *hammer*. Independent of color, a *karakasa* is a sell signal at a top and a buy signal at a bottom. Candlestick 47 is an example of a so-called *shooting star*, and candles 77, 57, 52, and 51 are all examples of doji candlesticks with varying shadow lengths. These doji variants all generate reliable alerts to possible changes in market trends. Taken as an ensemble, these indicators can be seen as signals of possible trend reversals and are consequently referred to as *reversal candles*.

The individual decimal codes obtained from a day's trading can be combined into an entity smoothed over a period of time and termed the *index of candle strength* (ICS), which is suitable for interpretations of market sentiment. We omit details here, but point out to the reader that ICS generates clear signals, often better than the usual technical indicators. In particular, ICS often signals turning points without the delays from which other indicators suffer.

COMMENTS

- The candlestick method we have described starts by translating market price fluctuations into a compact geometrical tool comprising four characteristic features: color, body length, and two shadow lengths of a candle. A visual inspection of these geometric features provides the analyst with an immediate *qualitative* sense of market trends. Thus, in general, long white candle bodies with long upper shadows are good news, while large black candle bodies with long lower shadows provide a strong bearish signal. These qualitative indicators were next quantified by first formulating a binary digital code, followed by a suitable translation into decimal code. These codes were somewhat arbitrary, but justifiably so, in that they put the greatest weight on color (i.e., the daily price change), followed by body length and upper/lower shadow lengths. The decimal code, which ranged from 0 to 127, gave bullish signals when high, bearish signals when low and provided a reliable signal of possible impending change in the intermediate range 35–80. The Japanese candlestick provides an excellent example of the application of simple and ingenious geometrical tools to illuminate what is essentially a highly complex process.

◆ EXAMPLE 3.4 Surveying: The Join Calculation and the Triangulation Intersection

Surveying is an activity concerned with the position fixing of objects on the ground. These positions are expressed in terms of several distinct parameters, displayed in Fig. 3.7.

- *Coordinates*. These are determined by an arbitrarily chosen reference point 0 and the ordinate N, which traditionally points *north*. It follows that the abscissa points east, and together the two axes define the coordinates of a point, commonly designated by N and E.
- *Bearing*: The bearing α is defined as the *clockwise* angle with respect to north. Thus the bearing for the line AB in Fig. 3.7 is designated α_{BA} at the point B and α_{AB} at the point A. α_{BA} is known as the *reverse* or *back bearing* and is given by

$$\begin{aligned} \alpha_{BA} &= \alpha_{AB} + 180^\circ \quad \text{if } \alpha_{AB} \leq 180^\circ \\ \alpha_{BA} &= \alpha_{AB} - 180^\circ \quad \text{if } \alpha_{AB} \geq 180^\circ \end{aligned} \tag{3.5}$$

- *Direction*: Direction θ is the angle between two lines, or two sighted targets. Both bearings and directions are determined in the field by means of instruments known as *theodolites*.
- *Distance D*: This quantity is defined in the usual way as the distance between two points.

We now turn to the application of these concepts to two important procedures used in surveying: the join calculation and the triangulation intersection.

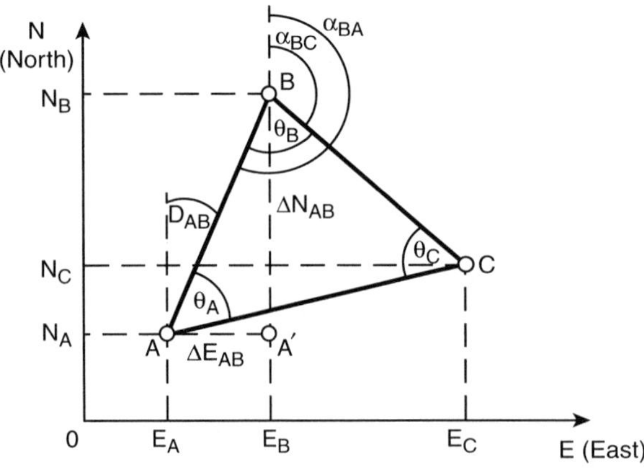

Figure 3.7 The rectangular coordinate system: α = bearing, θ = direction, D = distance, E, N = coordinates.

Join Calculation

The calculation of bearing and distance from the coordinates (N, E) of two points is known as a *join calculation* and is probably the most important and fundamental calculation in surveying. It is used frequently either to check existing data or to prepare data for more complex operations. The pertinent equations are based on the properties of the triangle AA′B of Fig. 3.7, from which we can write

$$\tan \alpha_{AB} = \frac{\Delta E_{AB}}{\Delta N_{AB}} \tag{3.6a}$$

$$\cot \alpha_{AB} = \frac{\Delta N_{AB}}{\Delta E_{AB}} \tag{3.6b}$$

and

$$D_{AB} = \left(\Delta E_{AB}^2 + \Delta N_{AB}^2\right)^{1/2} \tag{3.7a}$$

$$D_{AB} = \frac{\Delta E_{AB}}{\sin \alpha_{AB}} \tag{3.7b}$$

$$D_{AB} = \frac{\Delta N_{AB}}{\cos \alpha_{AB}} \tag{3.7c}$$

To minimize errors, one uses the formula that has the largest absolute difference as the denominator (set 3.6), or as the numerator (set 3.7). For the latter set it is common practice to use the first of the three formulae and to rely on the other two to check the result.

Let us see how this works out in practice. Suppose we are given Table 3.3, with the coordinates of two points, and we wish to calculate the joins of AB (i.e., the distance between them) and the bearing α_{AB}. We have for the coordinate differences

$$\Delta N_{AB} = N_B - N_A = 3912.56 - 4102.96 = -190.40\,\text{m}$$

$$\Delta E_{AB} = E_B - E_A = 5997.73 - 8246.18 = -2248.45\,\text{m}$$

To calculate bearing, we choose the relation (3.6b) with the largest absolute denominator and obtain

$$\cot \alpha_{AB} = \frac{\Delta N_{AB}}{\Delta E_{AB}} = \frac{-190.40}{-2248.45} = 0.084681$$

Hence $\alpha_{AB} = 85°09'35''$ or $265°09'35''$. Since both ΔN and ΔE are negative, the bearing is in the third quadrant, and we have

$$\boxed{\alpha_{AB} = 265°09'35''}$$

Table 3.3 Coordinates of Points A and B

	N (m)	E (m)
Point A	4102.96	8246.18
Point B	3912.56	5997.73

To calculate the distance D_{AB} we turn to the first relation in set (3.7) and obtain

$$D_{AB} = \left[\Delta N_{AB}^2 + \Delta E_{AB}^2\right]^{1/2} = [(-2248.45)^2 + (-190.40)^2]^{1/2}$$

$$\boxed{D_{AB} = 2256.497\ \text{m}}$$

We verify this value by using the second relation in the set. Thus

$$D_{AB} = \frac{\Delta E_{AB}}{\sin \alpha_{AB}} = \frac{-2248.45}{-0.996434}$$

$$\boxed{D_{AB} = 2256.497\ \text{m}}$$

This is the distance between the two points A and B with given coordinates.

Triangulation Intersection

In triangulation, the join calculations are in essence reversed. The task here is to establish unknown coordinates of a point C from the known coordinates of two other points A, B, and their measured bearings or directions. The situation is depicted in Fig. 3.8, which also outlines the essentials of the solution. It consists of the following steps:

1. A triangle with an apex angle θ_C is erected on the side AB. This can be done, for example, by intersecting the circle through A, B, and C (not shown) with the line CF.
2. The triangles BHD and AHC established in this construction are seen to be similar, since corresponding pairs of angles are identical. This implies proportionality of the relevant triangle sides, which in turn leads to the conclusion that the triangles AHB and CHD must likewise be similar. From this it follows that the angle HDC $= \theta_B$, and therefore BDF $= \theta_A$. [We recall for the benefit of the reader that two triangles are similar if (a) all three pairs of sides are proportional, (b) all three pairs of angles are identical, or (c) two pairs of sides are proportional and the angles in one pair are identical.]

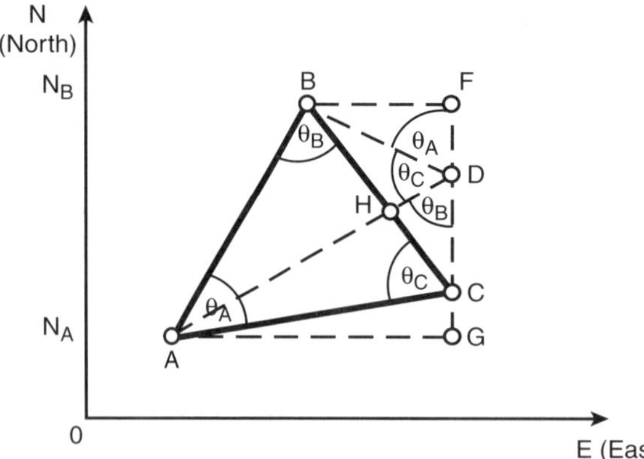

Figure 3.8 Triangulation intersection.

3. Focusing on the triangle BFD, the following relations are established:

$$\begin{aligned} FD &= BF\cot\theta_A = (E_C - E_B)\cot\theta \\ DG &= AG\cot\theta_B = (E_C - E_A)\cot\theta_B \end{aligned} \tag{3.8}$$

from which one obtains

$$N_B - N_A = FD + DG = E_C(\cot\theta_A + \cot\theta_B) - E_A\cot\theta_B - E_B\cot\theta_A \tag{3.9}$$

Solving equation (3.9) for E_C yields

$$\boxed{E_C = \frac{E_A\cot\theta_B + E_B\cot\theta_A - N_A + N_B}{\cot\theta_A + \cot\theta_B}} \tag{3.10}$$

and similarly

$$\boxed{N_C = \frac{N_A\cot\theta_B + N_B\cot\theta_A - E_A + E_B}{\cot\theta_A + \cot\theta_B}} \tag{3.11}$$

Equations (3.10) and (3.11) thus express the unknown coordinates of the point C in terms of the coordinates of two known points A, B, and the measured directions θ_A and θ_B.

COMMENTS

- It is not at all evident at first sight that unknown coordinates of a point can be easily computed from two known locations and a pair of measured angles. We know that one side and two angles define a triangle and thus fix the position of C, but it still requires the imaginative use of geometry to arrive at the coordinate equations (3.10) and (3.11). Note that only simple concepts involving triangles were invoked—similarity and basic trigonometry—and that these led to simple expressions allowing the direct evaluation of N_C and E_C.

◆ EXAMPLE 3.5 The Global Positioning System (GPS)

Example 3.4, on surveying, we addressed the problem of fixing the unknown position of an object located on a plane, given certain observations made at points with known coordinates. The present example in a sense inverts the problem. Here we are asked to find the coordinates of the observer himself, given certain measurements with respect to objects in space, whose coordinates are known. In the early days of nighttime aerial navigation, this was done by measuring the bearing of the plane with respect to certain stars or constellations. This practice was superseded by the use of radio and radar. These methods are not convenient for everyday use in determining the coordinates of an average observer located at some unknown position on earth.

The launching over the last three decades of so-called geosynchronous satellites (i.e., satellites located at a fixed point above the surface of the earth) has made it possible to determine positions anywhere on earth. This is done by measuring the time it takes for a signal to travel between the observer and any one satellite and translating this into a distance r_i between the two.

To extract the desired coordinates of the observer from these measurements, we construct a sphere of radius r_i about each of four satellites. The equations, describing these spheres are given by

$$(x - x_1)^2 + (y - y_1)^2 + (z - z_1)^2 = r_1^2 \tag{3.12a}$$

$$(x - x_2)^2 + (y - y_2)^2 + (z - z_2)^2 = r_2^2 \tag{3.12b}$$

$$(x - x_3)^2 + (y - y_3)^2 + (z - z_3)^2 = r_3^2 \tag{3.12c}$$

$$(x - x_4)^2 + (y - y_4)^2 + (z - z_4)^2 = r_4^2 \tag{3.12d}$$

where x_i, y_i, z_i are the known coordinates of the satellites in space, and x, y, z are the unknown coordinates of the observer on earth (Fig. 3.9).

We now subtract the first of these equations from each of the other three. This eliminates quadratic terms in x, y, z and leads to the following set of nonhomogeneous *linear* equations in x, y, and z:

$$2(x_2 - x_1)x + 2(y_2 - y_1)y + 2(z_2 - z_1)z = r_1^2 - x_1^2 - r_2^2 + x_2^2 \tag{3.13a}$$

$$2(x_3 - x_1)x + 2(y_3 - y_1)y + 2(z_3 - z_1)z = r_1^2 - x_1^2 - r_3^2 + x_3^2 \tag{3.13b}$$

$$2(x_4 - x_1)x + 2(y_4 - y_1)y + 2(z_4 - z_1)z = r_1^2 - x_1^2 - r_4^2 + x_4^2 \tag{3.13c}$$

This system can be solved by the method of Gaussian elimination outlined in Chapter 2.

Additional examples of this procedure and their numerical solution are taken up in Chapter 8.

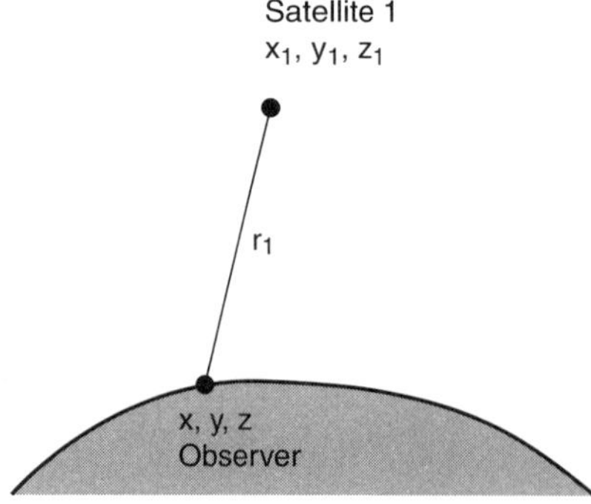

Figure 3.9 Satellite and observer in the global positioning system.

Let us consider an actual numerical example. Suppose the coordinates of the satellites with respect to a chosen point on earth are given by the following values (in meters).

$$x_1 = 2{,}088{,}202.299 \qquad x_3 = 35{,}606{,}984.591$$
$$y_1 = -11{,}757{,}191.370 \qquad y_3 = 94{,}447{,}027.239$$
$$z_1 = 25{,}391{,}471.881 \qquad z_3 = 9{,}101{,}378.572$$
$$x_2 = 11{,}092{,}568.240 \qquad x_4 = 3{,}966{,}929.048$$
$$y_2 = -14{,}198{,}201.090 \qquad y_4 = 7{,}362{,}851.831$$
$$z_2 = 21{,}471{,}165.950 \qquad z_4 = 26{,}388{,}447.172$$

Suppose that in addition the distances registered by the GPS device are given by

$$r_1 = 23{,}204{,}698.51$$
$$r_2 = 21{,}585{,}835.37$$
$$r_3 = 31{,}364{,}260.01$$
$$r_4 = 24{,}966{,}798.73$$

When these values are substituted into the set of equations (3.13), we obtain the numerical equations

$$(1.80887 \times 10^7)x - (4.88202 \times 10^6)y - (7.84061 \times 10^6)z$$
$$= 1.9119422563676106 \times 10^{14} \tag{3.14a}$$
$$(6.70376 \times 10^7)x + (1.46203 \times 10^7)y - (3.25802 \times 10^7)z$$
$$= 8.182379897872735 \times 10^{14} \tag{3.14b}$$
$$(3.75745 \times 10^6)x + (3.82401 \times 10^7)y + (1.99395 \times 10^6)z$$
$$= -7.3507068654016 \times 10^{13} \tag{3.14c}$$

We use Gaussian elimination to solve the system, obtaining

$$x = -3.0168 \times 10^6 \text{ m}$$
$$y = 7231.03 \text{ m}$$
$$z = -3.13188 \times 10^7 \text{ m}$$

These are the coordinates of the observer with respect to a fixed known origin on earth.

COMMENTS

- We have here the application of simple expressions drawn from the analytical geometry of a sphere and the elementary use of Gaussian elimination to arrive at the solution of a problem of some sophistication. The practical implementation of the GPS had to await the deployment of geosynchronous satellites and the hardware needed to produce compact GPS devices, which can now be installed in automobiles.

It is to be noted that three equations (i.e., the use of only three satellites) would have been sufficient to calculate the coordinates x, y, z of the observer. In Practice Problem 3.5, the reader is asked to explain why this was not done.

◆ EXAMPLE 3.6 The Orthocenter of a Triangle

This geometry problem would require a considerable effort if it were to be addressed by conventional geometrical procedures. We use instead an ingenious application of the scalar product of vector algebra, which provides the essentials of the solution in a single line.

The task is to prove that the altitudes of a triangle ABC intersect in a single point H called the *orthocenter* of the triangle (see Fig. 3.10). Use of the scalar product suggests itself here, since we must give expression to orthogonality of altitudes and the respective triangle sides. We can accomplish this by setting three appropriate scalar products equal to zero. (Recall that two orthogonal vectors have a zero scalar product.) These scalar products are composed of the position vectors we have sketched in Fig. 3.10.

Consider, for example, the altitude orthogonal to the side AB. We can express this orthogonality by writing

$$\overrightarrow{\mathbf{BA}} \cdot \overrightarrow{\mathbf{CH}} = 0 \tag{3.15}$$

or alternatively

$$(\mathbf{a} - \mathbf{b}) \cdot (\mathbf{h} - \mathbf{c}) = 0 \tag{3.16}$$

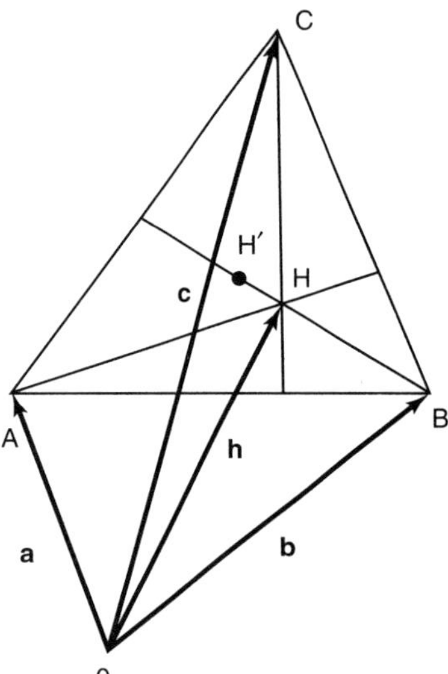

Figure 3.10 The orthocenter (H) of a triangle ABC.

Similarly we have for the other two altitudes

$$(\mathbf{b} - \mathbf{c}) \cdot (\mathbf{h} - \mathbf{a}) = 0 \tag{3.17}$$

$$(\mathbf{c} - \mathbf{a}) \cdot (\mathbf{h}' - \mathbf{b}) = 0 \tag{3.18}$$

where we have substituted a position vector $\mathbf{h}'$ for $\mathbf{h}$ to show that we have not as yet proven intersection at a single point. Addition of the equations (3.16) to (3.18) yields the identity

$$(\mathbf{a} - \mathbf{b}) \cdot (\mathbf{h} - \mathbf{c}) + (\mathbf{b} - \mathbf{c}) \cdot (\mathbf{h} - \mathbf{a}) + (\mathbf{c} - \mathbf{a}) \cdot (\mathbf{h}' - \mathbf{b}) = 0 \tag{3.19}$$

which can be dot-multiplied out to give, after cancellation of terms,

$$\mathbf{h} \cdot (\mathbf{a} - \mathbf{c}) + \mathbf{h}' \cdot (\mathbf{c} - \mathbf{a}) = 0 \tag{3.20}$$

But equation (3.20) can hold only if $\mathbf{h} = \mathbf{h}'$. Hence intersection must occur at a single common point.

COMMENTS

- The use of both scalar products and position vectors is crucial to the proof here, since they are needed to arrive at equation (3.20), which is the centerpiece of the procedure. One could have used the vectors represented by the sides and altitudes instead of the position vectors, but this would not have yielded a useful result. Some initial thought and trial-and-error is evidently needed to get off to a successful start.

◆ EXAMPLE 3.7 Relative Velocity and the Wind Triangle

In this example, we prepare the ground for Example 3.8, which deals with the interception of an aircraft and requires consideration of the motion of the pursuing plane with respect to a *fixed* reference frame (the ground) and a *moving* reference frame (the wind). We therefore begin by examining the velocities that arise in such systems.

Consider the particle P, which is moving with respect to a fixed reference frame F (Fig. 3.11). If its path has the equation $\overrightarrow{OP} = \mathbf{r} = \mathbf{r}(t)$ in terms of time as a parameter, then the velocity vector $\mathbf{v}$ is defined as the time derivative

$$\mathbf{v} = \frac{d\mathbf{r}}{dt} \tag{3.21}$$

which can be written in the form

$$\mathbf{v} = \frac{d\mathbf{r}}{ds}\frac{ds}{dt} = \mathbf{T}\frac{ds}{dt} = v\mathbf{T} \tag{3.22}$$

where $\mathbf{T}$ is the unit tangent vector to the path of the particle. The velocity of P is thus a vector tangent to its path in the direction of motion, and of length numerically equal to its speed. Proof that $d\mathbf{r}/ds = \mathbf{T}$ is left to Practice Problem 3.6.

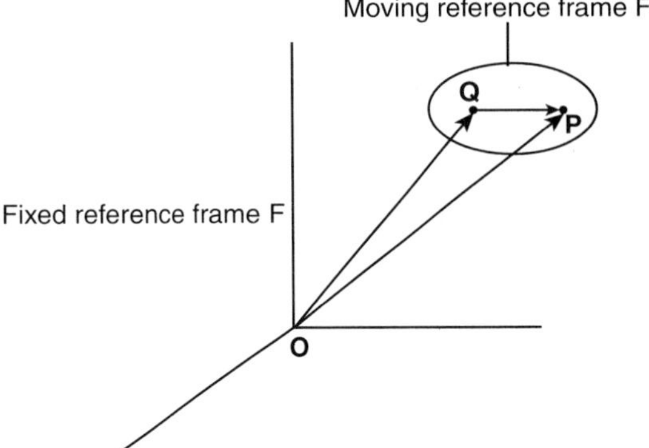

Figure 3.11 Relative velocity.

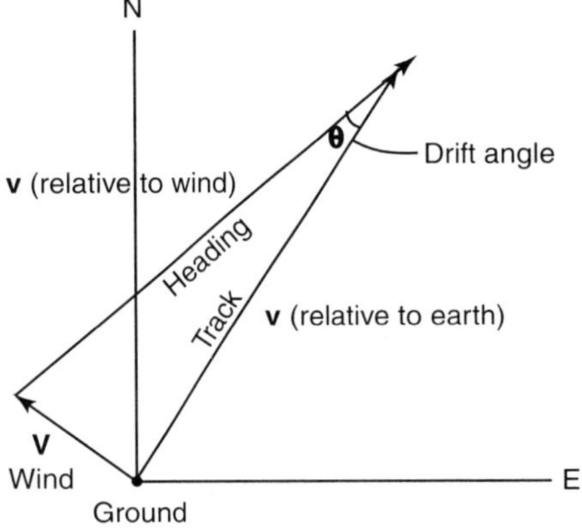

Figure 3.12 The wind triangle.

Consider next a point Q, which is fixed *within* the moving reference frame F′ (i.e., moving with it). It then follows from the construction shown in Fig. 3.11 that

$$\overrightarrow{\mathbf{OP}} = \overrightarrow{\mathbf{OQ}} + \overrightarrow{\mathbf{PQ}} \tag{3.23}$$

and that consequently

$$\frac{d}{dt}\overrightarrow{\mathbf{OP}} = \frac{d}{dt}\overrightarrow{\mathbf{OQ}} + \frac{d}{dt}\overrightarrow{\mathbf{PQ}} \tag{3.24}$$

or alternatively

$$\mathbf{v}_{PF} = \mathbf{v}_{QF} + \mathbf{v}_{PF'} \tag{3.25}$$

Here $\mathbf{v}_{PF}$ is the absolute velocity of P with respect to the fixed reference frame, $\mathbf{v}_{PF'}$ is the relative velocity of P with respect to the moving reference frame, and $\mathbf{v}_{QF}$ is the velocity of the moving reference frame itself. It follows that the absolute velocity of a particle is equal to the sum of the relative velocity and the velocity of the moving reference frame. Let us see how these concepts can be applied to the motion of a plane relative to the ground and the air. The situation is depicted in Fig. 3.12.

The two velocities of the plane are denoted by $\mathbf{v}$ and $\mathbf{v}'$, where $\mathbf{v}$ is the absolute velocity of the plane, whose magnitude is termed the *ground speed,* while $\mathbf{v}'$ is the relative velocity, whose magnitude is called the *air speed.* Air speed is registered by velocity-measuring equipment onboard the plane, usually a Pitot tube (see Fig. 4.12E). The *directions* of $\mathbf{v}$ and $\mathbf{v}'$ determine the so-called *heading* and *track* of the plane. Heading is the direction in which the plane is pointed, while the track is the path traced out over the ground. The angle ($\mathbf{v}$, $\mathbf{v}'$) from heading to track is the so-called *drift angle* θ. The resulting construction is termed the *wind triangle.*

With these notions in place, we are in a position to address the problem of the interception of a plane.

◆ EXAMPLE 3.8 Interception of an Airplane

In Example 3.9, we shall consider the interception of a moving object by a pursuer whose path is continually adjusted to allow interception in minimum time. The present example, which considers a simpler, straight-path interception of one object, an airplane, by another, is complicated by the need to consider the effect of wind. Example 3.7 prepared the ground for this task by examining the different relative velocities that arise in this case and by deriving the so-called wind triangle, which needs to be applied to establish the path (track) of the pursuing plane. The diagrams displayed in Fig. 3.13 show the position or locations of the pursuer and the pursued, as well as their associated velocities. The task will be to establish the track $\overrightarrow{\mathbf{eq}}$ of the pursuer and the time necessary to effect the interception. To do this, we proceed as follows.

Let us assume that the pursued plane p is flying over the track PX with the ground speed ep. As plane p passes the point P, the pursuing plane q departs from point Q to intercept the plane p, with an air speed wq. The wind velocity $\overrightarrow{\mathbf{eq}}$ is known. Thus, at the start of this problem, the following are known and can be entered in the diagrams:

- Track PX of the pursued plane p
- Starting point Q of the pursuing plane q
- *Direction* $\overrightarrow{\mathbf{QP}}$ of the velocity of q relative to p
- *Magnitude* of the velocity of q relative to the wind (i.e., the air speed wq)
- *Magnitude* and *direction* of the wind velocity $\overrightarrow{\mathbf{ew}}$

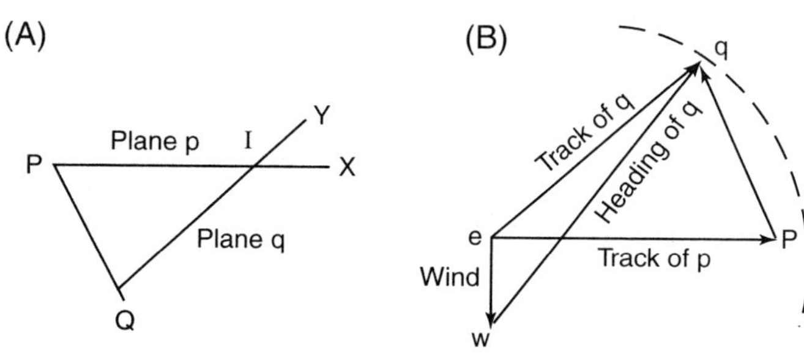

Figure 3.13 Interception of a plane: (A) positions and (B) velocities.

The unknown quantities are

- *Magnitude* of the velocity of q relative to p
- *Direction* of the velocity of q relative to the wind (i.e., the heading of the pursuer q).

These two items together fix point q in Fig. 3.13B and the wind triangle ewq. The remaining unknown pieces then easily fall into place. We proceed as follows.

1. Enter the known air velocity $\overrightarrow{\mathbf{ew}}$ and the ground velocity $\overrightarrow{\mathbf{ep}}$ of the pursued plane p in Fig. 3.13B. Both these velocities are with respect to the earth.
2. With w as a center, describe a circle having the known air speed wq as its radius. Note that the location of q is not known at this stage, but we do know that it must lie on the circle.
3. Draw a line through p parallel to the known direction $\overrightarrow{\mathbf{QP}}$. Intersection with the circle fixes point q.

With these items in place, the unknown quantities are quickly established as follows.

The track of the pursuing plane is obtained by connecting e and q in Fig. 3.13B. A line parallel to $\overrightarrow{\mathbf{eq}}$ drawn through Q of Fig. 3.13A establishes the location of the point of interception I. Interception time can then be derived by dividing distance covered to point I by the ground speed of either plane p or plane q. Thus

$$\boxed{\text{Flying time to interception} = \frac{\text{QI}}{\text{eq}} = \frac{\text{PI}}{\text{ep}}}$$

COMMENTS

- We have here an elegant application of simple vector concepts to a problem of some complexity. It is evident that some initial "doodling" involving the location and velocity diagrams is needed to establish known and unknown quantities and to map out a strategy for the solution of the problem. The concepts established in Example 3.7 and the wind triangle shown in Fig. 3.13 aid in clarifying the situation.
- Depending on the position of p, and the direction of $\overrightarrow{\mathbf{QP}}$, two interesting and notable subcases arise.

    1. The ray through p parallel to $\overrightarrow{\mathbf{QP}}$ does not intersect the circle. This nonintersection happens when $\overrightarrow{\mathbf{ep}}$ is large or $\overrightarrow{\mathbf{PQ}}$ is steep, and it corresponds to the situation of a pursued plane that is too fast or too far ahead to allow interception. The strategy here would be to raise the air speed of the pursuing plane, thus increasing the radius of the described circle to obtain intersection. There will be a critical minimum air speed at which this can be accomplished, and Fig. 3.13B helps us to establish this value in rapid fashion. One simply increases the radius of the circle until the ray through p parallel to PQ in Fig. 3.13A yields a first intersection. This determines the minimum air speed of q required to effect an interception.
    2. The ray through p parallel to $\overrightarrow{\mathbf{QP}}$ intersects the circle in *two* points, say q_1 and q_2. Plane q can then intercept p along two different tracks parallel to $\overrightarrow{\mathbf{eq}_1}$ and $\overrightarrow{\mathbf{eq}_2}$.

We next turn to a problem where the path of interception is curved, rather than straight as was the case in Example 3.8.

◆ EXAMPLE 3.9 Path of Pursuit

In this example, involving geometry as well as geometrical aspects of calculus, we consider a predator B, moving at a constant speed b, in pursuit of a prey A, moving in a straight line at a constant speed a (Fig. 3.14). The task will be to find the path of pursuit, $y = f(x)$, which the predator must follow to intercept the prey in minimum time.

To accommodate this situation, we position the prey A on the y axis, which is also the straight-line path taken by it. The predator B is initially placed on the x axis, from which it moves away as the pursuit develops. Since the predator will keep the prey in constant sight, a "line of sight" can be drawn connecting the two, which will perforce be tangent to the moving predator, hence tangent to the path of pursuit. This is shown in Fig. 3.14. Note that the variable position y of the prey has been related to elapsed time and velocity of the prey through the term at. This is the first of a series of ingenious moves to quantify the problem.

We start the modeling process by writing out the equation for the tangent. This serves to introduce the derivative of the function we ultimately seek. We obtain

$$\frac{dy}{dx} = \frac{y - y_1}{x - x_1} = \frac{y - at}{x - 0} \tag{3.26a}$$

or alternatively

$$xy' - y = -at \tag{3.26b}$$

The task is now to eliminate time t as a variable by expressing it in terms of the system coordinates x, y. This is done in a series of clever steps, which we outline next.

We start by differentiating equation (3.26b) with respect to x. This leads to the result

$$xy'' = -a\frac{dt}{dx} \tag{3.27}$$

We next decompose dt/dx along the arc s of the path, obtaining

$$\left(\frac{dt}{dx}\right) = \left(\frac{dt}{dx}\right)\left(\frac{ds}{dx}\right) \tag{3.28}$$

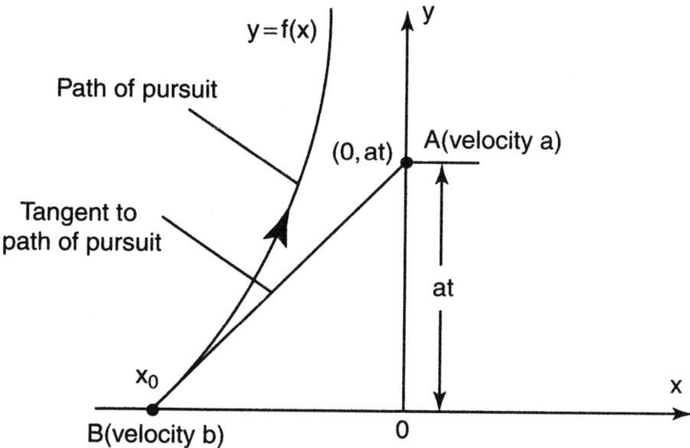

Figure 3.14 Path of pursuit.

This appears to be a retrograde step, since we have merely managed to introduce an additional variable, s. Closer examination shows, however, that the derivatives appearing on the right side of equation (3.28) can be expressed in terms of the system parameters and coordinates, devoid of either time t or arc length s. We first recognize dt/ds as the inverse of the speed of the predator. Thus

$$\frac{dt}{dx} = \frac{1}{b} \tag{3.29}$$

The second term in equation (3.28), ds/dx, will be recognized as the directional derivatives, which can be related to dy/dx by the standard formula known from elementary calculus. We have

$$\frac{ds}{dx} = [1 + (y')^2]^{1/2} \tag{3.30}$$

We have thus managed, at one and the same time, to eliminate the undesired variables t and s. It remains to introduce the relations (3.28) to (3.30) into equation (3.27). This gives the result

$$xy'' = -\frac{a}{b}[1 + (y')^2]^{1/2}$$

or equivalently,

$$1 + \left(\frac{dy}{dx}\right)^2 = \frac{b^2}{a^2}x^2\left(\frac{d^2y}{dx^2}\right)^2 \tag{3.31}$$

Solution of this nonlinear ODE, which is the object of Practice Problem 3.8, yields the desired path of pursuit $y = f(x)$.

COMMENTS

- One will note the ingenious use of several concepts and relations, drawn mainly from calculus, to arrive at the final model, equation (3.31). The procedure will to many appear to have been circuitous, starting with the expression for the location of the prey in terms of velocity and elapsed time, whereupon elapsed time had to be eliminated by introducing the directional derivatives ds/dx. The temptation here would have been to denote the position of the prey as y, but this would have brought the proceedings to a halt. The next step might have been to try to express prey location in terms of x, but this is clearly not possible because $x = 0$ along the path. This leaves as the only alternative an expression in t, and here it takes a good sense of the tools available from calculus to realize that a conversion of t to x, or dy/dx, not only is possible but can be accomplished with relative ease.
- The reader will have noted that the path presented here is only one of several alternative trajectories that present themselves. The result obtained from equation (3.31) is clearly an *optimum*, since interception is achieved in minimum time by continually adjusting the trajectory. A more straightforward but less efficient method is to use a path of constant slope, that is to proceed along a straight line. This type of trajectory was addressed in Example 3.8 in the context of airplane interception. The complicating factor there was the effect of wind, which calls for an adjustment in the trajectory and requires the use

of vector concepts. Finally, interception may be complicated by evasive action taken by the prey. This is clearly a more complex problem, which is not addressed here. The model equations we have established however, provide some guidelines for the solution of that more complicated case.

◆ EXAMPLE 3.10 Trilinear Coordinates: The Three-Jug Problem

The customary procedure for dealing with three functions x, y, z is to represent them in rectangular, cylindrical, or spherical coordinates, depending on the geometry involved. In many instances it becomes more convenient to carry out the representation in a graduated trilinear coordinate system, an example of which appears in Fig. 3.15. Such systems find many practical applications, among them operations such as solvent extraction, in which a substance C (e.g., acetic acid) dissolved in a liquid (e.g., water), is extracted with an immiscible solvent B (e.g., benzene). The associated triangular diagram is shown in Fig. 3.16 and an example of its application is to be found in Practice Problem 3.10.

This example uses trilinear coordinates to address the following problem: We are given three jugs A, B, and C, with respective capacities of 8, 5, and 3 L. The first jug is full of water. A procedure is to be found for using the two empty jugs to measure out exactly 4 L without spilling a drop. None of the jugs is graduated.

We start our deliberations by examining the sample trilinear diagram shown in Fig. 3.15. The graduations are given in three digits, which represent the respective contents of the jugs A, B, and C. Thus the apex A (800) denotes a full jug A (8 L) and jugs B and C completely empty. Pouring water from one jug into another is equivalent to moving along one of the lines. For example, if we consider three 8 L jugs and denote the initial state by a black circle and the final state by a white circle, we can identify four different paths for reaching the final point. If the jugs are not graduated, only one of these paths (shown by a thick line), consisting of either completely filling or completely emptying one of the jugs,

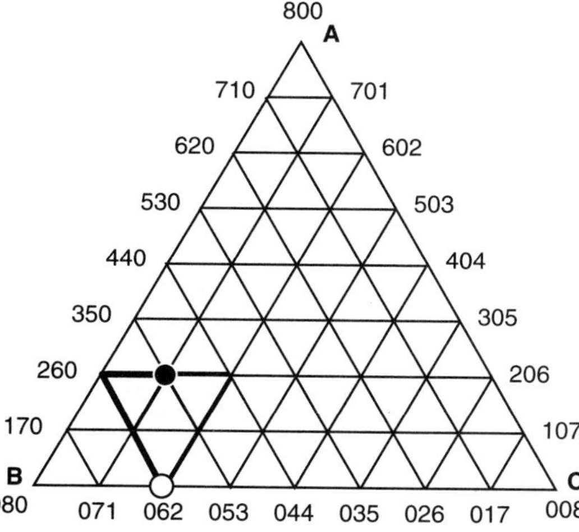

Figure 3.15 Trilinear coordinates showing four possible pathways to proceed from position 251 to position 062. All three jugs are of 8 L capacity. Heavy line shows the only possible pathway for ungraduated jugs.

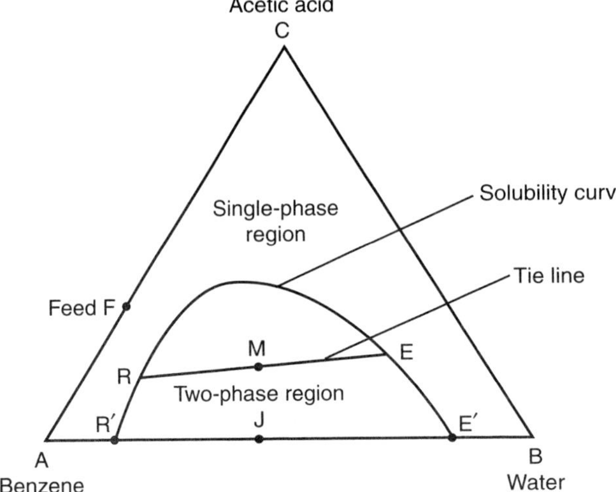

Figure 3.16 Use of trilinear coordinates to describe systems composed of three liquids, two of which (A and B) are partially soluble.

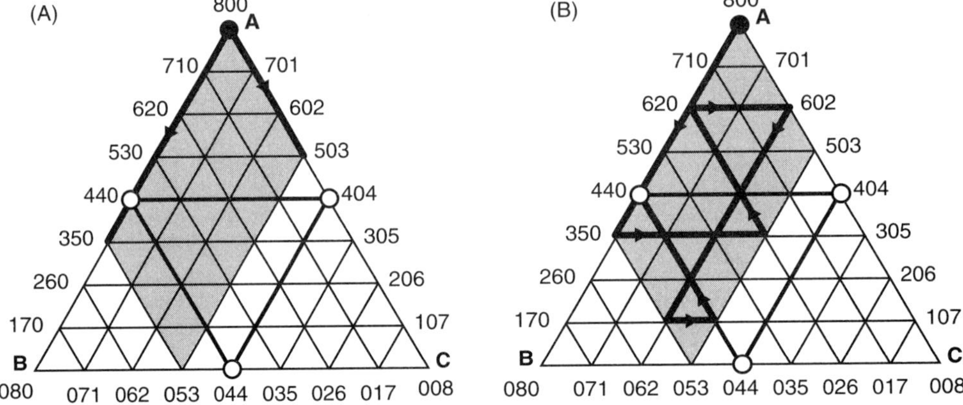

Figure 3.17 Trilinear coordinates and the three-jug problem: (A) feasible region and perimeters for jugs of 8, 5, and 3 L capacity, and (B) one of several pathways to fill two jugs with 4 L each.

is possible. At the start of the operation the three jugs are respectively filled with 2, 5, and 1 L. Emptying jug C (1 L) into jug B (5 L), we fill the latter to 6 L while retaining 2 L in jug A. We thus reach the coordinates 260. If we next empty jug A (2 L) into the empty jug C, we arrive at the final state 062.

We return to our original jug problem and start with an examination of Fig. 3.17A. We observe that since jugs B and C only hold 5 and 3 L, respectively, not every point in the diagram will be a feasible state. To take these reduced capacities into account we draw lines through 053 and 350 to denote the 5 L limit of jug B and through 053 and 503 to denote the 3 L limit of jug C. The resulting shaded parallelogram represents the feasible region. Also shown in the diagram is the triangular perimeter on which at least one jug holds 4 L. There are seven such points within the feasible region and we choose one of them, the point with coordinates 440, as the desired final state.

There are only two ways in which the trajectory can be started. We fill either jug B or jug C with part of the water from jug A. We choose the first of the two, leaving the second path to the exercises. The starting point is the apex A.

To establish the subsequent steps of the trajectory we make use of the following important rule: each step must start and end on the perimeter of the shaded parallelogram. This follows from the requirement that we must in the course of a single step either empty or fill one of the jugs completely. With this rule in hand, the number of choices is vastly reduced and one can, with some imagination and skill, trace out the complete trajectory by inspection. It takes six intermediate steps to do this, which are as follows (Fig. 3.17B):

1. Fill 5 L from jug A into jug B. This lands us on the coordinates 350.
2. We next fill 3 L from jug B into jug C, reaching the coordinates 323.
3. Then 3 L are emptied from jug C into A, yielding the coordinates 620.
4. The next step is to empty 2 L from B into C. This puts us at the coordinates 602.
5. We follow this up by emptying 5 L from A into B, reaching the coordinates 152.
6. Now 1 L is poured from B into C, leading to the coordinates143. We could at this point terminate the procedure because one of the jugs now contains exactly 4 L. However, since the more ambitious coordinates 440 were chosen as the end point, one more step is required to complete the trajectory.
7. We empty 3 L from jug C into jug A. This yields the desired coordinates 440. We have thus in seven steps, using six intermediate stages, split the original 8 L of jug A into two equal portions of 4 L each.

COMMENTS

- Although it could be argued that the trajectory could have been deduced by inspired guesswork, the use of the trilinear coordinates and the delineated feasible region vastly simplify and rationalize the entire procedure. Furthermore it enables us, by simple inspection, to establish both the existence and number of alternative routes. For example, a quick examination of Fig. 3.17B shows that using the same first step leading from the coordinates 800 to 530, two additional pathways can be traced to arrive at the coordinates 440. In one of these, the first four steps leading to the coordinates 602 remain the same. Instead of then moving to the coordinates 152, we proceed to 503, cross over to 530, and finally descend to the end point 440. The second alternative retains the first five steps (up to the coordinates 152), but then crosses over to 350 and finally moves along the side of the triangle to the end point 440. We reiterate that all three pathways can be found, once the feasible perimeter has been established, by a cursory inspection of the triangular diagram. This is a considerable achievement and demonstrates the power of this particular geometric concept.

◆ EXAMPLE 3.11 Inflecting Production Rates and Multiple Steady States: The van Heerden Diagram

The observation has been made in a number of disciplines that if one plots the production rate of a physical entity, such as heat or mass, against a pertinent state variable (temperature or concentration), an inflection may occur, resulting in an S-shaped curve. For example,

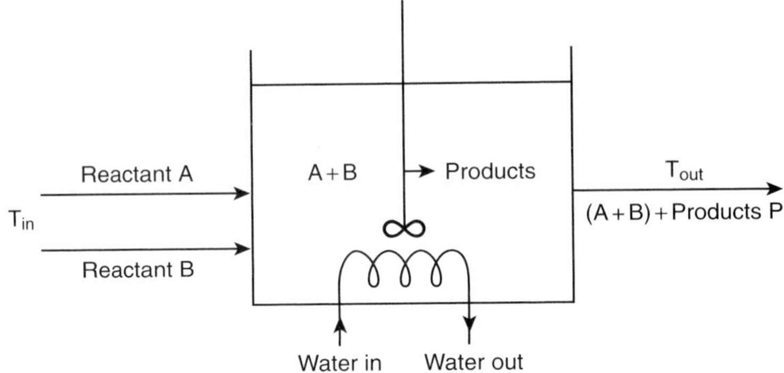

Figure 3.18 Exothermic reaction in a stirred tank.

if an exothermic or heat-producing chemical reaction is carried out in a tank receiving a constant inflow of reactants A and B, while products P are continuously removed (Fig. 3.18), the rate of heat production plotted against temperature will exhibit an inflection, as shown in Fig. 3.19A. The reason for this lies in the existence of two opposing factors. The first is the strong exponential dependence on temperature of the reaction rate constant k, expressed through the well-known Arrhenius equation

$$k_r(T) = A \exp\left(\frac{-E_A}{RT}\right) \tag{3.32}$$

which is linked to the overall reaction rate r through the expression

$$r = k_r(T) C_A^m C_B^n \tag{3.33}$$

Here r is expressed in units of moles of A or B reacted per volume × time, and C_A, C_B are the reactant concentrations in units of moles per volume; E_a is known as the activation energy (J/mol), and R is the universal gas constant = 8.314 J/mol K.

The second and opposing factor lurks in the term $C_A^m C_B^n$ of equation (3.33), which indicates that the reaction rate r will *diminish* as reactants are consumed to form product. This causes the production rate curve to inflect and to approach a constant asymptotic value, which corresponds to the rate of heat produced by complete conversion to products.

It is customary in such exothermic reactions to provide external cooling (Fig. 3.18) to prevent undesirable side effects due to excessive temperature levels (evaporation, decomposition, degradation, and extraneous reactions). One can then argue that for a steady-state operation, the heat generated by the reaction will be exactly balanced by the heat removed by the cooling medium and the heat content (or enthalpy) difference between entering and exiting streams. We can express this formally by writing the following energy balance:

$$\text{Rate of energy in} - \quad \text{Rate of energy out} \quad = 0$$

$$\begin{bmatrix} -\Delta H_r V \\ +Q\rho H_{in} \end{bmatrix} - \begin{bmatrix} Q\rho H_{out} \\ +UA(T_{out} - T_{ext}) \end{bmatrix} = 0 \tag{3.34}$$

where ΔH_r = heat of reaction (J/mol) (negative for exothermic reactions), V = volume of tank contents (m^3), Q = flow rate (m^3/s), ρ = density (kg/m^3), H = enthalpy

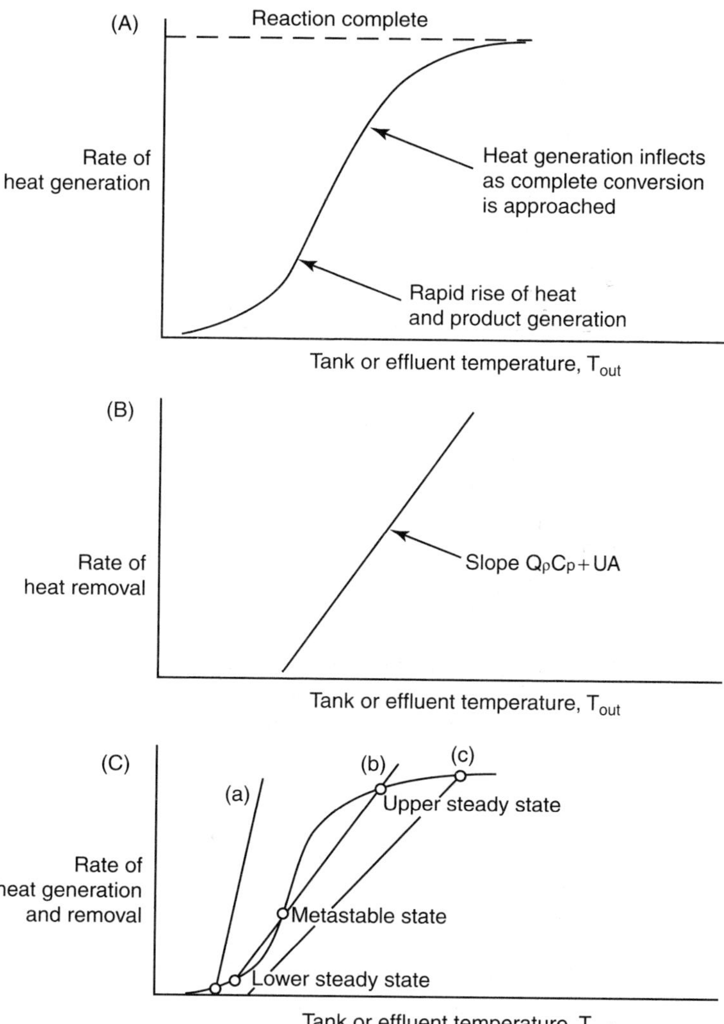

Figure 3.19 Genesis of the van Heerden diagram: (A) rate of heat production for an exothermic irreversible reaction, (B) rate of heat removal, and (C) superposition of A and B leading to the van Heerden diagram.

(J/mol), U = overall heat transfer coefficient (J/m^2 s), A = heat transfer area (m^2), and T = temperature (K).

For practical purposes one expresses enthalpy in terms of temperature, using the thermodynamic relation $H_{out} - H_{in} = \Delta H = C_p(T_{out} - T_{in})$, where C_p = heat capacity (J/kg K) [equation (1.21) of Table 1.3]. Rearrangement of equation (3.34) then leads to the following convenient form:

$$\begin{aligned} &\text{Rate of heat generation} = && \text{Rate of heat removal} \\ &-\Delta H_r r V && = (-Q_\rho C_p T_{in} - UAT_{ext}) + (Q_\rho C_p + UA)T_{out} \end{aligned} \tag{3.35}$$

The left side represents the heat generation curve shown in Fig. 3.19A. The right side can be plotted as a straight line of rate vs T_{out} and is shown in Fig. 3.19B. We now proceed to superpose the two plots and arrive at the composite shown in Fig. 3.19C, often referred to as a van Heerden diagram. Several features of this diagram are noted.

We start with the observation that depending on the location of the rate-of-removal line, one can have *one* or *three* intersections with the rate of production curve, that is, one or three steady-state tank or effluent temperatures. At very low or very high cooling rates (low or high values of the product UA), or at very low or high flow rates Q, only one steady state results. It is at intermediate values of these variables that one encounters *several* steady states. Of these, only the upper and lower states are stable, while the middle state is metastable (i.e., does not materialize spontaneously). This can be shown as follows.

Suppose the middle state in Fig. 3.19C is slightly disturbed upward by a small temperature rise. Heat produced will then exceed the heat removed, leading to a further increase in temperature. The process thus repeats itself until the upper steady state is reached. That point is *stable*, for a further increase in temperature would result in heat removal exceeding heat production and a consequent return to the point C. By similar reasoning, if a slight disturbance *lowers* the temperature of the middle state, that process will continue until the lower (and stable) steady state is reached.

The existence of three steady states, two stable and one metastable, is fairly common for highly exothermic reactions in stirred tanks. Perhaps even more common is the existence of only one steady state. In the production of polystyrene from styrene, a highly exothermic reaction carried out extensively in the plastics industry, all three steady states exist only for a limited range of the feed temperature T_{in}. If T_{in} is sufficiently high, only the upper "runaway" condition can be realized. For intermediate values of T_{in}, all three steady states are possible.

Let us consider the example of styrene polymerization in more detail, using a feed temperature $T_{in} = 300$ and *no* external cooling (UA = 0). Cooling is instead provided by the incoming feed and can still result in three steady states. In addition, the following information is provided:

$$\text{Residence time } \tau = \frac{\text{Reactor volume}}{\text{Flow rate}} = 2\text{h}$$

$$\text{Reaction rate } r = k_r C_{out} = 10^{10} \exp\left(-\frac{10^4}{T_{out}}\right) \text{ mol/h L and } C_{in}\frac{\Delta H_r}{\rho C_p} = 400\text{ K}$$

The problem to be addressed is the calculation of the resulting steady-state temperatures T_{out} and the associated fractional conversion X of the reactant (styrene) to product (polystyrene). That conversion is given by the expression

$$X = \frac{\text{moles reacted}}{\text{moles in}} = \frac{\text{moles in} - \text{moles out}}{\text{moles in}} = 1 - \frac{C_{out}}{C_{in}} \tag{3.36}$$

where C = moles of styrene per liter.

We start by noting that the energy balance (3.35) *alone* does not suffice to solve the problem because it contains two variables, temperature T_{out} and rate r or concentration C_{out}. We must supplement it with a second equation, for which we draw on a styrene *mass balance* about the reactor. We obtain

$$\text{Rate of styrene in} - \text{Rate of styrene out} = 0$$

$$QC_{in} \quad - \quad \begin{bmatrix} QC_{out} \\ +k_r C_{out} V \end{bmatrix} = 0 \tag{3.37}$$

Equations (3.35) and (3.37) comprise the model for the required solution. We rearrange them by dividing them by flow rate Q and solving or C_{out}/C_{in}. This yields the following set

$$\frac{C_{out}}{C_{in}} = \frac{1}{1 + \tau k_r} = \frac{1}{1 + 2 \times 10^{10} \exp(-10^4/T_{out})} \tag{3.38}$$

and

$$\frac{C_{out}}{C_{in}} = \frac{T_{out} - T_{in}}{-\dfrac{C_{in}\Delta H_r}{\rho C_p} r} = \frac{T_{out} - 300}{400 \times 2 \times 10^{10} \exp(-10^4/T_{out})} \tag{3.39}$$

These two nonlinear equations can be solved for C_{out} and T_{out} by the Newton–Raphson method (see Section 8.2.5) and yield the results given in Table 3.4, so that there are three steady-state solutions. In the low temperature solution, the reactor acts merely as a storage vessel, and no significant reaction occurs. The high temperature solution represents an upper, runaway condition where the reaction goes to near completion. The middle steady state is metastable. Surprisingly, industrial practice is to operate at this metastable state rather than at stable upper state, since the latter produces a lower grade product brought about by the high operating temperature and excessive reaction rate. The control necessary to maintain the system in this inherently unstable condition is achieved through autorefrigeration (i.e., cooling by boiling). The reactor pressure is set so that the styrene, which is a liquid, boils when the desired operating temperature is exceeded. The latent heat of vaporization rapidly reduces the temperature of the reactor below the boiling point. Boiling then stops until the heat of reaction returns the temperature to the boiling point. Steady state is thus never attained. Instead, the temperature cycles about the metastable state, causing a cyclic variation in the outlet product concentration. Although this is not generally considered to be a desirable condition, the need to ensure the required quality makes this type of operation necessary.

The existence of multiple steady states is not confined to exothermic reactions. It has become increasingly evident in recent years that such multiplicities may be at work in a host of naturally occurring systems, where they perform the task of a "biological switch." We have sketched its mode of operation in Fig. 3.20. As an illustration of its control action, we consider the process of blood coagulation, which exhibits a proven set of multiple steady states.

Table 3.4 Values of C_{out} and X for Equation (3.39)

T_{out}	$X = 1 - C_{out}/C_{in}$
300.03	0.00007
404	0.262
699.97	0.99992

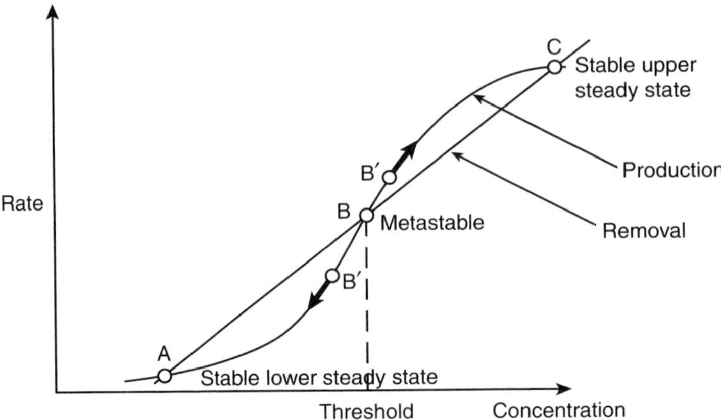

Figure 3.20 Example of a biological switch.

Blood coagulates when it comes in contact with an injured site or a foreign surface. The event triggers a cascade of enzyme-catalyzed reactions that culminates in the production of the clotting protein thrombin. Thrombin has the ability to form aggregates of platelets and strands of a high molecular weight protein, fibrin. These two components together form a blood clot that adheres to the vessel wall.

Examination of the reaction network that leads to thrombin production shows that it gives rise to an S-shaped curve when rate of production is plotted against thrombin concentration. This rate of production is balanced by the rate at which thrombin is swept away from the vessel wall by the flowing blood. That rate is a linear function of thrombin concentration and can thus be represented by a straight line, as shown in Fig. 3.20.

Suppose now that thrombin is produced at a rate corresponding to the point B′ in Fig. 3.20. Since the removal rate there is higher than the production rate, concentration will diminish until the stable lower steady state of A is reached. By similar reasoning, a production rate B″ will cause a continuous increase in concentration until the stable upper steady state, at C is attained. It follows that the metastable state at B represents a *threshold concentration*. Any concentration above it will turn the switch ON to give the high level of thrombin concentration needed for clotting to proceed. Conversely, concentrations below point B will turn the switch OFF and cause thrombin concentration to subside to a "standby" level represented by point A. This low level is required as an initiator for the production of the high concentrations required in case of acute need.

COMMENTS

- One notes here the power of simple geometric constructions and arguments to unravel complex problems such as the stability of a system. This question of stability arises in a host of situations and is usually resolved by examining characteristic values ("eigenvalues") of the underlying model. The geometric procedure given here dispenses with that need and at the same time conveys in graphical fashion the genesis of instabilities and its consequences.

We now turn our attention from questions of multiplicity and stability to a problem in optimization, using a geometric version of a technique known as linear programming.

◆ EXAMPLE 3.12 Linear Programming: A Geometrical Construction

Linear programming deals with the problem of optimizing a linear function f of the form

$$f = a_1x_1 + a_2x_2 + \cdots + a_nx_n + b_1 \tag{3.40}$$

That is, the requirement is to find a combination of values for $x_1, \ldots, x_n$ and b_1 that will either minimize or maximize the function f subject to a series of constraints. Such problems arise generally in decision-making processes and more specifically in management science in connection with production schedules, optimum inventories, shipping schedules, and a host of related problems. This example demonstrates the use of simple geometrical concepts to arrive at the solution of a rather elementary linear programming problem.

Consider two factories that manufacture three different grades of paper. There is some demand for each grade. The company that controls the factories has contracts to supply to customers 16 tons of low grade, 5 tons of medium grade, and 20 tons of high grade paper, all of which it draws from the two plants under its control.

It costs \$10,000 per day to operate the first factory and \$20,000 per day to operate the second factory. Factory number one produces 8 tons of low grade, 1 ton of medium grade, and 2 tons of high grade paper in one day's operation. Factory number two produces 2 tons of low grade, 1 ton of medium grade, and 7 tons of high grade paper per day. We have summarized these details in Table 3.5.

The problem to be addressed is the following. Given that x_1 is the number of days that factory 1 operates and x_2 the number of days that factory 2 operates to produce the required amounts, how many days should each factory be operated to fill the order most economically?

We can cast the problem into the following algebraic inequalities.

At least 16 tons of low grade paper are required:

$$8x_1 + 2x_2 \geq 16 \tag{3.41}$$

At least 5 tons of medium grade paper are required:

$$x_1 + x_2 \geq 5 \tag{3.42}$$

At least 20 tons of high grade paper are required:

$$2x_1 + 7x_2 \geq 20 \tag{3.43}$$

Table 3.5 Operation of Paper Plants

	Production (tons/day)		
Grade	Factory 1	Factory 2	Tons Needed
Low	8	2	16
Medium	1	1	5
High	2	7	20
Daily cost	\$10,000	\$20,000	

The number of days operated must be nonnegative:

$$x_1 \geq 0 \tag{3.44}$$

$$x_2 \geq 0 \tag{3.45}$$

These five inequalities represent five restrictions on our variables x_1 and x_2. We add to this an algebraic statement of the cost function we wish to minimize, namely:

$$f(x_1, x_2) = 10{,}000x_1 + 20{,}000x_2 \tag{3.46}$$

The geometrical construction begins by noting that the solution must be located in the first quadrant $x_1 > 0$, $x_2 > 0$ (see Fig. 3.21). Next the inequalities (3.41) to (3.43) are entered into the diagram. Using the inequality (3.41) as an example, we have

$$8x_1 + 2x_2 \geq 16 \tag{3.47}$$

or equivalently

$$4x_1 + x_2 \geq 8 \tag{3.48}$$

The expression (3.48) represents all points *on or above* the line $4x_1 + x_2 = 8$. That line passes through the points (0, 8) and (1, 4) and is shown in Fig. 3.21. In similar fashion, the lines

$$x_1 + x_2 = 5 \tag{3.49}$$

and

$$2x_1 + 7x_2 = 20 \tag{3.50}$$

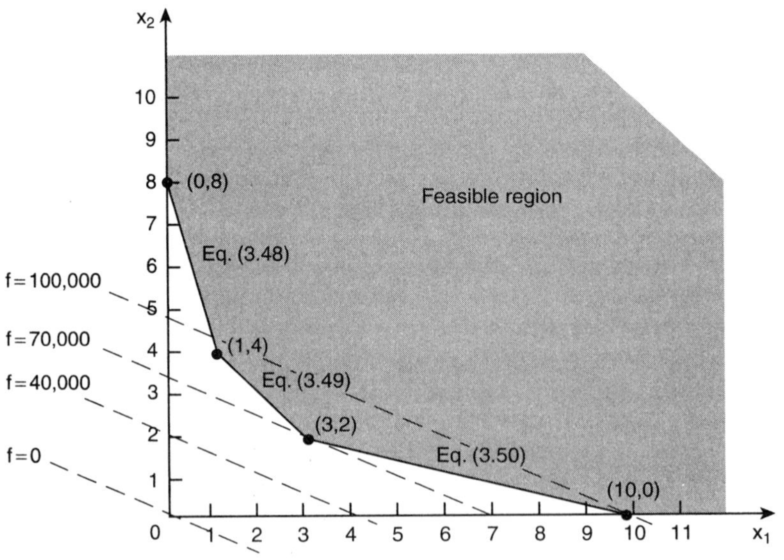

Figure 3.21 Geometrical solution of a linear programming problem.

are established and entered in the diagram. Intersections occur at points (1, 4) and (3, 2).

We next turn to the cost function, equation (3.46). If f, the total production cost, is assigned arbitrary constant values, the equation in each case represents a straight line of slope −1/2. In other words, with f as a parameter, equation (3.46) represents a family of parallel lines (dashed lines in Fig. 3.21). They show increasing values of f, the total cost, as one moves up from the origin. The process is continued until a first intersection with the feasible region is obtained. That intersection occurs at point (3, 2). Any lower line does not intersect the feasible region, while higher lines represent larger cost values of f. Hence point (3, 2), which is the intersection of lines (3.41) and (3.42), yields the minimum total production cost, given by

$$f = 10{,}000(3) + 20{,}000(2) = \$70{,}000$$

Thus if factory 1 operates 3 days, and factory 2 for 2 days, the requirements will be met at lower cost than with any other combination.

COMMENTS

- This is a striking example of the application of simple geometrical concepts to solve an optimization problem of some complexity. It consisted essentially of representing the "equation part" of the expressions (3.41) to (3.43) as straight lines in an x_1–x_2 diagram, yielding the polygonal line shown in Fig. 3.21. The "inequality part" of the same expressions was then used to identify the feasible region, which lay above the polygon. The cost function, equation (3.46), was next plotted for constant values of f until intersection with the feasible region occurred at the lowest value of f. This gave us the required minimum production cost. We note that if more than two factories had been involved, the simple construction shown would not have been possible. In these more complex situations, the problem has to be solved algebraically, by means of matrix methods.
- A further point of note is that if the optimum operating schedule, $x_1 = 3$ days, $x_2 = 2$ days is adhered to, the low grade paper production will be given by

$$8x_1 + 2x_2 = 8 \times 3 + 2 \times 2 = 28 \text{ tons}$$

This is almost twice the minimum requirement of 16 tons, but any other combination of operating days (including fractions) will result in a greater cost. We thus have the paradoxical situation that a *surplus* is obtained while a factory is operating at minimum cost.

◆ EXAMPLE 3.13 Stagewise Adsorption Purification of Liquids: The Operating Diagram

Liquids containing either valuable or undesirable dissolved "solutes" are often treated by contacting them with adsorbents such as activated carbon that exhibit a strong affinity for the substances in question. In a typical operation, the solution is contacted with the adsorbent in a tank, usually under stirring, and allowed to come to equilibrium. At that point, the solid is saturated with solute, no further adsorption takes place, and the solution,

(A)

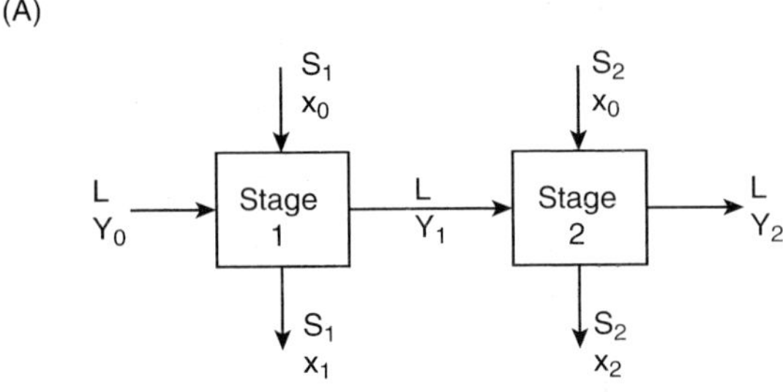

(B)

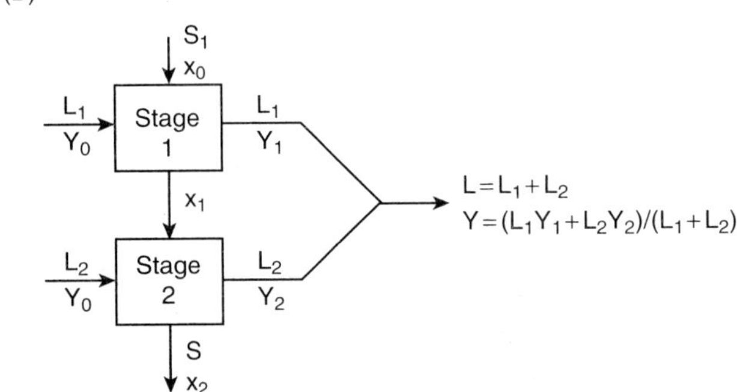

(C)

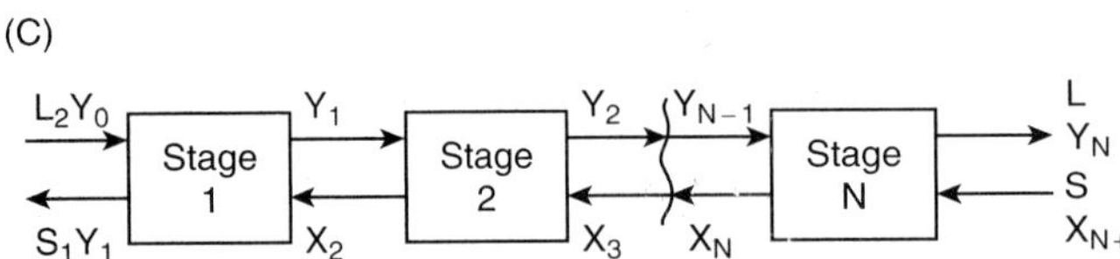

Figure 3.22 Various contacting configurations: (A) cocurrent, (B) cross current, and (C) countercurrent. L = kg solvent, S = kg adsorbent, X = kg solute/kg adsorbent, Y = kg solute/kg solvent.

which still contains some residual material, is either drained or filtered and then subjected to a second treatment. The process is repeated until the desired recovery or purification has been achieved.

The operation can be carried out in a variety of ways or configurations, some of which are displayed in Fig. 3.22. In the so-called *cocurrent* mode, *fresh* adsorbent is used in each tank or "stage" to ensure a maximum degree of solute removal. In the *cross-current configuration*, on the other hand, the same adsorbent is used repeatedly. This is also the case in the *countercurrent* mode, where the flow of material is in parallel but opposite directions. The latter two operations evidently have a lower adsorbent inventory, but this advantage is bought at the cost of reduced removal efficiency or a higher number of stages.

Choosing the most economical operation is evidently one of the major concerns of the design engineer. This a complex task that requires balancing adsorbent and equipment costs against the value of recovered solute or the benefit of impurity removal. Instead of doing this here, we will address a number of subsidiary but important questions that prepare the ground for a more thorough analysis. In particular, we wish to find answers to the following:

- Given a prescribed final purity level, and with the amount of *adsorbent fixed,* what is the number of stages required to achieve the desired result?

- Conversely, with the *number of stages fixed*, how much adsorbent is needed to achieve the required solution purity?
- For a given operation, is there a minimum amount of adsorbent that will reduce solute concentration to the desired level?

Let us consider the tools we shall require. They are two in number. We will first need a statement of solute inventory (i.e., a mass balance). For example, for stage (tank) 1 of the cocurrent mode (Fig. 3.22A), this leads to the equation

$$\begin{aligned} \text{Impurity entering (kg)} &= \text{Impurity leaving (kg)} \\ LY_0 + S_1X_0 &= LY_1 + S_1X_1 \end{aligned} \tag{3.51}$$

Similarly, for the second stage, we write

$$LY_1 + S_2X_0 = LY_2 + S_2X_2 \tag{3.52}$$

Each term is made up of the product of kilograms of solvent L or of adsorbent S, neither of which change, during passage through the stage, and the concentrations X (kg solute/kg adsorbent) and Y (kg solute/kg solvent), which do change. Numbered subscripts refer to the stage the stream is leaving, with 0 denoting a hypothetical zeroth stage.

The second tool is a statement that the concentrations X_i, Y_i leaving a particular stage are in equilibrium with each other. These values are obtained experimentally and are formally expressed by the relation

$$Y_i^* = f(X_i) \tag{3.53}$$

where the asterisk denotes equilibrium conditions.

Laboratory data of X_i, Y_i are not always easily expressed in the analytical form of equation (3.53), and this has led to the development of a geometrical construction termed an *operating diagram*, shown for the cocurrent and countercurrent modes in Fig. 3.23. These plots are in essence graphical representations of both the material balance equations and the equilibrium relation (3.53). They display in vivid form the interrelation of the operational variables and enable the analyst to make various calculations of interest in rapid fashion. We now take up these two cases in some detail.

Cocurrent Operation

Consider the two-stage purification process depicted by the "operating diagram" shown in Fig. 3.23A. The first step in its construction is to plot the experimental equilibrium data obtained in the laboratory in the form of an "equilibrium curve," formally given by equation (3.53). The second step is to represent the mass balances, equations (3.51) and (3.52), which is best achieved by rearranging them in the form

$$-\frac{S}{L} = \frac{\Delta Y}{\Delta X} = \frac{Y_0 - Y_i}{X_0 - X_i} \tag{3.54}$$

This is the equation of a line termed the *operating line* which has a slope of $-S/L$. In this equation, both the amount of solvent L and the initial or "feed" concentrations X_0, Y_0 are known variables, the others being either prescribed or unknown.

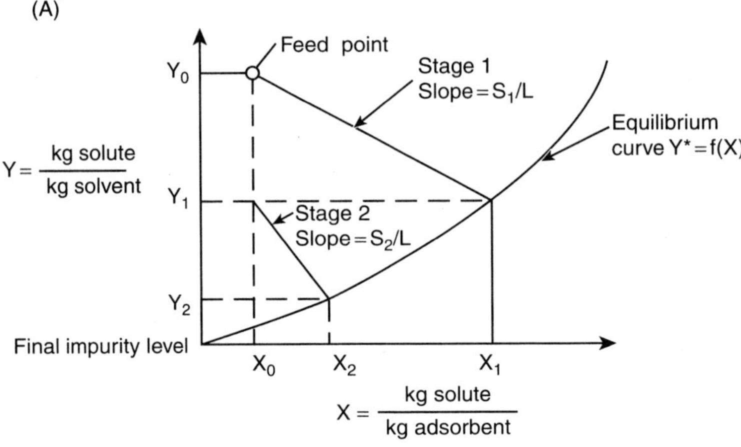

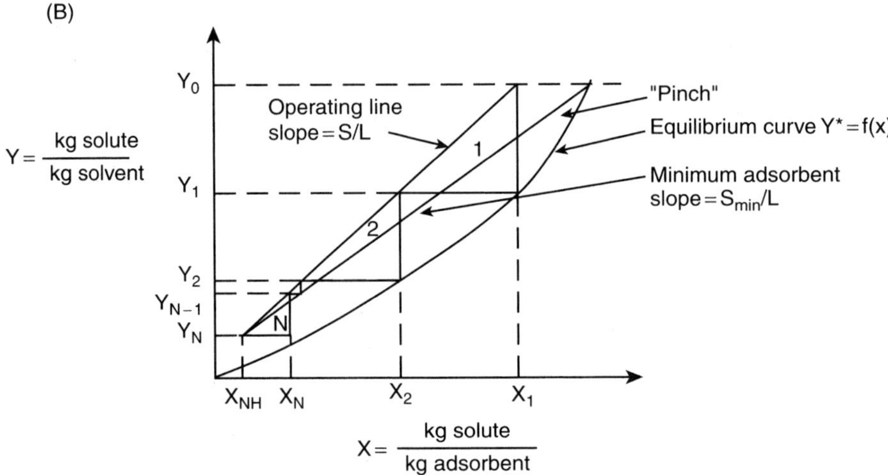

Figure 3.23 Operating diagrams for (A) cocurrent and (B) countercurrent configurations.

Suppose, for example, that one wishes to calculate the final impurity level Y_2 that results in a two-stage process when a solution with known L, X_0 and Y_0 is treated with *prescribed* amounts S_1 and S_2 of adsorbent. To construct the operating diagram, one proceeds as follows.

- Drop a line of slope $-S_1/L$ through the known feed point (X_0, Y_0). Its intersection with the equilibrium curve fixes the concentration (X_1, Y_1) leaving the first stage.
- Drop a second operating line of slope $-S_2/L$ through the point (X_0, Y_1). Its intersection with the equilibrium curve defines the final impurity level Y_2.

It is clear from this construction that there is an infinite number of combinations of S_1 and S_2, which will reduce the feed impurity level Y_0 to Y_2. It is also seen that if either S_1 or S_2 is set at 0 (i.e., a single-stage contact is used), the resulting operating line, which extends from (X_0, Y_0) to (X_2, Y_2), will always have a slope $-S/L$ greater than the combined slopes $(S_1/L + S_2/L)$ so that

$$S_{\text{single stage}} > S_1 + S_2 \tag{3.55}$$

We deduce from this that between the two extreme values of S for single-stage operation, there will probably exist a combination $S_1 + S_2$ that represents a *minimum* total adsorbent inventory $(S_T)_{min}$. This minimum amount can be determined by simple calculus. Suppose, for example, that the equilibrium is linear, a condition that holds at low impurity levels. We then have

$$X_i = HY_i \qquad H > 1 \tag{3.56}$$

where H is the so-called Henry's constant. Assume further that the adsorbent is initially clean, that is, $X_0 = 0$ (this is frequently the case). Substitution of equation (3.56) into the mass balances (3.51) and (3.52), followed by addition, yields

$$S_T = S_1 + S_2 = HL\left(\frac{Y_0}{Y_1} + \frac{Y_1}{Y_2} - 2\right) \tag{3.57}$$

The condition $dS_T/dY_1 = 0$ will then establish the value of the optimum intermediate concentration Y_1, which yields the minimum value of S_T. We obtain

$$\frac{dS_T}{dY_1} = -Y_0Y_1^{-2} + Y_2^{-1} = 0 \tag{3.58}$$

and hence

$$(Y_1)_{opt} = (Y_0Y_2)^{1/2} \tag{3.59}$$

In other words, the optimum impurity level that will minimize S_T is the geometric mean of the impurities entering and leaving the cascade. Let us see how this affects the amount of adsorbent to be used.

Back-substitution into the material balances (3.51) and (3.52) yields

$$\frac{S_1}{L} = H\left[\frac{Y_0}{(Y_0Y_2)^{1/2}} - 1\right] \tag{3.60}$$

$$\frac{S_2}{L} = H\left[\frac{(Y_0Y_2)^{1/2}}{Y_2} - 1\right] \tag{3.61}$$

But we also have

$$\frac{Y_0}{(Y_0Y_2)^{1/2}} = \frac{Y_0(Y_0Y_2)^{1/2}}{Y_0Y_2} = \frac{(Y_0Y_2)^{1/2}}{Y_2} \tag{3.62}$$

so that the bracketed terms in equations (3.60) and (3.61) are identical, hence $S_1 = S_2$.

Hence it is seen that in this case (linear equilibrium) optimum operation calls for the use of *equal amounts of adsorbent* in each stage, given by either equation (3.60) or (3.61).

For nonlinear equilibria, unequal dosages will be required in each state, which can be established in similar fashion. The optimum intermediate concentration will however in general appear in a nonlinear algebraic equation that must be solved numerically (see Practice Problem 3.15).

Countercurrent Operation

The operating diagram for the case of countercurrent operation is shown in Fig. 3.23B and shows an operating line, this time with a positive slope S/L, and a staircase construction representing the various stages and spanning the interval between operating line and equilibrium curve. Let us see how this comes about.

A material balance over the entire process or cascade leads to the equation

$$\text{kg impurity/time in} = \text{kg impurity/time out}$$
$$LY_0 + SX_{N+1} = LY_N + SX_1 \tag{3.63}$$

Rearrangement then gives

$$\frac{S}{L} = \frac{Y_0 - Y_N}{X_1 + X_{N+1}} \tag{3.64}$$

Usually in the design of such countercurrent systems, the liquid feed parameters L and Y_0 are known, and the desired exiting impurity level in the solution Y_N is prescribed. If in addition one chooses a certain adsorbent input S (kg/s), of known impurity content X_{N+1}, equation (3.64) can be plotted in the operating diagram of Fig. 3.23B by drawing a line of slope S/L through the known point (X_{N+1}, Y_N).

We next perform an impurity balance on a *single* stage, for example, stage 1. We obtain

$$\text{kg impurity/time in} = \text{kg impurity/time out} \tag{3.65}$$
$$LY_0 + SX_2 = LY_1 + SX_1$$

This equation can be represented in the operating diagram as follows. We locate the point (Y_0, X_1), where the abscissa value X_1 is obtained from the intersection of the operating line with the horizontal ordinate value $Y = Y_0$. We next draw a *vertical* line through (Y_0, X_1) to the equilibrium curve $Y^* = f(X)$. The point of intersection will have the coordinates (Y_1, X_1), since these values represent the impurity levels leaving stage 1 and are known to be in equilibrium with each other. This is followed by drawing a *horizontal* line through the point (Y_1, X_1), all the way to the operating line. The point of intersection will have the coordinates (Y_1, X_2). This is shown to be the case by rearranging equation (3.65) to read

$$\frac{S}{L} = \frac{Y_0 - Y_1}{X_1 - X_2} \tag{3.66}$$

and noting that this represents a straight line of slope S/L passing through the point (Y_0, X_1); that is, it is identical to the operating line equation (3.64) established earlier. This construction of alternating vertical and horizontal lines between operating line and equilibrium curve is continued until (Y_N, X_{N+1}), the known concentrations at the outlet, are reached. A count is then made of the steps between the entrance and outlet impurity levels. That number represents the number of stage contacts required to reduce the feed impurity level from Y_0 to the prescribed value Y_N, using a fixed adsorbent flow rate S. A fractional step at the outlet end of the staircase is usually rounded off to one stage.

We now note a number of features of this construction that are of use in both engineering calculations and analysis.

- The operating line in essence represents impurity mass balances around a single stage or an aggregate of stages. Its coordinate points (Y_j, X_{j+1}) establish the relation between concentration *entering* a stage, while the equilibrium curve relates concentrations *leaving* a stage (Y_j, X_j).
- Reducing the amount of adsorbent used (S) *lowers* the slope of the operating line and simultaneously *increases* the number of stages required for a prescribed purification. This process can be continued until the operating line intersects the equilibrium curve. At that point a "pinch" results at the high concentration end (Fig. 3.23B) yielding an *infinite number of stages* and a corresponding *minimum amount of adsorbent* S_{in}. This is the least quantity of adsorbent that will achieve the desired purification and is a useful limiting value to establish. Below that value the prescribed exit impurity Y_N can no longer be attained, even if an infinite number of stages were used.
- An increase in adsorbent flow will reduce the number of contact stages, hence will diminish the capital cost of the plant. This advantage is earned at the expense of an increase in operating costs occasioned by the higher adsorbent inventory. There consequently exists an *optimum* adsorbent flow that will *minimize* the combined capital and operating costs. That optimum can be established only by a detailed economic analysis of the process. It has been found in such studies, however, that S_{opt} usually lies in the range 1.5 to 2 times S_{min}. That range is commonly used to establish S, hence the operating line in preliminary designs of a process.
- The reverse task to the design problem just considered (i.e., the prediction of the effluent impurity level in an *existing* plant, N_P, S, and X_{N+1} fixed) cannot be achieved in the same direct fashion. One must resort instead to a trial-and-error procedure by drawing a series of parallel operating lines of slope S/L until the staircase construction accommodating N_P stages reaches the known adsorbent impurity level X_{N+1}. This presents no undue difficulties and can be accomplished quite rapidly.

COMMENTS

- The operating diagrams we have discussed not only are elegant in their simplicity but also are quite universal in their application to stagewise operations in general. They can be applied to processes in which the impurity is extracted by a *liquid* (liquid–liquid or solvent extraction) or where the impurity resides in a *gas stream* and is removed by contacting it with a liquid solvent (gas absorption or scrubbing). It is also used to analyze the design of *distillation* processes where the corresponding operating diagrams are known as *McCabe–Thiele diagrams*, after their inventors.

We leave the preceding example drawn from chemical and environmental technology to address a problem in a totally different domain, biology.

◆ EXAMPLE 3.14 Supercoiled DNA

Deoxyribonucleic acid, or DNA, is the primary genetic material of most living organisms. Its structure, as originally elucidated by Crick and Watson in the 1950s, was visualized as consisting of an open-ended double helix winding around a straight common axis (Fig. 3.24A). The backbones of the two strands are made up of sugar–phosphate units that carry, attached to them at right angles, four base units called nucleotides, whose sequence determines the genetic message. In the late 1950s and early 1960s, it gradually became

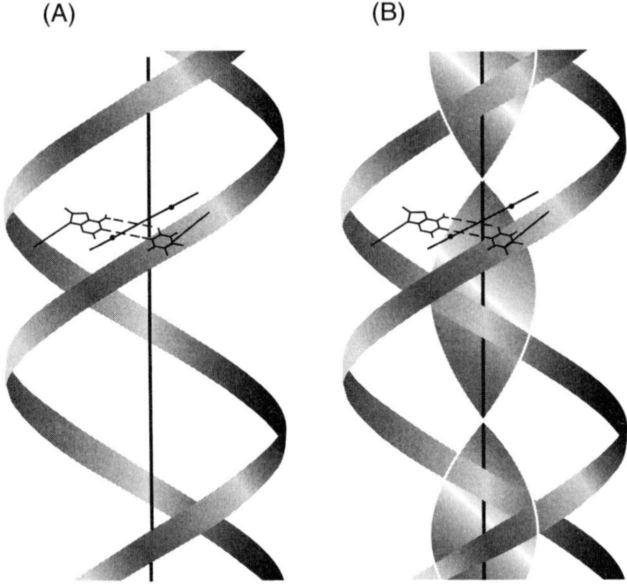

Figure 3.24 (A) The DNA double helix and (B) simulation by a thin twisted ribbon.

clear that often the axis of the double helix is not straight and open ended but can wind in space and even close into a twisted loop. The term "supercoiled" was coined to describe such structures.

To study supercoiling mathematically, it is most convenient to choose a model in which the structure is represented as a narrow twisted ribbon of infinitesimal thickness. The edges of this ribbon represent the sugar–phosphate backbones, and its axis coincides with that of the double helix (Fig. 3.24B). This ribbon model follows the axis of the DNA double helix and twists as the two chains of the molecule twist around that axis. When the ends of the ribbon are joined, each edge of the ribbon will describe a closed curve in three-dimensional space, and these edges will be twisted with respect to each other.

To describe this configuration mathematically, use is made of three mathematical concepts.

The *linking number* Lk, which expresses the number of times a curve loops through another, the *twisting number* Tw, which is a measure of the degree of twisting in the ribbon, and the *writhing number* Wr, which depends exclusively on the curvature of the axis of the ribbon. The linking number expresses a topological property of the structure, while the twisting and writhing numbers are purely geometrical concepts. Let us now examine these three entities in greater detail.

A reasonable way of defining a linking number is to set it equal to zero for a pair of unlinked curves, or curves that do not loop through each other, such as two straight parallel lines. It is equal to 1 for a curve that loops through another just once, 2 for a curve that loops through another twice, and so on. A further distinction is made by assigning a plus or minus sign to the number, depending on the orientation of the two curves and to give each curve a direction. The convention adopted is that each crossing point is given the index number of $+1$ or -1 according to the direction in which the top piece must be rotated to coincide with the bottom piece. If the rotation is clockwise, the index is positive, and if it is counterclockwise, the index is negative. Adding up the index number of all the crossing points and dividing by 2 (the number of linked curves) gives the linking

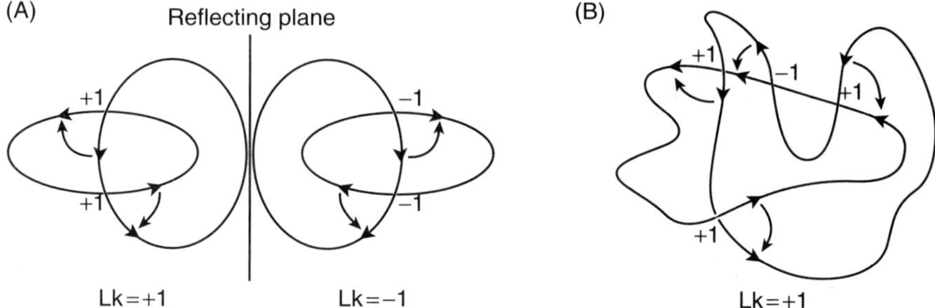

Figure 3.25 The linking number Lk is an integer indicating the number of crossings of the loops. It is positive if superposition of the top onto the bottom loop is by clockwise rotation and negative for a counterclockwise rotation.

number Lk. We show this in Fig. 3.25, where the top piece is fully drawn while the bottom piece is represented by a broken curve. In Fig. 3.25A, the two circles to the right of the reflecting plane have two crossing points, each of which is given a negative index because a counterclockwise rotation is required to bring about coincidence of the two curves. The total linking number is then given by $Lk = [(-1) + (-1)]/2 = -1$. Note that when the structure is reflected through a plane, the sign of the linking number changes. Similar considerations apply to the structure shown in Fig. 3.25B, which has a total of four crossing points, three of which are positive. The linking number is consequently $Lk = [3(+1) + (-1)]/2 = -1$. To summarize, the linking number is a positive or negative integer that expresses the number of crossing points of two linked curves. It is a topological property: that is, no matter how the curves are deformed by pulling, twisting, and so on, as long as neither one is broken, they will remain linked in exactly the same way, and the linking number will remain constant.

The twisting number differs from the linking number in two important ways: it is based on geometrical, not topological considerations, and it can assume fractional values. If a ribbon is deformed, the twist may be altered, but the linking number does not change. Twist need not be an integer and can be considered locally, in which case it is often a fraction. The twist values of individual sections can be summed to obtain the total for a ribbon. Note that twisting increases the energy of a structure and that a relaxed molecule has very little twist.

The convention adopted is that for a right-handed (i.e., clockwise) twist of 360 degrees, the twist number has a value of +1, while a left-handed or counterclockwise twist has a value of −1. A good way of understanding this concept is to imagine a small arrow placed perpendicular to the axis of the ribbon. As the arrow is moved along the twisting ribbon, it rotates about the axis, and the twist of the ribbon can be defined as the integral of the arrow's angular rate of rotation with respect to the arc length of the axis curve. In the three cases shown in Fig. 3.26, the straight ribbon, case A, has no twist at all ($Tw = 0$), case B has two half-twists, each of which carry a twist number $Tw = 1/2$, while the one full twist of case C has the twist number of +1 (right-handed rotation). In the special case of the axis of the ribbon being confined to a plane, the twist number can be measured directly as the number of rotations the arrow completes about the axis as it is moved along the ribbon for a complete loop. For example, when the ribbon represents a closed circular strand of DNA with 1000 base pairs, the arrow makes one complete rotation for every turn of the double helix, each turn containing 10 base units. Hence the total twist Tw equals +100, the plus sign indicating that the double helix is right-handed. Since the 100 twists

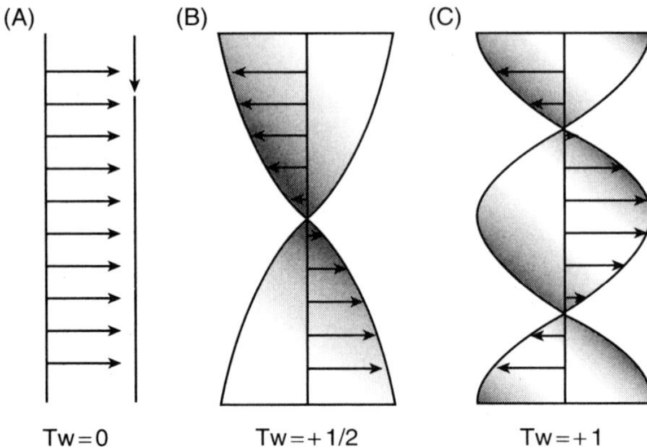

Figure 3.26 The twist number Tw: Tw = 1 for a full 360° twist, and assumes fractional values for lesser degrees of rotation.

will involve 100 crossing points of the two edges of the ribbon representing the sugar–phosphate backbones, the linking number Lk will likewise be 100. Note that the twisting number is not divided by 2, since the entire ribbon is used in defining and determining it.

That the linking and twisting numbers in this example, were numerically identical turns out to be a coincidence, as indeed it should be, since these two entities describe two distinctly different types of behavior. This raises the question of whether there is a geometrical significance to the linking number of a ribbon and its total twist. At about the same time that biochemists were first studying circular DNA structures, several groups of mathematicians were addressing the problem of the relation between linking and twist. It was ultimately shown (1968) that the linking number of a ribbon and its total twist differ by a quantity that depends exclusively on the curve of the axis of the ribbon and was termed the writhing number Wr. Thus

$$\mathrm{Wr} = \mathrm{Lk} - \mathrm{Tw} \tag{3.67}$$

It follows from this relation that if the axes of two closed ribbons trace out the same curve in space, then even if the ribbons themselves turn and twist in entirely dissimilar ways, the difference between linking number and total twist will in both cases be exactly the same.

Mathematically, the writhing number equals the Gauss integral of the axis curve, that is,

$$\mathrm{Wr} = \frac{2}{\sqrt{\pi}} \int_0^S e^{-\lambda^2} d\lambda \tag{3.68}$$

but it is usually more convenient to compute it from the relation (3.67). Unlike the linking and twist numbers, Wr cannot be estimated visually or in an intuitive way. Thus a ribbon that writhes in the ordinary sense of the word may have an overall writhing number of zero. On the other hand, when the writhing number is made to change by deforming the axis of the ribbon in space, the linking number remains the same, while Wr, and with it Tw necessarily undergo a change in accordance with equation (3.67). A further property of the writhing number, which can be deduced by evaluation of the Gauss integral (3.68), is that if the axis of a ribbon lies entirely in a plane, or entirely on the surface of a sphere, Wr is zero so that Lk = Tw. This was the case for the circular DNA molecule we considered previously, when we found by inspection that Lk = Tw = 100.

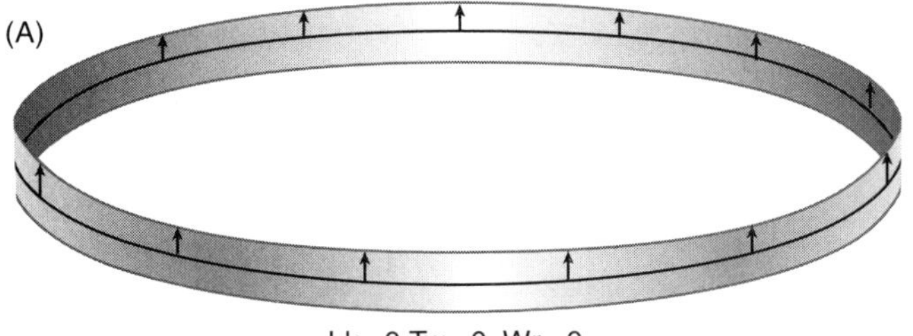

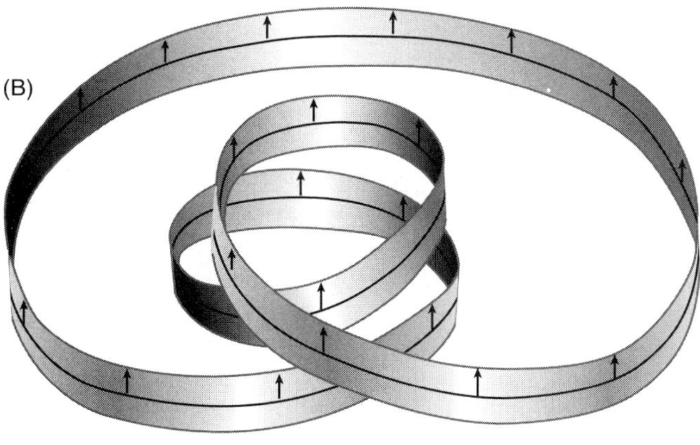

Figure 3.27 Relation between link, twist, and writhe with the numbers Wr = Lk − Tw.

We give two additional examples of the interrelation between writhing, linking, and twisting numbers. Figure 3.27A shows an untwisted circular ribbon for which we find Lk = Tw = Wr = 0. In Figure 3.27B there are four crossing points between the upper and lower edges of the ribbon so that Lk = +2. There is no apparent twist or very little of it so that Tw ≈ 0. It then follows from the relation (3.67) that Wr $\approx +2$.

COMMENTS

- This example introduces the reader at an elementary level to the fascinating world of microbiology. The concepts of linking, twist, and writhing, introduced in the 1960s, have by now become everyday tools in the study of DNA structures. They have also found their way increasingly into the field of genetic engineering. Experiments in this area alter the supercoiling structure of DNA molecules by the action of enzymes, thus creating new, "genetically engineered" DNA molecules. For example, an enzyme known as ω-protein is capable of making a nick in one of the backbones of a DNA double helix, relaxing the molecule by a few turns and then closing the nick. Such "nick-and-closing" enzymes have proved to be invaluable tools for studying the physical chemistry of DNA.
- The concepts of writhing, linking, and twist, which we introduced at an elementary level, have their deeper roots in the advanced mathematical fields of topology and differential geometry. We were nevertheless able to reduce their complexity to a point of rendering then understandable and both visually and intuitively acceptable. This demonstrates the power of even the simplest mathematical models.

PRACTICE PROBLEMS

3.1 **A navigational problem: distance from a lighthouse** A ship is sailing due east when light from a lighthouse is observed bearing N 62°10′ E. After the ship has traveled 2250 m, the light bears N 48°25′ E. If the course is continued, how close will the ship approach the lighthouse?

ANSWER 2934 m

3.2 **The decimal code of a karakasa** Establish the decimal code of the karakasa shown in Fig. 3.6C.

3.3 **Triangulation from bearings** Show that the unknown coordinates of a point C are given in terms of the coordinates of two known points and their bearings by the formulae

$$E_C = \frac{E_A \cot \alpha_{AC} - E_B \cot \alpha_{BC} - N_A + N_B}{\cot \alpha_{AC} - \cot \alpha_{BC}}$$

$$N_C = N_A + (E_C - E_A) \cot \alpha_{AC}$$

(*Hint*: Use equation [3.6b].)

3.4 **Trilateration intersection** The trilateration problem arises when distances rather than angles have been measured from points with known coordinates. Given in Table P3.4, derive the coordinates of P. (*Hint*: Use the cosine formula $\cos \alpha = (b^2 + c^2 - a^2)/2\,bc$.)

ANSWER $N_P = 4096.265$ m

$E_P = 7103.332$ m

3.5 **The use of three satellites in the global positioning system** In principle, a set of three equations (i.e., the use of only three satellites) should suffice to solve for the unknown coordinates x, y, and z. Why, then, is this not done?

3.6 **The unit vector tangent** Consider a position vector **r** that traces out a curve **r** = **r**(s), where s is the arc of the curve. Show that the derivative d**r**/dx is a vector tangent to the curve and has unit length; that is, it is the unit tangent **T** given in equation (3.22).

Table P3.4 Data for Trilateration Problem

Point	Coordinates (m) N	E	Line	Distance (m)
A	3912.56	5997.73	AP	1120.76
B	4643.10	7279.56	BP	574.53

3.7 **Airplane returning to a carrier** An airplane leaves a carrier s at the positon 0 and patrols along the track 0Y, while the carrier follows the course 0X with a constant speed v. If the fuel in the tank allows the plane T hours of flying time at a given air speed, at what point B on the track 0Y must the plane turn back to rejoin the carrier at the point A on the track 0X, T hours after departure? What will be the position C of the carrier at the moment the plane turns back? What is the return track of the plane? (*Hint*: Start by drawing a location diagram with $0A = T/v$. Enter the carrier velocity $\overrightarrow{\mathbf{es}}$ and the wind velocity $\overrightarrow{\mathbf{ew}}$ in a velocity diagram. Draw the air speed circle with w as the center and follow this up by drawing a ray through e parallel to 0Y cutting the circle at p_1. Draw a second line through p_1 and s, intersecting the circle at a second point p_2.)

3.8 **Path of pursuit—the solution** Derive the solution to equation (3.31), using the boundary conditions $y = dy/dx = 0$ at $x = x_0$. (*Hint*: Use a p-substitution [Item 9 of Table 2.8].)

ANSWER $$y(x) = \frac{1}{2}x_0\left[\frac{k}{1-k}\left(\frac{x}{x_0}\right)^{(k-1)/k} + \frac{k}{k+1}\left(\frac{x}{x_0}\right)^{(k+1)/k} + \frac{k}{k^2-1}\right]$$

3.9 **The three-jug problem revisited** Using the procedure outlined in Example 3.10, trace out the trajectory for the three-jug problem stated there, using an alternative route that starts by filling the 3 L jug C from the 8 L jug A, which is full of water at the outset. Show that this leads to an eight-step pathway with seven intermediate stages.

3.10 **Use of trilinear coordinates in a single-stage extraction process** A triangular diagram composed of trilinear coordinates that describe a system of three liquids, two of which are only partially miscible was shown in Fig. 3.16. We use benzene (A) and water (B) as an example of the two partially miscible solvents, while the third component, acetic acid (C), is completely miscible with either benzene or water. Systems of this type give rise to two regions, one in which only a single phase can exist, and a second one in which the system consists of two phases. The two regions are separated by the so-called solubility curve, which envelopes the two phase region. Any three-component mixture that lies within that curve, such as M, will form two insoluble phases whose equilibrium compositions are indicated by R (A-rich) and E (B-rich). The line RE joining these two compositions is called a tie line. Such tie-line compositions are determined experimentally and then entered in the triangular diagram as a family of tie lines. We note that the point M describing the overall composition of the two phases must lie on the tie line. In general, any mixture that separates into two other mixtures of different composition, and any two mixtures that are joined to give a third mixture, must lie on the same line. Proof of this important property is omitted here, but follows from some simple geometrical considerations based on material balances.

Consider now a solution of acetic acid in benzene of known total mass F (kg) and compositions x_{Fi}, which is contacted with a prescribed amount of water B (kg). Sketch a procedure by which the composition x of R and y of E can be established. How many kilograms of R, known as the raffinate, and E, termed the extract, are formed in this process? Assume that the relevant tie-line compositions have been entered in the diagram as mass fractions.

Hint: Perform appropriate total and component balances.

3.11 Effect of system variables on multiplicity

(a) Analyze the effect of inlet and coolant temperatures on multiplicity in a chemical reactor in which an exothermic reaction is taking place.

(b) What would be the effect of flow rate on multiplicity in a biological switch related to blood coagulation?

3.12 Linear programming: minimizing transportation costs A trucking company owns trucks of two types. Type A has $20\,\text{m}^3$ of refrigerated space and $30\,\text{m}^3$ of nonrefrigerated space. Type B also has $20\,\text{m}^3$ of refrigerated space but only $10\,\text{m}^3$ of nonrefrigerated space. A customer wants to haul some produce a certain distance and will require $160\,\text{m}^3$ of refrigerated space and $120\,\text{m}^3$ of nonrefrigerated space. The trucking company figures it will take 300 gallons of gas for the type A truck to make the trip and 200 gallons of gas for the type B truck. Find the number of trucks of each type that the company should allow for the job to minimize gas consumption.

ANSWER Two trucks A and six trucks B

3.13 Linear programming: maximizing the number of television viewers A local television network is faced with a problem: it has been found that program A, with 20 min of music and 1 min of advertisement, draws 30,000 viewers, while program B with 10 min of music and 1 min of advertisement, draws 10,000 viewers. The advertiser insists that at least 6 min be devoted to his advertisement, and the network can afford no more than 80 min of music. To obtain the maximum number of viewers, how many times per week should each program be given?

ANSWER Program A twice a week, program B four times a week

3.14 Stage efficiency Suppose that in a given adsorption process, equilibrium between solid and solution cannot be achieved within a reasonable time and exiting adsorbent is loaded to only a fraction of its equilibrium value, say 70% of Y*. How would you accommodate this fact in the operating diagram? Use countercurrent operation as an example.

3.15 The Freundlich equation: minimum adsorbent inventory for the nonlinear case Two common equilibrium relations, or adsorption isotherms as they are known, are the Langmuir and Freundlich isotherms:

Langmuir isotherm

$$X = \frac{HY}{1 + BY}$$

Freundlich isotherm

$$X = \left(\frac{Y}{m}\right)^{1/n}$$

where H = Henry's constant, and m, n, B are arbitrary constants. The Langmuir isotherm is based on a well-defined theoretical model of adsorption and meets two important criteria: it reduces to the linear case $Y = HX$ at low concentrations and converges, as it should, to a constant saturation concentration $X_S = H/B$ as $Y \to \infty$. The Freundlich equation fulfills neither of these criteria but often provides a good empirical representation at low and intermediate concentrations.

Consider a two-stage cocurrent adsorption process as shown in Fig. 3.22A, using an adsorbent that satisfies the Freundlich isotherm, just given. Show that for adsorbent inventory to be a minimum, the solution concentration Y_1 leaving the first stage must satisfy the following nonlinear relation:

$$\left(\frac{Y_1}{Y_2}\right)^{1/n} - \frac{1}{n}\frac{Y_0}{Y_1} = 1 - \frac{1}{n}$$

Assume that the adsorbent is initially clean (i.e., $X_0 = 0$). Do not solve the equation.

3.16 **The cross-current cascade** Derive an operating diagram for the cross-current cascade shown in Fig. 3.22B. Assume that the adsorbent is initially clean and that its quantity is known. The solution split is prescribed as well, leaving Y_2 as the only unknown.

3.17 **The coiled telephone wire** Consider a coiled telephone wire lying on the floor that someone picks up and stretches. Give the magnitude (large or small) of its writhing and twist in the relaxed and stretched states. How is the linking number affected?

CHAPTER 4

The Effect of Forces

4.1 Introduction

It seems at first sight odd to have chosen the subject of forces, and their effect, as a unifying theme for an entire chapter on modeling. A superficial argument can be made that a single example, or perhaps two or three examples, would suffice to demonstrate the use of Newton's law. That law is commonly, and somewhat, restrictively stated as Force = Mass × Acceleration. This ignores two important facets of this relation. The first is the *variety* of forces or changes in velocity that can arise under different physical circumstances. Among forces, those due to gravitation, electrical fields, or magnetic fields immediately spring to mind. Pressure and stress, which are defined as force per unit area, are additional sources of expressions involving force. So is work, which involves the application of a force over a given distance. Changes in velocity, which is a vector, can come about as a result of a change in its magnitude (speed) or because of a change in direction. The second facet is that Newton's law also applies to static or stationary systems, in which case one uses the special form: Sum of Forces = 0. The Examples and Practice Problems we present in this chapter reflect this variety of situations.

Gravitational forces in dynamic systems are at the core of Examples 4.5 (Path of a Projectile) and 4.6 (The Law of Universal Gravitation). In its static form, gravity is used to calculate the lift capacity of a hot air balloon in Example 4.8. Electrical and magnetic forces make their appearance in Example 4.3, where they form the basis for Thomson's experiment to determine the ratio of charge to mass of an electron. Pressure forces arise in three different contexts: In Example 4.4, they appear as the pressure of a gas caused by impacts of the moving molecules with a container wall, while Example 4.7 considers pressures that arise in moving fluids. Example 4.9 involves pressure in the context of the compression of a gas, which is essentially a static system. The work concept makes its appearance in Example 4.6, where it is used to calculate escape velocity form the earth's gravitational field, and again in Example 4.9, where we derive the expression for compression work done on a gas. This is complemented by the amusing example of the power output of a hovering bumblebee. Let us now consider these examples in somewhat greater detail.

We begin with two examples drawn from the field of solid mechanics (statics). Example 4.1 introduces the reader to the concepts of stress and strain and uses them to calculate the elongation of a beam caused by the impact of a falling mass. Example 4.2 considers an axially loaded beam and derives the famous Euler formula for the onset of buckling. This so-called catastrophic event has counterparts in other physical systems. The mathematics here leads to a second-order ODE, which is solved by the D-operator method. In Example 4.3 we examine the experimental setup and the underlying model used by Thomson to determine the ratio of charge to mass e/m of an electron. Both the electrical and magnetic field concepts are brought in, and we show how Thomson used them ingeniously, both singly and in combination, to measure and derive e/m. Example 4.4 turns to the field of physical chemistry by considering the pressure exerted by an ideal gas as a result of the impact of gas molecules on a confining wall. This leads to the derivation of Boyle's law in terms of molecular properties of the gas and enables us to derive explicit expressions for the molecular speed of a gas and other quantities of interest. The mathematics here involves the choice of an appropriate element in the gas space and integrating the result obtained for a differential segment of that element. Example 4.5 examines the trajectory of a projectile under the influence of gravity by using the differential form of Newton's law $F = m(d^2x/dt^2)$. The resulting differential equations are solved by separation of variables. Gravitational forces also enter in Example 4.6, which deals with the escape velocity from the gravitational field of the earth and computes as well the height to which a geosynchronous satellite has to be lifted. Forces that arise in a flowing fluid are the topic of Example 4.7, and we use them to derive the classical Bernoulli equation for frictionless or inviscid flow. That equation interrelates pressure, velocity, and elevation of a flowing fluid and is applied to a number of practical physical examples. Example 4.8 examines the popular hot air balloon, which acquires its lift from the hot combustion gases of propane. The basic equation is a static force balance of weight and lift; but the important calculation is that of the temperature of the combustion gases, for which we use some elementary concepts drawn from physical chemistry. A static balance is also used in the last example of this chapter: Example 4.9 deals with the work and power associated with the compression of a gas and a hovering bumblebee.

Table 4.1 gives some variables that apply to rotational motion in circular coordinates. These are needed in some of the practice problems given at the end of the chapter.

Table 4.1 Rotational Variables

Angular velocity	$\omega = \frac{d\theta}{dt} = \frac{2\pi}{T}$
	$\omega = \frac{v}{r}$
Angular acceleration	$a = \frac{d\omega}{dt}$
	$a = \frac{v^2}{r} = \omega^2 r$
Angular force (centripetal or centrifugal force)	$F = \frac{mv^2}{r} = m\omega^2 r$

EXAMPLE 4.1 The Stress–Strain Relation: Stored Strain Energy and Stress Due to the Impact of a Falling Mass

A mass falling from a height h causes stress and displacement upon impact with a solid structure. Stress is defined as the force per unit area of surface and is commonly decomposed into a component normal to the surface (*normal stress* σ), and one that acts tangentially upon it (*tensile stress* τ). Thus

$$\sigma = \frac{F_n}{A} \tag{4.1}$$

$$\tau = \frac{F_t}{A} \tag{4.2}$$

where F_n, F_t are the normal and tangential force components.

Displacement or extension δ is commonly expressed in terms of the so-called *strain* ε, defined as

$$\varepsilon = \frac{\text{Displacement}}{\text{Original length}} = \frac{\delta}{L} \tag{4.3}$$

The determination of stresses and displacement caused by either static or moving loads form the core subject matter of the topic of solid mechanics. The particular example chosen here is important in itself and allows us as well to address several basic notions of solid mechanics.

Stress–Strain Relation

The relationship between stress σ and strain ε is exemplified in the diagram Fig. 4.1. The plot exhibits several critical points associated with a change in material behavior.

- Point A denotes the transition from linear to nonlinear elastic behavior. The limit of the latter occurs at B. Thus up to point B, the material will always recover its original shape upon removal of the load or stress.

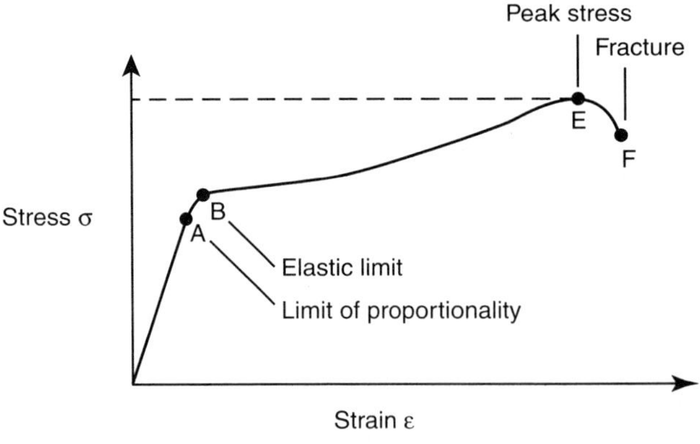

Figure 4.1 Stress–strain relation.

Table 4.2 Tensile Strength of Some Materials

Material	Tensile Strength (MPa)
Cast iron	100–400
Standard stainless steel	450–800
Copper	220
Aluminum	80
Glass fibers	3400
Carbon fibers	5200

Table 4.3 Young's Modulus of Some Materials

Material	Young's Modulus (GPa)
Standard steel	210
Copper	120
Fresh concrete	19
Old concrete	36
Oak	12

- Point E signals the attainment of the peak stress, known as the *tensile strength* of the material. This value is used to characterize strength of construction materials, and typically falls in the range 100 to 1000 MPa (see Table 4.2). Note that while the material will not fail under stress up to this point, it will nevertheless be permanently deformed.
- Point F denotes fracture or failure of the material.

In the design of structures, the stress is normally kept within the range of linear proportionality whose slope is known as *Young's modulus E*.

Thus

$$\frac{\text{Stress } \sigma}{\text{Strain } \varepsilon} = E \tag{4.4}$$

Young's modulus is an important material characteristic, which makes its appearance in many elasticity problems, including the one to be undertaken here. Large values of E correspond to a high resistance to elongation under stress, while low values are characteristic of weak materials. Typical values for several materials are compiled in Table 4.3. An additional application of Young's modulus appears later, in Example 8.2.

Stored Strain Energy

With these basic notions of stress and strain in place, we are in a better position to map a strategy for solving the problem at hand. We argue as follows.

During impact of the falling mass on the beam, the initial potential energy of the mass (or its final kinetic energy) is converted into what we term *stored strain energy* U_s, that is, the work required to elongate the beam by the distance δ (see Fig. 4.2A). This quantity, once known, can then be equated to the potential energy of the mass to obtain the resultant displacement δ. The initial task, then, is to derive an expression for U_s from elementary principles. We proceed as follows.

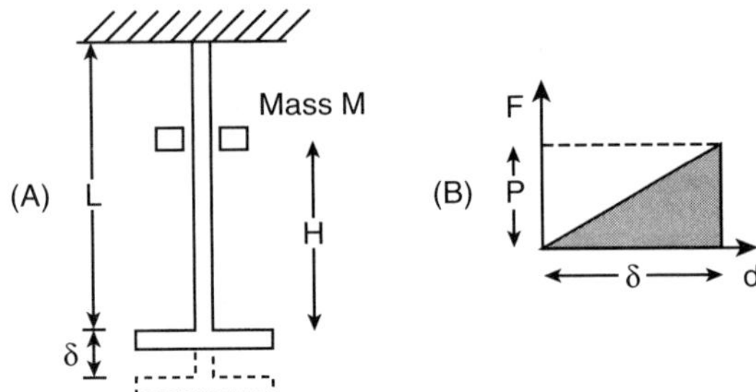

Figure 4.2 Rod under impact: (A) system of falling mass and rod and (B) force–distance diagram.

Figure 4.2B shows the relation between applied force and displacement. For $d = 0$, $F = 0$, and $d = \delta$, $F = P$, so that the *average* force for a displacement from 0 to σ is P/2. The attendant work or stored strain energy is then given by

$$U_s = \tfrac{1}{2}P\delta \tag{4.5}$$

that is, the shaded area of Fig. 4.2B. We now express P and δ in terms of system properties. From equations (4.3) and (4.4) we have

$$\delta = \varepsilon L = \frac{\sigma L}{E} \tag{4.6}$$

and from equation (4.1)

$$P = \frac{\sigma}{A} \tag{4.7}$$

so that U_s becomes

$$U_s = \frac{\sigma^2 LA}{2E} = \frac{\sigma^2}{2E} \times \text{Volume of the rod} \tag{4.8}$$

This is the expression required to complete our calculations.

Impact of a Falling Mass

We start by equating potential energy of the mass to the stored strain energy given by equation (4.8) and obtain

$$mg(H + \delta) = \frac{\sigma^2}{2E} AL \tag{4.9}$$

or by substituting for δ from equation (4.6)

$$mg\left(H + \frac{\sigma L}{E}\right) = \frac{\sigma^2}{2E} AL \tag{4.10}$$

Equation (4.10) can be rearranged into the quadratic equation

$$\sigma^2 - \frac{2mg}{A}\sigma - \frac{2mgHE}{AL} = 0 \tag{4.11}$$

which has the positive root

$$\boxed{\sigma = \frac{mg}{A} + \left(\frac{m^2g^2}{A^2} + \frac{2mgHE}{AL}\right)^{1/2}} \tag{4.12}$$

The corresponding displacement is obtained from equation (4.6).

Let us consider the following numerical example in which M = 1 kg, A = 2 cm × 2 cm, L = 1 m, H = 0.5 m, and E(steel) = 2×10^{11} Pa. Then

$$\sigma = \frac{1 \times 9.81}{4 \times 10^{-4}} + \left[\frac{1^2 \times 9.81^2}{(4 \times 10^{-4})^2} + \frac{2 \times 1 \times 9.81 \times 0.5 \times 2 \times 10^{11}}{4 \times 10^{-4} \times 1}\right]^{1/2}$$

$$\boxed{\sigma = 70.07\ \text{MPa}}$$

By comparison, the corresponding stress due to a resting load of 1 kg would be

$$\sigma = \frac{mg}{A} = \frac{1 \times 9.81}{4 \times 10^{-4}} = 0.025\ \text{MPa}$$

The impact stress is therefore about 2800 times greater than the static stress. Note that since the former is only about 10% below the tensile strength of aluminum (Table 4.2), relatively minor increases in mass or height would lead to rupture.

COMMENTS

- This problem, which at first sight appears rather challenging, yields to a simple solution based on elementary concepts of solid mechanics. The increase of impact stress over static stress was of course anticipated, but its magnitude, for a short fall of a 1 kg mass, nevertheless, is surprising. Another interesting result, taken up in Practice Problem 4.1, is that a load applied suddenly at zero velocity causes *twice* the stress that would be due to a load applied gradually. What is remarkable here is that no time constraints are placed on the two motions, which are vague and ill defined; yet the resulting stresses always differ by a factor of 2.

◆ EXAMPLE 4.2 Bending of Beams: Euler's Formula for the Buckling of a Strut

As a second example of the application of simple solid mechanics principles, we consider the bending of beams in response to an applied load. Two such cases are depicted in Fig. 4.3: the application of a *lateral* load to a beam clamped at one end (cantilever) and the application of an *axial* load to a beam hinged or pinned at both ends. A beam under axial compression in general is termed a *strut*.

The principal difference between the two cases is that application of the end load to a cantilever causes it to bend, the deflection δ increasing gradually with an increase in

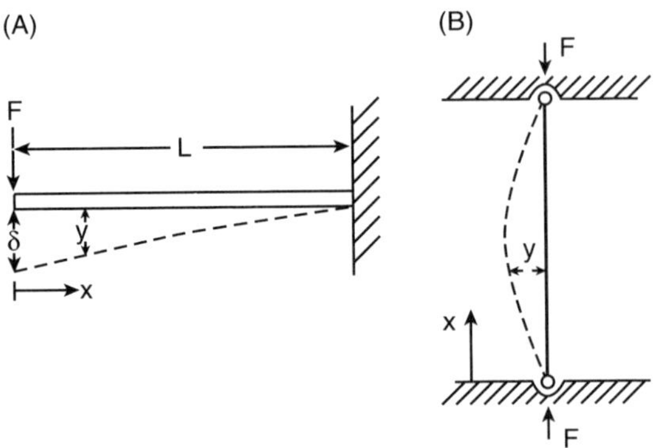

Figure 4.3 Bending of beams due to lateral and axial loads: (A) cantilever with a lateral end load; and (B) hinged strut.

load (Fig. 4.3A). Ultimately, if the load is raised to the level of the tensile strength of the material, the beam will rupture or crack, but only after it has gone through a gradual process of increasing deflection. In the case of an increase in the load on a strut, there is at first no lateral movement at all; a point may be reached, however, at which the resistance to bending has become so small that the slightest increase causes the strut to suddenly buckle. This is often termed a *catastrophic failure* or *instability*. Note again that the failure is sudden and occurs at a particular critical load F_C given by Euler's celebrated formula. It is the derivation of this formula we wish to address. The gradual bending of a cantilever, which makes use of the same basic formulae but under different conditions (lateral rather than axial application of the load), is taken up in Practice Problem 4.2.

It is not clear at first sight how this derivation is to be accomplished, a common occurrence in modeling. Individuals with exceptional vision might argue that the sinusoidal form of the buckling curve arises from a second-order differential equation of the type listed in Table 2.8 (item 4), and this ultimately proves to be the case. To arrive at that equation, however, requires a relation between the bending moment of the applied load and the deflection y caused by it (Fig. 4.3A), and this in turn calls for a consideration of the stresses and strains within the material.

Let us start with a diagram of the system under consideration. This is always a useful first step in modeling. We have for this purpose depicted in Fig. 4.4A a longitudinal element of the bent beam. It is evident from that diagram that the fibers at the top are stretched, causing tensile stresses in those fibers, whereas the bottom fibers will decrease in length, leading to compressive stresses. Somewhere in between these will be what is termed a *neutral layer*, which will be in neither compression nor tension. The intersection of the neutral layer with the cross section of the beam is referred to as the neutral axis (NA). Practice Problem 4.2 asks the reader to show that NA always coincides with the *centroid or first area moment of the cross section*, (i.e., with the location where $\int r\,dA = 0$). Thus for a beam of circular cross section, NA is the diameter of the circle; for a T section, it is shifted toward the horizontal member of the beam. We point this out because we shall shortly make use of the *second area moment* $\int r^2\,dA$, and it is therefore of importance to know the location of the axis of these moments.

We now proceed to establish a relation between applied moment M, caused by either a lateral or an axial load, and the stress and strain within the bent beam. We have from Fig. 4.4A

$$AB = R\theta \qquad \text{and} \qquad CD = (R + r)\theta$$

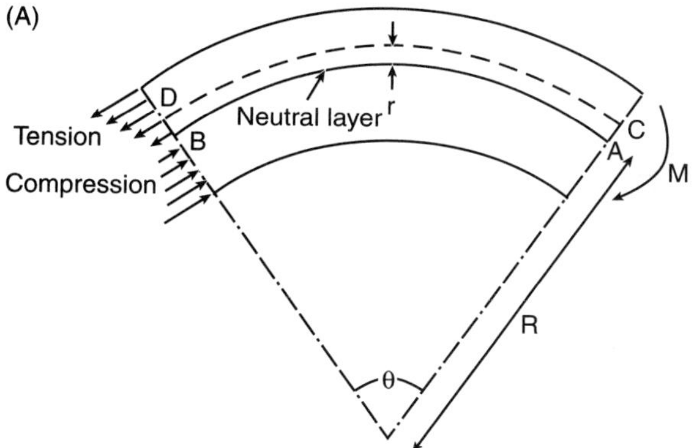

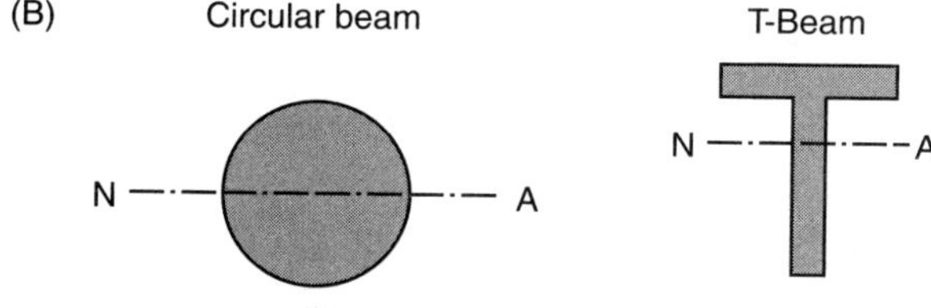

Figure 4.4 Parameters in the bending of beams: (A) longitudinal beam element and (B) the neutral axis.

so that the strain ε [equation (4.3)] is given by

$$\varepsilon = \frac{CD - AB}{AB} = \frac{(R + r)\theta - R\theta}{R\theta}$$

and hence [equation (4.4)]

$$\sigma = E\frac{r}{R} \tag{4.13}$$

where R = radius of curvature of the bent beam.

Now it is seen from Fig. 4.4A that the bending stresses cause a moment about the neutral axis, which from equilibrium considerations must be equal and opposite to the externally applied moment M, so that

$$M = \int \sigma r \, dA \tag{4.14}$$

Substituting equation (4.13) into equation (4.14), we obtain

$$\boxed{M = \frac{E}{R}\int r^2 \, dA = \frac{E}{R} I} \tag{4.15}$$

where $I = \int r^2 \, dA$ = second area moment. To aid the reader in computations that flow from this relation, we have listed in Fig. 4.5 the second moment of some common cross-sectional areas. We note further that equation (4.15) is the fundamental relation to be applied in bending problems involving either axial or lateral loads. Thus for the lateral

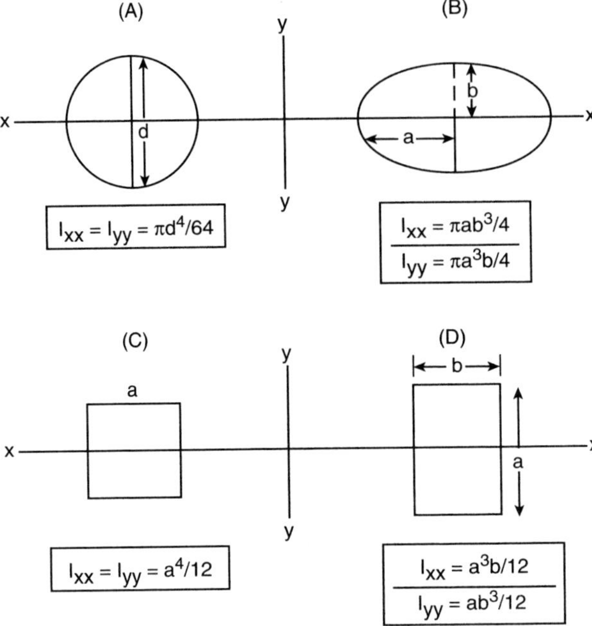

Figure 4.5 Second moments of area for some common cross sections: (A) circular, (B) elliptical, (C) square, and (D) rectangular.

load of Fig. 4.3A, M = FL, and for the axial load, M = Fy, given that moment equals force times lever arm. This establishes the relation between applied load F and the properties of the beam.

Although we have not established the desired relation between load F (contained in M) and the deflection y, one senses that the latter lurks in the radius of curvature R. To quantify the relation, we reach into elementary calculus and apply the expression for the radius of curvature of a curve y = f(x), that is,

$$\frac{1}{R} = \frac{d^2y/dx^2}{[1 + (dy/dx)^2]^{3/2}} \tag{4.16}$$

If the deflections are small, which is the usual requirement in structural design, then $(dy/dx)^2$ is negligible compared with unity so that

$$\frac{1}{R} = \frac{d^2y}{dx^2} \tag{4.17}$$

Upon introduction of equation (4.15), we then obtain the following *fundamental differential equation for the bending of beams or struts*

$$\boxed{EI\frac{d^2y}{dx^2} = M} \tag{4.18}$$

Let us now apply this equation to the axially loaded beam of Fig. 4.3B. We obtain the ODE

$$EI\frac{d^2y}{dx^2} = -Fy \tag{4.19}$$

which has the solution (see Table 2.9)

$$y = A\cos\alpha x + B\sin\alpha x \tag{4.20}$$

with the boundary conditions set by the requirement of zero deflection at both ends. Thus

$$y(0) = y(L) = 0 \tag{4.21}$$

Introduction of equation (4.21) into (4.20) leads to the following expressions for the integration constants:

$$A = 0 \qquad \text{and} \qquad B\sin\alpha L = 0 \tag{4.22}$$

Since B cannot be zero, we must have

$$\sin\alpha L = 0 \qquad \text{or} \qquad \alpha L = 0, \pi, 2\pi, \ldots, n\pi \tag{4.23}$$

where

$$\alpha = \left(\frac{F}{EI}\right)^{1/2} \tag{4.24}$$

From equations (4.23) and (4.24), it follows that the relation between load F and the system properties E, I, and L is given by

$$F = \frac{n^2\pi^2EI}{L^2} \tag{4.25}$$

Since we are only interested in the *smallest* load F_b that will cause buckling, we set $n = 1$ and obtain

$$F_b = \frac{\pi^2EI}{L^2} \tag{4.26}$$

This is the buckling expression due to Euler.

Let us see what magnitudes are obtained in practice. We choose a steel rod of diameter $d = 1$ cm, $L = 1$ m and obtain

$$E = 2.1 \times 10^{11}\ \text{N/m}^2$$

and

$$I = \frac{\pi d^4}{64} = \pi\frac{10^{-8}}{64} = 4.9 \times 10^{-10}\ \text{m}^4$$

from Table 4.3 and Fig. 4.5A, respectively.

Hence from equation (4.26) we write

$$F_b = \frac{\pi^2 \times 2.1 \times 10^{11} \times 4.9 \times 10^{-10}}{1^2} = 1.01 \times 10^3 \text{ N}$$

This corresponds to a mass of 104 kg applied to the ends of the beam.

COMMENTS

- We start by noting that this striking result was obtained by the sole application of relatively simple concepts of solid mechanics and elementary calculus and differential equations. The theory is admittedly applicable only to long and slender struts, since we have tacitly assumed coincidence of the neutral axis and the end of the radius of curvature. It has nevertheless proven itself to be a powerful tool both for obtaining limiting first estimates and as a starting point for more elaborate calculations.
- The boundary conditions to be applied in the solution of equation (4.19) turn out to be of importance. It was already pointed out in connection with Fig. 4.3 that beams can be mounted in different ways: clamped, hinged, or with one end free. For a strut that is clamped at both ends, for example, the buckling load turns out to be *four times* that of a hinged beam. Proof of this is left to Practice Problem 4.4.
- The buckling of a strut is but one example of a wide range of *catastrophic events* or *catastrophes*. The subject has been unified in the so-called *catastrophe theory*, which has been superseded by the wider ranging *chaos theory*, of which it forms a part. We bring to the attention of the reader another example of a catastrophic event that occurs at the so-called critical temperature T_C of a substance. The situation, sketched in Fig. 4.6A, involves a liquid in equilibrium with its vapor in a sealed vessel. Upon heating, the liquid expands, causing its density to decrease, while that of the vapor increases as a result of evaporation (Fig. 4.6B). At a particular point termed the critical temperature T_C, the two densities become equal, the liquid phase vanishes, and one is left with a single (vapor)

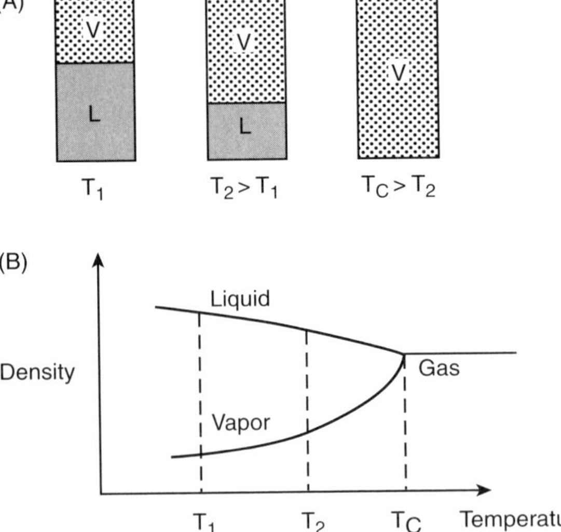

Figure 4.6 Vapor–liquid equilibrium and the critical temperature: (A) the system and (B) bifurcation diagram.

phase that obeys the gas laws. Such "catastrophic" transitions occur in other similar contexts. The graph depicting them (Fig. 4.6B) is referred to as a *bifurcation*.

EXAMPLE 4.3 Electrical and Magnetic Forces: Thomson's Determination of e/m

We consider here two forces that differ from those due to electrical and magnetic fields. These forces played a crucial role in the pathbreaking measurement in 1897 by J. J. Thomson of the ratio of charge e to mass m of an electron. The experiment, which amounted to the discovery of the electron, consisted of observations of its deflection in combined electrical and magnetic fields. To prepare the ground for an examination of the underlying model, we introduce the notions of electrical and magnetic fields and consider as well the deflection experienced by an electron in a purely electrical field. The three items together provide the tools for deducing the ratio e/m from experimental observations. Some of this material was presented earlier (Example 1.4) and is repeated here for convenience.

The Electrical Field

If we place a test charge of magnitude q_0 into the space near a charged rod, an electrostatic force will act on the charge. We speak of an *electric field E* in this space and define it as the *force per unit charge*, that is,

$$\mathbf{E} = \frac{\mathbf{F}}{q_0} \tag{4.27}$$

Since force is a vector, the electric field, which has units of newtons per coulomb (N/C), must also be a vector. Their magnitudes are related by the scalar equation

$$E = \frac{F}{q_0} \tag{4.28}$$

We note that neither of these two equations is used to measure E. One makes use instead of the *potential difference* ΔV between two points in the field, defined as the *work* required to move a unit charge from one point to the other against the electrical field force. Thus

$$\Delta V = F\Delta L = -q_0 E \Delta L \tag{4.29}$$

In the limit, and since $q_0 = 1$, we have

$$\frac{dV}{dL} = -E \tag{4.30}$$

Thus E can be determined by mapping the potential, or voltage, of an electrical field and applying equation (4.30) to obtain E at a particular point.

The Magnetic Field

The definition of the magnetic field, denoted by the vector **B**, is somewhat more circuitous. It involves consideration of the force acting not on a charge at rest, as in electrostatics, but on one that moves through the magnetic field with a velocity **v**. Let us assume that the field is due to a magnet and that the magnetic force lines have been rendered visible (e.g., through the use of iron filings). We would then find the following by experiment:

- No force acts on a charge at rest.
- No force acts on a charge moving tangentially to the field lines.
- A sideways force at right angles to **v** will act in all other cases.
- A maximum force will be exerted when the charge moves at right angles to the field lines.

We use these observations to make the following definition of the vector **B**:

- The *magnitude* of **B** is proportional to the *maximum* force (F_{max}) as follows:

$$B = \frac{F_{max}}{q_0 v} \tag{4.31}$$

It has the units newtons per ampere-meter (N/Am), also termed tesla (T).

- The *direction* of **B** is along the lines of *zero* force (i.e., tangentially to the field lines).

These findings and definitions are combined in the following vector equation

$$\mathbf{F} = q_0 \mathbf{v} \times \mathbf{B} \tag{4.32}$$

Note that for opposite charges, the deflection is in opposite directions.

Let us see how this expression reflects the previous findings and definitions. We use for this purpose the elements of vector algebra we had compiled in Table 2.1.

- For a charge at rest, $\mathbf{v} = 0$, hence $\mathbf{F} = 0$, in agreement with experiment.
- For a charge moving tangentially to the field lines, that is, parallel to **B**, we have (see Table 2.1)

$$F = q_0 vB \sin(v, B) = 0 \tag{4.33}$$

again in agreement with experimental observations.

- For F_{max}, **v** and **B** are orthogonal, hence

$$F_{max} = q_0 vB \sin(v, B) = q_0 vB \tag{4.34}$$

which agrees with the definition of equation (4.31).

- From the definition of the cross product **F** is orthogonal to **v**, in agreement with observations.

For a combined electrical and magnetic field, the total force is given by the sum of equations (4.27) and (4.32)

$$\mathbf{F} = q_0 \mathbf{E} + q_0 \mathbf{v} \times \mathbf{B} \tag{4.35}$$

This is the expression Thomson used as a basis for his model.

We note in closing that the foregoing equations are not used for measuring B. This is done, for example, by placing a coil carrying an electric current and suspended from a balance into the magnetic field. The weight necessary to counterbalance the magnetic field force is then registered and used in appropriate equations. Thus values for both E and B were available to interpret Thomson's results.

Deflection of a Charge in an Electrical Field

The final item needed in the derivation of e/m, the deflection of a charge in an electrical field, constitutes one of several steps in Thomson's experiment. It involves projecting an electron with initial velocity v_0 into a uniform electrical field at right angles. A parabolic path results, much like that of a projectile fired horizontally in the earth's gravitational field (see Example 4.5). Upon emerging from the electrical field, the electron continues in a straight path because there is no longer any force acting on it. One can let it fall on a fluorescent screen, where together with other electrons it makes itself visible as a small luminous spot. This is the principle of the *cathode-ray oscilloscope* (Fig. 4.7).

To derive an equation $y = f(x)$ for the path within the electrical field, we first note that there is no acceleration in the x direction, so that

$$x = v_0 t \tag{4.36}$$

In the y direction there is an acceleration a, and we have from elementary mechanics

$$\frac{d^2y}{dt^2} = a \Rightarrow y = \frac{1}{2}at^2 \tag{4.37}$$

which we can write in the form

$$y = \frac{1}{2}at^2 = \frac{1}{2}\frac{F}{m}t^2 = \frac{1}{2}\frac{eE}{m}t^2 \tag{4.38}$$

by drawing on equation (4.28). Finally, we eliminate t between equations (4.36) and (4.38) and obtain

$$y = \frac{1}{2}\frac{eE}{mv_0^2}x^2 \tag{4.39}$$

This describes the parabolic path of the electron within the electrical field.

We now have the tools to address Thomson's experiment.

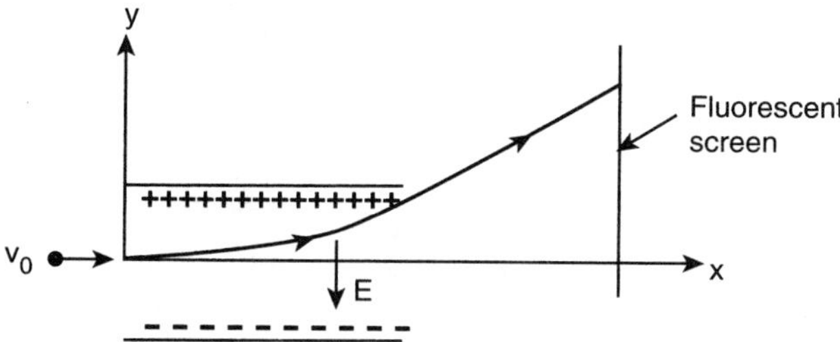

Figure 4.7 Electron moving in a uniform electrical field.

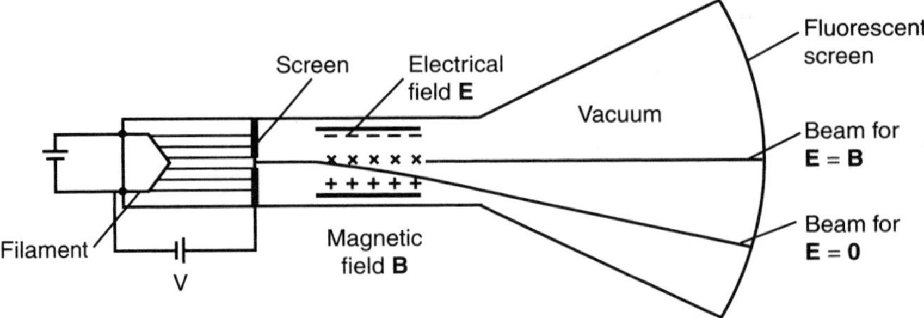

Figure 4.8 Thomson's determination of e/m.

Thomson's Experiment

In Fig. 4.8, which is a modernized version of Thomson's apparatus, electrons are emitted from a hot filament and accelerated by a potential V while passing through a hole in a screen. They then enter a region in which they move at right angles to an electric field **E** and a magnetic field **B**. The two fields themselves are arranged in such a way that they cause deflections in opposite directions.

Thomson's threefold procedure was (a) to note the position of the undeflected beam spot with **E** = **B** = 0, (b) to expose the beam to a purely electrical field and measure the attendant deflection, and (c) to apply a magnetic field and adjust its value until the beam deflection is restored to zero. For step b, equation (4.39) applies. In this expression x (length of deflecting plates) and E are known, and y (deflection at far edge of the field) can be calculated from the observed beam displacement and the known geometry of the apparatus. This leaves v_0 as an extra unknown. The problem is resolved by the ingenious addition of step c, for under its conditions the deflecting force **F** becomes zero, and the vector equation (4.35) now reads, in terms of its magnitudes,

$$eE = ev_0 B \sin(v_0, B) = ev_0 B \times 1 \tag{4.40}$$

or equivalently

$$v_0 = \frac{E}{B} \tag{4.41}$$

[Note that equation (4.40) makes use of the fact that **E** and **B** act in opposite directions.] It remains to substitute equation (4.41) into (4.39), and one obtains

$$\boxed{\frac{e}{m} = \frac{2yE}{B^2x^2}} \tag{4.42}$$

Thomson's value for e/m was 1.7×10^{11} C/kg, in almost exact agreement with the 1977 value of 1.758803×10^{11} C/kg.

COMMENTS

- It is clear here that the underlying model played a crucial role in the design of Thomson's experiment. Equally important was the ingenious introduction of step c, which eliminated v_0 as an unknown.

- The determination of e (hence m) had to await the famous oil drop experiments of Robert Milliken in 1911, in which tiny charged droplets were placed in an electric field **E**. Then **E** was adjusted until the electrical force was equal and opposite to its weight, whereupon e was deduced from the drop density and diameter and the electrical field strength **E**, since the number of charges e had to be arrived at as the lowest common denominator in a series of experiments.

◆ EXAMPLE 4.4 Pressure of a Gas in Terms of Its Molecular Properties: Boyle's Law and the Ideal Gas Law, Velocity of Gas Molecules

Initial work on the behavior of gases dealt principally with their *macroscopic* properties, that is, the interrelation between pressure p, volume V, temperature T, and the total mass m of the gas, or its derivative, the so-called number of moles n

$$n = \frac{m}{M} \tag{4.43}$$

Here M is the mass of a *single* gaseous particle relative to the mass of a carbon atom set at 12 and is termed molar mass or molecular weight. Thus n is a measure of the number of particles in the gas. For one mole of a substance (n = 1), that number, called Avogadro's number N_{Av} was later found to be 6.02217×10^{23} particles.

The early investigations, all experimental in nature, led to the proposal by Robert Boyle in 1662 that at sufficiently low pressure and high temperature, the product PV was constant for a fixed temperature and mass of gas. Thus

Boyle's Law

$$pV = f(m, T) \tag{4.44}$$

This was later generalized into an explicit relation, termed the ideal gas law:

Ideal Gas Law

$$pV = nRT \tag{4.45}$$

$$R = \text{Gas constant} = 8.314 \text{ J/mol K}$$

Some 100 years later, attempts were begun to relate these macroscopic entities to the *microscopic* or *molecular* properties of the gas. This was a more formidable and entirely theoretical task, since experiments at the molecular level could not at that time be carried out. The initial model, which we take up here, was based on the following assumptions:

1. The particles are taken to be *point* masses. This avoids the complication of having to take the finite particle volume into account and is valid in the limit of low pressure.
2. The particles undergo elastic collisions with one another and with the container; that is, the particles have the same *speed* but opposite *directions* before and after the collision. This allows us to calculate the *force* due to impact on the container

walls in simple fashion but ignores the complex attractive and repulsive forces that come into play during the collision process. We justify this by arguing that speed cannot increase or diminish indefinitely and must therefore, on the average, remain constant.

3. The particles spend the preponderant part of their time *in flight* between collisions, which implies that the average distance between particles, the so-called *mean free path* λ, is large in comparison to their size. This reinforces the first assumption and, like it, becomes valid in the limit of low pressures. We use the concept of the mean free path to address the problem of calculating the pressure exerted by a gas on the container walls. To do this, we proceed as follows.

Consider the imaginary oblique cylinder of gas shown in Fig. 4.9A. The area of its base is A, its slant height $v\Delta t$ is taken to be less than the mean free path λ, and Δt is arbitrary. In other words, the particles entering at the top will undergo no further collisions or change of direction before reaching the wall. The axis of the cylinder is located by the angle θ from the normal to the wall, and the angle ϕ. Consequently, molecules moving parallel to the cylinder axis have a velocity component perpendicular to the wall of $v \cos\theta$, and upon striking the wall acquire a new perpendicular component, $v \cos\theta$. Note that we have here made use of assumption (2) of an *elastic* collision. The momentum imparted to the wall in one such collision is then

$$\Delta(\text{momentum}) = 2mv\cos\theta \tag{4.46}$$

or, drawing on Newton's law

$$\text{Force F} = \frac{\Delta(\text{momentum})}{\Delta t} \tag{4.47}$$

and consequently

$$\text{Pressure p} = \frac{F}{A} = \frac{2mv\cos\theta}{A\Delta t} \tag{4.48}$$

This is the result of a *single* collision. We must now first find the number N_θ of all molecules that move at an angle θ from the perpendicular and then integrate it over all

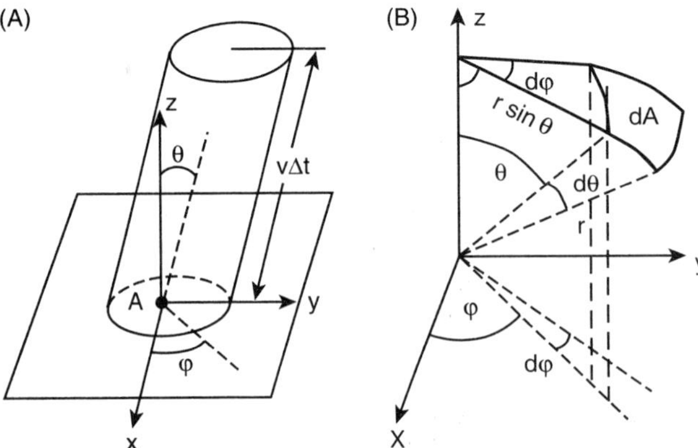

Figure 4.9 Derivation of gas pressure in terms of molecular properties: (A) imaginary cylinder of gas and (B) spherical area element.

possible angles θ and ϕ to obtain the total pressure. To accomplish the first task, we cleverly decompose N_θ into the following product

Volume of cylinder $Av\Delta t\cos\theta$

$\times$ Number of particles per unit cylinder volume N/V

$\times$ Fraction of molecules moving parallel to the cylinder axis

Note that we have tacitly assumed the entire cylinder volume to be available for particle motion, as postulated under item 1 of the list that begins soon after equation (4.45).

The second task, that of integration, requires that we start with the differential form of the desired fraction of molecules. That form is given by the differential spherical surface area shown in Fig. 4.9B, divided by the total surface area of a sphere. The requirement for a spherical geometry comes from the need to integrate over both θ and φ. This can be visualized in Fig. 4.9A by rotating the cylinder around its axis, thus varying φ, and repeating this process for various inclinations, $0 \le \theta \le \pi/2$, thus varying θ. The fraction is then given by

$$\frac{dA_s}{A_s} = \frac{(r\sin\theta\, d\varphi)\times(r\, d\theta)}{4\pi r^2} = \frac{\sin\theta\, d\theta\, d\varphi}{4\pi} \tag{4.49}$$

and hence

$$dp = \left(\frac{2mv\cos\theta}{A\Delta t}\right)(Av\Delta t\cos\theta)\left(\frac{N}{V}\right)\left(\frac{\sin\theta\, d\theta\, d\varphi}{4\pi}\right) \tag{4.50}$$

Integrating over φ and θ we obtain

$$p = \frac{Nmv^2}{2\pi}\int_0^{2\pi} d\varphi \int_0^{\pi/2} \cos^2\theta \sin\theta\, d\theta \tag{4.51}$$

with

$$\int_0^{\pi/2} \cos^2\theta\sin\theta\, d\theta = -\int_0^{\pi/2} \cos^2\theta\, d(\cos\theta) = \frac{1}{3} \tag{4.52}$$

The expression for the total pressure is then

$$p = \frac{1}{3}\frac{N}{V}mv^2 \tag{4.53}$$

Before further use is made of this expression, an examination of the particle velocity v is required. Elementary considerations show that v cannot be the same for all particles. For suppose that two such particles approach each other with the same speed but from different directions. Collision will then cause them to deflect in new directions with an attendant change in speed. This process results in a *distribution* of speeds having the shape of a skewed bell-shaped curve, (Fig. 4.10). In Fig. 4.10A, v_{mp} is the *most probable speed* (maximum N/N_{tot}), v_{av} the *average speed*, and v_{rms} the root-mean-square speed $(v^2)^{1/2}$. The three speeds are in the ratio 1:1.13:1.22, and for many purposes they can be

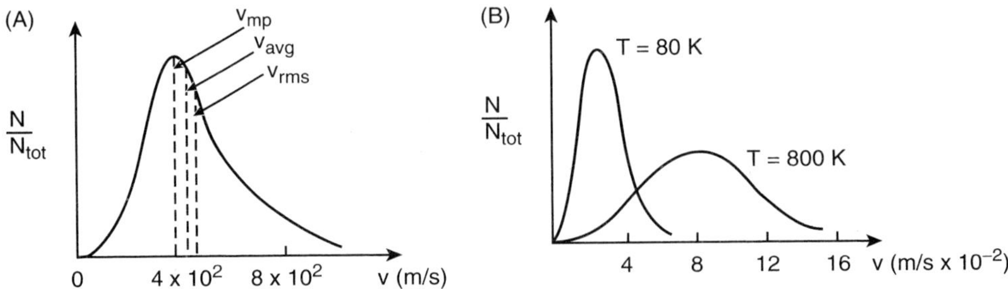

Figure 4.10 Molecular speed distribution for oxygen: (A) at 273 K, and (B) at two other temperatures.

considered to be identical. We note that speed is independent of pressure or volume but does depend on temperature (Fig. 4.10B). This is because thermal energy supplied to the gas will increase its kinetic energy, hence particle velocity. We use these considerations to rewrite equation (4.53) in the form

$$pV = \frac{2}{3}N\frac{m\overline{v^2}}{2} = f(m, T) \tag{4.54}$$

This equation in essence confirms Boyle's law, since the product pV depends only on the mass Nm of the gas and its temperature, as postulated by Boyle. There are three additional important deductions to be made from this equation, which we take up now.

Kinetic Energy of a Gaseous Particle

The reader will have noted that within equation (4.54) lurks the term $m\overline{v^2}/2$, which represents the kinetic energy E_K of a single gas particle. To obtain an explicit expression for E_K in terms of gas temperature, we first note that the number of particles N can be expressed as $N = nN_{Av}$ where n = number of moles and N_{Av} = number of particles in one mole.

By using this expression and the (experimental) ideal gas law (4.45) in equation (4.54), we obtain

$$\text{Kinetic energy of a mole of gas} = N_{Av}\frac{mC^2}{2} = \frac{3}{2}RT \tag{4.55}$$

Note that this expression applies to *translational* kinetic energy only and is therefore restricted to *monoatomic* particles. Polyatomic particles possess two additional energy modes, *rotational* and interatomic *vibrational* energy, neither of which is included in equation (4.55).

Heat Capacity of a Gas

Heat capacity C of a substance is defined as the amount of heat q required to raise the temperature of one mole of the substance by one degree centigrade. The heat added must appear as kinetic energy E_K of the particles, as expansion work done by the gas, or as both. Therefore, any energy q we add to the gas at *constant volume* must appear as an increase in E_K only. Using equation (4.55) we can then write, for a mole of monoatomic particles,

$$\Delta q = \Delta E_k = \frac{3}{2} R \Delta T \qquad (4.56)$$

Hence we obtain from the definition of heat capacity C

$$\boxed{C_V = \frac{\Delta q}{\Delta T} = \frac{\Delta E_k}{\Delta T} = \frac{3}{2} R = 12.47 \text{ J/mol K}} \qquad (4.57)$$

where the subscript V denotes constant volume.

Let us see what this number signifies in practice. Suppose that it is desired to heat one mole of helium, or 4 g, by 100 °C. We obtain

$$\Delta q = nC_V \Delta T = 1 \times 12.47 \times 100 = 1.247 \text{ kJ}$$

Suppose next that we wish to heat the same amount of mercury vapor by 100 °C. Mercury vapor is monoatomic and has a molar mass $M = 201$, so that from equation (4.43) $n = 4/201 = 0.02$. We obtain

$$\Delta q = n \times C_V \times \Delta T = 0.02 \times 12.47 \times 100 = 24.9 \text{ J}$$

Thus the heat requirement is approximately 50 times less than that for heating helium because of the higher molar mass of mercury. Put another way, mercury, while having the same (molar) *heat capacity* (J/mol K) as helium, has a 50 times lower *specific heat* (J/g K), which is based on unit mass. This difference is reflected in the difference in the required heat load.

Velocity of Gaseous Particles

We now return to equation (4.55) to derive an expression for particle velocity in terms of known or observable quantities. Noting that $N_{Av}m$ = mass of one mole = M, we obtain

$$\boxed{\text{Root-mean-square velocity } (\overline{c^2})^{1/2} = \left(\frac{3RT}{M}\right)^{1/2}} \qquad (4.58)$$

For hydrogen, for example, which has a molar mass $M = 2$, we obtain at $T = 298$ K

$$(\overline{c^2})^{1/2} = \left(\frac{3 \times 8314 \text{ J/kg mol K} \times 298\text{K}}{2 \text{ kg/kg mol}}\right)^{1/2} = 1930 \text{ m/s}$$

For heavier particles, the velocity is correspondingly lower, as can be seen in the tabulation of Table 4.4.

Table 4.4 Root-Mean-Square Velocities of Gas Particles at 298 K

Gas	Velocity (m/s)
Argon	431
Carbon dioxide	411
Chlorine	323
Helium	1360
Hydrogen	1930
Oxygen	482
Water	642
Xenon	238

COMMENTS

- The remarkable feature of this model, apart from its simplicity, is that it was able not only to provide a theoretical basis for Boyle's experimental observations but also to deduce properties (e.g., particle velocity) that were not easily accessible to direct observation. It is admittedly a *limiting* model, since it applies at low pressures only. It has nevertheless been successful in deducing a host of important properties, including the heat capacities already mentioned, as well as the thermal conductivity, viscosity, and diffusivity of low density gases.
- The reader should not overlook or dismiss the ingenious simplifications used to arrive at the model. Thus by limiting ourselves to low pressures, we were able to eliminate the complex interparticle collision forces from consideration, since they now come into play only during a very brief interval of a particle's existence. The same complications arise in collisions with the container wall, but these were cleverly avoided by approximating them as *elastic* collisions, with the simple result that the momentum change equaled 2 mv.
- The geometrical constructions used in the model derivation also deserve some attention. Here we made use of the mean free path concept to simplify the trajectory of the particles. The angle of approach to the container wall was taken into account through the use of the oblique cylinder of Fig. 4.10A and the contribution of *all* collisions arrived at by integration over a hemispherical domain. All this is done with seeming ease, but the model is in fact the result of trial and error, accompanied by considerable thought and many false starts.
- One important result of the model was to provide the first evidence that the noble gases (helium, neon, argon, krypton, and xenon) are monoatomic elements. This came from measurements of their heat capacities, which all converged to the value given by equation (4.57) at low pressures. Additional vindication of the model came from an unexpected quarter, the velocity of sound in gases. We have listed some typical values in Table 4.5, and note that they lie remarkably close to those given by particle velocities shown in Table 4.4. The reason for this lies in the nature of sound, which is known to cause local increases in gas pressure or density. That increased pressure can travel or be transmitted only by the actual motion of the gas molecule. It follows therefore that the speed of sound propagation is of the same order as the speed of the particles themselves. The reader is referred to Chapter 8,

Table 4.5 Velocity of Sound in Various Media

Medium	Temperature (°C)	Speed (m/s)
Air	0	331
Hydrogen	0	1286
Oxygen	0	317
Water	15	1450
Iron	20	5130
Rubber	0	54

Example 8.5, for another interesting example of the propagation of pressure due to an explosion.

◆ EXAMPLE 4.5 Path of a Projectile

When a shell is fired from a gun, two forces come into play that determine the path of the projectile:

1. The gravitational force, expressed as $F_G = mg$.
2. The force due to the air resistance, also termed the drag force F_D. This force is commonly expressed in the form

$$F_D = C_D A_C \rho \frac{v^2}{2} = Kv^2 \tag{4.59}$$

where ρ = air density, v = instantaneous velocity, A_C = maximum cross-sectional area of the projectile normal to the path of flight, and C_D is an empirical factor termed the drag coefficient. For streamlined projectiles C_D is approximately unity; for blunt-nosed objects it is, as expected, considerably higher. Equation (4.59) can be derived by dimensional analysis, a topic taken up in Chapter 8.

As one senses intuitively, the trajectory will also depend on the angle of elevation α (Fig. 4.11) and the muzzle velocity v_0, which must therefore appear in any proposed model for the path of flight.

The questions we wish to address are the following:

1. How high does the projectile rise vertically?
2. How far does the projectile travel horizontally, that is, what is its range r (Fig. 4.11)?
3. What angle of elevation produces the maximum range?
4. What is the equation for the trajectory $y = F(x)$?

Evidently, answers to questions 2 and 3 are of interest in the use of ground artillery while question 1 has its importance in the use of antiaircraft guns.

We start by noting that the problem calls for the use of Newton's law, which is a vector equation. Since questions 1, 2, and 4 involve the coordinate directions x and y, that vector

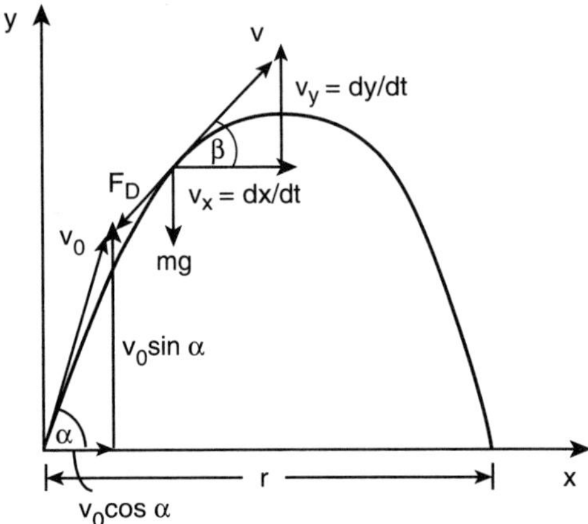

Figure 4.11 Path of a projectile.

equation will have to be broken up into its scalar components in x and y. We further note that inclusion of the air resistance, equation (4.59), will lead to nonlinear ODEs that are likely coupled and not amenable to analytical treatment. We therefore limit ourselves, in the first instance, to the asymptotic case of negligible air resistance. The solution is not a trivial one, since it sets *upper limits* to both height and range attained by the projectile and is thus of considerable usefulness in quickly assessing the potential of a given device. The more complex case of added air resistance is taken up in the exercises. The differential equations to be solved are then given by

$$0 = m\frac{d^2x}{dt^2} \tag{4.60}$$

$$-mg = m\frac{d^2y}{dt^2} \tag{4.61}$$

which yield, upon double integration,

$$x = C_1t + C_2 \tag{4.62}$$

$$y = -\tfrac{1}{2}gt^2 + C_3t + C_4 \tag{4.63}$$

Evaluation of the integration constants is accomplished by using the following four initial conditions at $t = 0$ (see Fig. 4.11):

$$v_x = \frac{dx}{dt} = v_0 \cos\alpha \tag{4.64a}$$

$$v_y = \frac{dy}{dt} = v_0 \sin\alpha \tag{4.64b}$$

$$x = 0 \tag{4.64c}$$

$$y = 0 \tag{4.64d}$$

The result is the following

$$C_1 = v_0 \cos\alpha, \qquad C_2 = 0, \qquad C_3 = v_0 \sin\alpha, \qquad C_4 = 0$$

so that the solutions (4.62) and (4.63) become

Horizontal Component

$$x = (v_0 \cos\alpha)t \tag{4.65}$$

Vertical Component

$$y = -\tfrac{1}{2}gt^2 + (v_0 \sin\alpha)t \tag{4.66}$$

Some thought will then lead us to the following strategy for answering questions 1 to 4.

- Set the vertical velocity component equal to zero, $dy/dt = 0$. This is the condition at which the maximum height is attained and yields y_{max}, answering question 1.
- Find the time t for which $y = 0$. This is the time at which the projectile strikes the ground and yields, after substitution into equation (4.64a), the range r sought in question 2.
- The r derived just will contain the angle of elevation α, which can be optimized to give r_{max}, thus answering question 3.
- Eliminate time t between equations (4.65) and (4.66). The result is the equation for the projectile trajectory $y = F(x)$ required in question 4.

We now take up each of these steps in turn.

1. We set $dy/dt = 0$ and obtain

$$\frac{dy}{dt} = -gt + v_0 \sin\alpha = 0 \tag{4.67}$$

from which it follows

$$t_{y_{max}} = \frac{v_0 \sin\alpha}{g} \tag{4.68}$$

and consequently

$$\boxed{y_{max} = -\frac{1}{2}gt_{max}^2 + (v_0 \sin\alpha)t_{max} = \frac{(v_0 \sin\alpha)^2}{2g}} \tag{4.69}$$

2. Here we set $y = 0$, so that from equation (4.66) we have

$$y = t\left[-\frac{1}{2}gt + v_0 \sin\alpha = 0\right] \tag{4.70}$$

and

$$t = 0 \tag{4.71a}$$

$$t = \frac{2v_0 \sin\alpha}{g} \tag{4.71b}$$

where equation (4.71a) corresponds to the instant when the shell is fired while equation (4.71b) represents the time it strikes the ground. Note that the latter is twice the time t_{max}, equation (4.68). Substitution of equation (4.71b) into equation (4.65) yields the desired expression for the range r:

$$r = v_0 \cos\alpha \frac{2v_0 \sin\alpha}{g} = \frac{v_0^2}{g} \sin 2\alpha \tag{4.72}$$

3. It follows immediately from equation (4.72) that r_{max} results when $\sin 2\alpha = 1$, so that

$$\alpha_{opt} = 45° \qquad r_{max} = \frac{v_0^2}{g} \tag{4.73}$$

4. We substitute $t = x/v_0 \cos\alpha$ from equation (4.65) into (4.66), obtaining

$$y = -\left(\frac{g}{2v_0^2 \cos^2\alpha}\right) x^2 + (\tan\alpha)x \tag{4.74}$$

Since this equation is linear in y and quadratic in x, it represents a *parabola*.

COMMENTS

Let us note some of the expected, as well as unexpected, results.

- Both y_{max} and r increase with the muzzle velocity, as expected. Somewhat less anticipated is the fact that they increase with the velocity *squared*, that is, doubling the muzzle velocity *quadruples* both the height and range attained. When air resistance F_D is included, the effect will be reduced, since F_D itself increases with v^2.
- The value of y_{max} increases with the angle α and reaches a maximum when the gun is pointed vertically upward, as expected. The effect of α on *range* r is more subtle: since r is by necessity zero for $\alpha = 0$ and near zero for $\alpha = 90°$, the existence of an optimum α was anticipated. It was not at all clear, however, that this angle should be exactly midway between the two extremes, at 45°. This fact could be revealed only by modeling the process. We shall see in Example 8.9 that when drag is included, the optimum angle is below 45°.
- The parabolic trajectory, is intuitively appealing, as is $t_{x=r} = 2t_{y_{max}}$, verification is required. It could be argued, for example, that the falling portion, which is accelerated by gravity, takes less time than the rising part of the trajectory, which is slowed down. However, the rise starts with velocity v_0, while the fall starts from rest. The two effects exactly cancel each other, a fact nicely brought out by the model.

  All the foregoing variables (i.e., x and y as well as y_{max} and r) are independent of *mass*. This is a startling result, since one would intuitively expect a heavier projectile to fall short of the range of a lighter shell fired at the same muzzle velocity. The reason for this is revealed in one of the model equations [equation (4.61)], where both gravity and acceleration are seen to be proportional to mass, the two thus canceling each other. When air resistance is included, this will no longer be the case, since F_D must now be added to the force balance. This, too, is nicely revealed by the model.

- During the French Revolution and the Napoleonic wars, famous mathematicians such as Pierre Simon de Laplace and Joseph Fourier were employed by the military to assist with gunnery problems. This practice was a novelty in the eighteenth century but is regarded as commonplace today. A contemporary example is the involvement of such physicists and mathematicians as Robert Oppenheimer, Enrico Fermi, Edward Teller, and John von Neumann in the development of the atom and hydrogen bombs.

◆ EXAMPLE 4.6 The Law of Universal Gravitation: Escape Velocity and Geosynchronous Satellites

In Example 4.5, we used Newton's second law to make some statements about the trajectory of a projectile. The present example draws on yet another proposition by Newton, that of the law of universal gravitation. It can be stated as follows: the force between any two particles having masses m_1 and m_2 separated by a distance r is an attraction acting along the line joining the particles and has the magnitude

$$F = G\frac{m_1 m_2}{r^2} \tag{4.75}$$

where G is a universal constant having the same value for all pairs of particles. This is the expression we saw Table 1.2 (item 4).

Lord Cavendish first determined G in 1798 by measuring the attractive displacement of two suspended dumbbells. The present accepted value of G is

$$G = 6.6720 \times 10^{-11}\,\mathrm{N\,m^2/kg^2}$$

We illustrate the use of equation (4.75) with the following.

Minimum Escape Velocity

The problem addressed here is to find the initial speed v_0 that m be imparted to a projectile such as a rocket, so that it can escape from the earth. We neglect for this purpose the effect of air resistance, which plays a secondary role during the initial part of the trajectory. The value we obtain as a result will therefore be a *minimum* escape velocity.

To model the system, we argue that the initial kinetic energy of the projectile must equal or exceed the potential energy required to move it from the surface of the earth to infinity, for example.

$$E_K = \tfrac{1}{2}mv_0^2 = \Delta E_p \tag{4.76}$$

In turn, E_p is given by the work necessary to overcome the gravitational force over the distance (R_e, ∞) where R_e = (mean) radius of the earth = 6.37×10^3 km. We have

$$\Delta E_p = W = \int_{R_e}^{\infty} F\,dr = \int_{R_e}^{\infty} G\frac{mM_e}{r^2}\,dr \tag{4.77}$$

or

$$\Delta E_p = -\frac{GM_e m}{r}\bigg|_{R_e}^{\infty} = \frac{GM_e m}{R_e} \tag{4.78}$$

where M_e = mass of the earth = 5.98×10^{24} kg.

Substituting the expression (4.78) into equation (4.76) and solving for v_0, we obtain

$$v_0 = \left(2\frac{GM_e}{R_e}\right)^{1/2} = \left[\frac{2(6.67 \times 10^{-11})(5.98 \times 10^{24})}{6.37 \times 10^6}\right]^{1/2}$$

$$v_0 = 11.2 \text{ km/s}$$

The lighter gas molecules in our atmosphere have translational speeds high enough to enable them to escape into outer space. Hydrogen, for example, which must have been present in the earth's atmosphere a long time ago, has now completely disappeared from it. Helium escapes at a steady rate, but much of it is resupplied by the radioactive decay from the earth's crust. The escape speed for the sun is much too great to allow hydrogen to escape from its atmosphere. On the other hand, the speed of escape on the moon is so small that it can hardly keep any atmosphere at all. To obtain a sense of the magnitude of the escape velocity from these three disparate bodies, let us calculate v at a distance of 10 radii of each of the three:

For the earth ($R_e = 6.37 \times 10^6$ m, $M_e = 5.98 \times 10^{24}$ kg)

$$v = \left(\frac{2GM_e}{R_e}\right)^{1/2} = \left[\frac{2(6.67 \times 10^{-11})(5.98 \times 10^{24})}{6.37 \times 10^7}\right]^{1/2} = 3540 \text{ m/s}$$

For the sun ($R_s = 6.96 \times 10^8$ m, $M_s = 1.99 \times 10^{30}$ kg)

$$v = \left(\frac{2GM_s}{10R_s}\right)^{1/2} = \left[\frac{2(6.67 \times 10^{-11})(1.99 \times 10^{30})}{6.96 \times 10^9}\right]^{1/2} = 195{,}000 \text{ m/s}$$

For the moon ($R_m = 1.74 \times 10^6$ m, $M_m = 7.36 \times 10^{22}$ kg)

$$v = \left(\frac{2GM_m}{10R_m}\right)^{1/2} = \left[\frac{2(6.67 \times 10^{-11})(7.36 \times 10^{22})}{1.74 \times 10}\right]^{1/2} = 750 \text{ m/s}$$

Now Table 4.4 in Example 4.4 gives a value of 1930 m/s for the root-mean-square velocity of hydrogen at 298 K. This value has to be corrected for the temperature at the locations in question, which we take to be 150 K for the moon and earth, and 1000 K for the sun (surface temperature 6000 K). Applying the $T^{1/2}$ correction of equation (4.58), we obtain adjusted values of 1370 and 3536 m/s, respectively. The numbers indicate that the moon will not be able to retain hydrogen at the distance in question, while the required value of v near the sun is some 55 times higher than the translational speed of hydrogen molecules.

Geosynchronous Satellites

Satellites used in the transmission of signals or for use in GPS devices (see Example 3.5) are usually required to revolve at a speed that will allow them to remain above the same point on the earth. A basic piece of information needed in the launch of such satellites is the *height* to which they must be lifted to be synchronous with the revolving earth. To model such a system, we must consider the forces acting on the orbiting satellite, which will be composed of the gravitational force at the required distance and the countervailing centrifugal force, given by $m_s\omega^2 r$ (see Table 4.1). At steady state we must therefore have

$$\underbrace{\frac{GM_e m_s}{r^2}}_{\text{Gravitational force}} = \underbrace{m_s \omega^2 r}_{\text{Centrifugal force}} \tag{4.79}$$

where ω = circular frequency = $2\pi/T$ and T = period of one revolution.

This gives us, for the orbital radius of the satellite

$$r = \left(\frac{GM_e}{\omega^2}\right)^{1/3} = \left(\frac{TGM_e}{4\pi^2}\right)^{1/3} \tag{4.80}$$

or

$$r = \left[\frac{(3600 \times 24)(6.67 \times 10^{-11})(5.98 \times 10^{24})}{4\pi^2}\right]^{1/3}$$

$$r = 4.23 \times 10^4 \text{ km}$$

The height h above the surface of the earth is then given by

$$H = h = r - R_e = 4.23 \times 10^4 - 6.37 \times 10^3 = 3.59 \times 10^4 \text{ km}$$

so that the satellite will revolve at an approximate height of 36,000 km.

◆ EXAMPLE 4.7 Fluid Forces: Bernoulli's Equation and the Continuity Equation

We have considered forces that arise as a result of magnetic and electrical fields (Example 4.3), gravitational fields (Examples 4.5 and 4.6), and change in momentum (Example 4.4), as well as those induced by stresses applied to solid bodies (Examples 4.1 and 4.2). We turn here to yet another category of forces, those associated with flowing fluids. Qualitatively, we can break down these forces into four categories: gravitational forces, forces due to a change in momentum, pressure forces, and frictional forces.

Gravitational forces arise when the direction of flow deviates from the horizontal.

Forces due to a change in momentum follow from Newton's law, which can be written in the form

$$\mathbf{F} = \frac{d\mathbf{M}}{dt} = F\Delta\mathbf{v} = \rho Q\Delta\mathbf{v} \tag{4.81}$$

where F and Q are mass and volumetric flow rates and **F** the vectorial force associated with momentum change d**M**/dt or velocity **v**. In steady flow such forces must be considered whenever there is a change in either magnitude or direction of the fluid velocity.

Pressure forces exist at any point in the flow field and are due to fluid pressure.

Frictional forces can be viewed as arising from friction or momentum transfer between adjacent fluid layers or the fluid and its containing wall s.

Early work in fluid mechanics, which took place in the eighteenth century, was confined to a consideration of the first three forces, which lead to what is often referred to as *inviscid flow*. Quantification of the frictional forces occurred only later. We will focus on the inviscid cases. We consider for this purpose the differential tubular element shown in Fig. 4.12 and take up each of the operative forces in turn.

1. *Gravity*. To take the component of gravity or weight in the vertical direction, we use an average cross-sectional area of (A + dA/2). Thus

$$dF_g = -\left(A + \frac{dA}{2}\right) ds\rho g \sin\theta \tag{4.82}$$

or, neglecting second-order differentials, and noting that $ds \sin\theta = dz$,

$$dF_g = A\rho g\, dz \tag{4.83}$$

2. *Momentum*. Since the flow is steady and there is no change in direction, we obtain from equation (4.81), noting that volumetric flow rate Q equals the product of cross-sectional area A and velocity v

$$dF_m = -\rho Q\, dv = -\rho A v\, dv \tag{4.84}$$

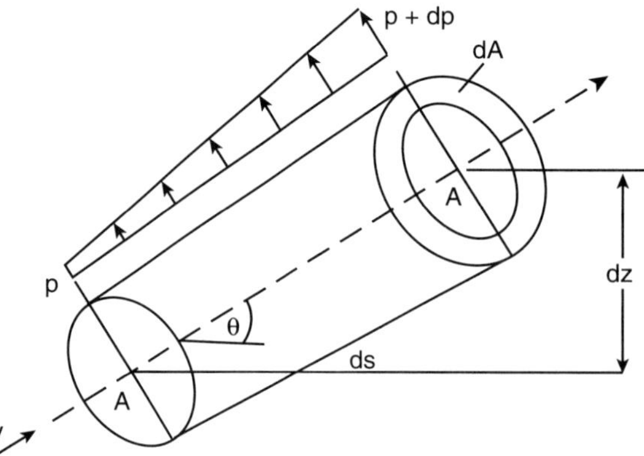

Figure 4.12 Derivation of Bernoulli's equation.

3. *Pressure*. A subtlety enters the picture here. In addition to the pressure forces acting on each cross-sectional area, there is a pressure force acting on the tube wall, which has a streamwise component $(p + dp/2)\,dA$ (see Fig. 4.12). The net pressure force is then given by

$$dF_p = Ap + \left(p + \frac{dp}{2}\right)A - (p + dp)\,dA \tag{4.85}$$

which, upon neglecting second-order differentials becomes

$$dF_P = -A\,dp \tag{4.86}$$

Setting the sum of forces equal to zero followed by division by $A\rho$, we obtain

$$\frac{dp}{\rho} + v\,dv + g\,dz = 0 \tag{4.87}$$

This is the differential form of the celebrated Bernoulli equation. For incompressible fluids, or those that are nearly so, $\rho \approx$ constant, and equation (4.87) can be integrated directly to yield

$$\frac{p_2 - p_1}{\rho} + \frac{v_2^2 - v_1^2}{2} + g(z_2 - z_1) = 0 \tag{4.88}$$

This expression is sometimes cast in the form

$$\frac{p}{\rho} + \frac{v^2}{2} + gz = \text{constant} \tag{4.89}$$

which states that the pressure, kinetic energy, and potential energy per unit mass are constant along a specified stream tube.

Another important result of this expression is that an increase in elevation z (at constant v) or an increase in velocity v (at constant z) causes a *decrease* in pressure. Similarly, a decrease in elevation at constant pressure causes an increase in velocity. These qualitative trends, sometimes referred to as the *Bernoulli effect*, have many applications, some of which we have sketched in Fig. 4.13. We now take up each one of these in turn.

The Siphon

This simple device, known from everyday life, is used to transfer liquids from containers that cannot be easily moved. We apply equation (4.88) to locations 1 and 2 in Fig. 4.13A, noting that $p_1 = p_2 = p_{atm}$ and that $v_1^2/2$ is negligible compared with $v_2^2/2$. We obtain

$$z_1 - z_2 = \frac{v_2^2}{2g} = \frac{(Q/A)^2}{2g} \tag{4.90}$$

where A = cross-sectional area of the tube.

This result shows that to have a positive flow rate $Q > 0$, the difference in elevation must be positive (i.e., $z_1 > z_2$). This is of course well known to anyone who has successfully used a siphon. The preceding text provides the scientific explanation for the phenomenon.

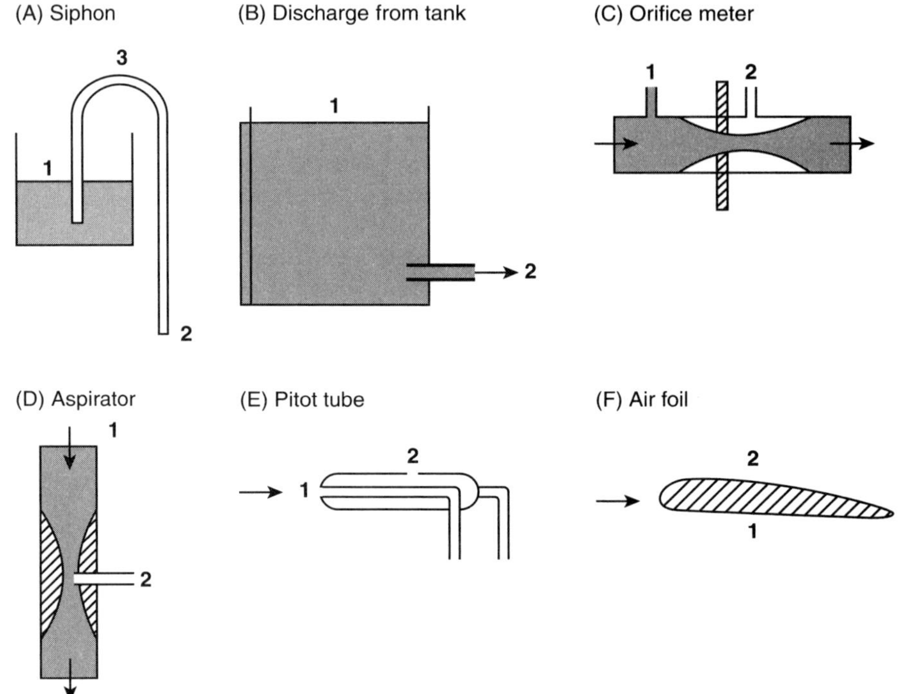

Figure 4.13 Applications of Bernoulli's equation.

Discharge from a Tank

We consider here the discharge of a liquid through a nozzle attached to the side of a tank, as shown in Fig. 4.13B. If the ratio of nozzle and tank diameters is small, which is usually the case, one can neglect the velocity term at location 1, as was done with the siphon. Similarly, one can again set $p_1 = p_2 = p_{atm}$, so that the only terms remaining in the Bernoulli equation are those due to the exit velocity v_2 and the change in elevation Δz. The liquid level is assumed to be constant. One then obtains

$$v_2 = (2g\Delta z)^{1/2} \tag{4.91}$$

This expression can be viewed as stating that the potential energy of the surface fluid is entirely converted into kinetic energy at the exit, consistent with the absence of friction and pressure work.

Orifice Meter and Aspirator

In these devices (Figs. 4.13C, 4.13D), a drop in pressure is artificially induced by constricting the conduit carrying the fluid, which serves to increase its velocity. Applying Bernoulli's equation to the two points in question we have, neglecting a change in elevation,

$$\frac{v_2^2 - v_1^2}{2} = \frac{p_1 - p_2}{\rho} \tag{4.92}$$

or, using flow rate $Q = vA$ and solving for p_2,

$$p_2 = p_1 - \frac{\rho Q^2}{2A_1}\left(\frac{d_1^4}{d_2^4} - 1\right) \tag{4.93}$$

This equation shows that by increasing flow rate or the diameter ratio d_1/d_2, one can, in principle, reduce the pressure to the limit of the vapor pressure of the fluid. In the orifice meter, this drop in pressure is registered on a pressure-measuring device (e.g., a U-tube manometer), to obtain the unknown flow rate Q of a fluid. Since the diameter d_1 of the constriction is usually unknown, the device must be calibrated with known flow rates. In the aspirator, on the other hand, no measurements are involved, the device is used merely to produce a vacuum for laboratory purposes.

The Pitot Tube

The Pitot tube (Fig. 4.13E) is a device that uses the difference in pressure at the nose of the tube (location 1) and that of the free stream (location 2) to measure gas velocity. Applying equation (4.92) to this case, and noting that at the nose, velocity $v_1 = 0$, we obtain

$$v_2 = \left(\frac{2\Delta p}{\rho}\right)^{1/2} \tag{4.94}$$

Thus, by measuring the pressure difference Δp between locations 1 and 2, the free stream velocity v_2 can be calculated.

The Airfoil

The airfoil (Fig. 4.13F) represents a simple, yet immensely beneficial application of the Bernoulli effect. That effect arises here from the asymmetric geometry of the wing, which has a curved upper portion bulging into the flow and a relatively level lower surface. The effect of the upper bulge is to constrict the air flow passing over it, thus increasing its velocity while the flow along the bottom remains essentially undisturbed. But we know from Bernoulli's equation that an increase in velocity is associated with a drop in pressure and that consequently, the top of the air wing will be exposed to a lower fluid pressure than the bottom. This pressure deficit results in a net *upward* force on the airfoil. In other words, it produces the "lift" necessary to maintain an aircraft in flight.

COMMENTS

- The striking feature of Bernoulli's equation is that it serves as a model for a host of seemingly disparate physical systems and phenomena. This is not an uncommon feature of modeling. We have, after all, made repeated use of Newton's law to model, in our examples, quite dissimilar systems whose only common feature was the requirement of a force balance. This is also the case in the present example, although the connection is not at all apparent from a first and casual inspection of the devices shown in Fig. 4.13. It is only upon deeper analysis that the common thread of a single model, the Bernoulli equation, is revealed.
- The question arises whether inclusion of friction forces would have significantly altered the results obtained. The answer here is no. The Bernoulli equation yields results in close agreement with observed values whenever the locations to which it is applied are in close proximity to each other, or friction can for some other reason be neglected. This is the case in all the devices shown in Fig. 4.13.

- In equation (4.90) we tacitly made use of what is termed the continuity equation. That relation is in essence an expression of the law of conservation of mass, which states that if no fluid is added or removed between points 1 and 2 of a flow system, the mass flow rate F (kg/s) is constant and equal to the product of density ρ, velocity v, and cross-sectional area A of flow. Thus

$$F = \rho_1 v_1 A_1 = \rho_2 v_2 A_2 = \text{constant} \tag{4.95}$$

or alternatively

$$F = \rho_1 Q_1 = \rho_2 Q_2 = \text{constant} \tag{4.96}$$

where Q = volumetric flow rate (m^3/s) as before.

For incompressible flow, such as flow of liquids, density is constant and we obtain

$$Q = v_1 A_1 = v_2 A_2 = \text{constant} \tag{4.97}$$

which is the relation used in equation (4.90).

◆ EXAMPLE 4.8 Lift Capacity of a Hot Air Balloon

The principle of a hot air balloon resides in the simple fact that a hot gas has a lower density than a cold gas. As a consequence, the weight of the displaced cold air outside the balloon (i.e., the buoyancy force) will exceed the weight of the hot internal gas and will impart a net upward force to the balloon. The effect has been enshrined in the popular saying: "hot air rises" (Fig. 4.14).

Typically in the operation of such balloons the hot gas is generated by the combustion of propane (C_3H_8). The product gases are then directed into the balloon. At launch, with the velocity still at zero and an absence of drag, a simple force balance leads to the result

$$\begin{aligned} \text{Buoyancy force} &= \text{Weight of balloon} \\ m_{air} g &= m_{gas} g + m_{lift} g \end{aligned} \tag{4.98}$$

or alternatively

$$m_{lift} = (\rho_{air} - \rho_{gas}) V_{balloon} \tag{4.99}$$

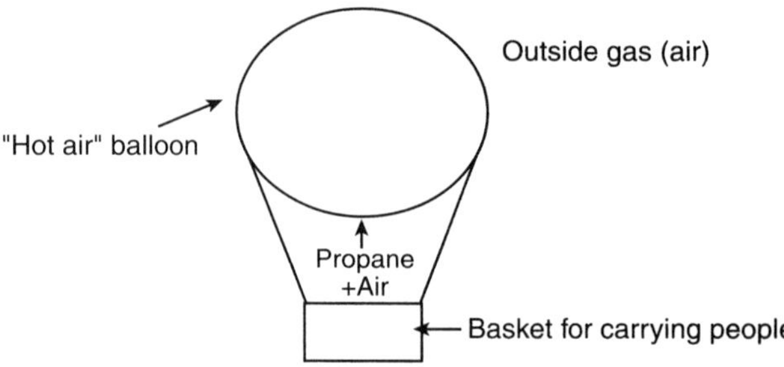

Figure 4.14 The hot air balloon.

where m_{lift} is the net mass of the balloon and its contents, that is, the maximum mass that can be lifted at launch or, more precisely, the mass that will keep the balloon in a stationary position. Calculation of m_{lift}, which is the task we set ourselves here, hinges on obtaining a value for the density ρ_{gas} of the hot combustion gases. Evidently this will require consideration of the heat of reaction and its effect on the temperature of the product gases. Calculation of that temperature is the first task we must address. Once its value is known, we can make use of gas laws to calculate the density of the hot combustion gases insides the balloon. On the way, to keep the calculations at a manageable level, we shall have to introduce some simplifying assumptions. Let us see how this works out in detail.

We start by noting that heats or enthalpies of combustion $\Delta H_c{}^\circ$ are usually tabulated for a standard temperature of 25 °C, or deduced from other enthalpies, those of formation, $\Delta H_f{}^\circ$, which are likewise tabulated for a temperature of 25 °C. The actual combustion is a complex process, which takes place over a range of temperatures, and the question then arises how this process can be described with the available data. We resolve this problem by considering a two-step hypothetical pathway having the same starting and end points as the actual process, hence being equivalent to it, that nevertheless makes use of the given standard enthalpy of combustion $\Delta H_c{}^\circ$. The stages in question are as follows:

1. Combustion of the propane with air at 25 °C. The energy change involved is given by $\Delta H_c{}^\circ$ kilojoules per moles of propane, the value for which is tabulated (Table 4.6).
2. Heating of the product gases to their ultimate temperature T_a. This quantity is given by the expression $\Sigma\, n_i C_{pi}\, (T_a - 25)$, where n_i represents the moles of the combustion gases and C_{pi} their molar heat capacities (J/mol °C) (see Table 4.6).

We now introduce a crucial assumption. We postulate that the process is adiabatic, that is, that during the launch no heat is lost to the surroundings, including the balloon fabric. Since that fabric is a poor conductor and the ultimate volume of the balloon quite large, this is not an unreasonable assumption. Furthermore, it enables us to state that the entire energy liberated by the process goes toward heating the combustion gases from 25 °C to their ultimate temperature T_a. We express this by writing

$$-\Delta H_c{}^\circ = \sum_i n_i C_{pi}(T_a - 25) \tag{4.100}$$

We follow this up by solving for the unknown quantity T_a, which is referred to as the *adiabatic flame temperature*.

The second task is to calculate the density of the combustion gases, for which we draw on the ideal gas law. We write

$$pV = nRT_a \tag{4.101}$$

where n is the total number of moles of combustion product. Multiplying each side by the molar mass M and dividing by volume V, we obtain for the density ρ_{gas} of the gases

$$\rho_{gas} = \frac{M_{avg}p}{RT_a} \tag{4.102}$$

Table 4.6 Molar Mass and Specific Heat Data for Some Common Gases

Compound	M (kg/kmol)	Cp (J/mol °C)
C_3H_8 (g)	44	52
CO_2 (g)	44	40
H_2O (g)	18	35
N_2 (g)	28	30
O_2	32	30

$\Delta H_c^\circ = -2146$ kJ/mol propane.

where M_{avg} is the *average* molar mass of the combustion products. The latter equals the weighted molar masses of the gases and is given by the expression

$$M_{avg} = \sum_i y_i M_i \tag{4.103}$$

where y_i is the mole fraction of the ith gaseous component.

We now have, in equations (4.100) and (4.102), the necessary expressions to calculate the lift capacity m_{lift} from equation (4.99). The data of Table 4.6, obtained from standard reference works, are available for this purpose.

Let us proceed with the specifics of our example. We postulate a balloon volume of 1000 m^3, external air density of 1.17 kg/m^3, and external pressure of 100 kPa. A 10% excess of air is to be used in the combustion of propane. We start by composing the reaction equation and proceed to calculate the moles n_i and mole fractions y_i that result in the combustion process. These quantities are required in the calculation of the adiabatic flame temperature T_a [equation (4.100)] and the average molar mass M_{avg} of the combustion products. We write

$$C_3H_8 + 5O_2 = 3CO_2 + 4H_2O$$

from which we obtain

$$n_{CO_2} = 3, \quad n_{H_2O} = 4, \quad n_{O_2}(\text{excess}) = 0.5, \quad n_{N_2} = 5.5 \times \frac{79}{21} = 20.7, \quad n_{tot} = 28.2$$

and

$$y_{CO_2} = \frac{3}{28.2} = 0.106$$

$$y_{H_2O} = \frac{4}{28.2} = 0.142$$

$$y_{O_2} = \frac{0.5}{28.2} = 0.010$$

$$y_{N_2} = \frac{20.7}{28.2} = 0.734$$

Turning first to the calculation of the adiabatic flame temperature, we have from equation (4.100)

$$\left[(nC_p)_{CO_2} + (nC_P)_{H_2O} + (nC_p)_{O_2} + (nC_p)_{N_2}\right](T_a - 25) = -\Delta H_c^\circ$$

or

$$10^{-3}[3 \times 40 + 4 \times 35 + 0.5 \times 30 + 20.7 \times 30](T_a - 25) = 2146$$

from which we obtain

$$T_a = 2421\,^\circ C = 2694\,K$$

Drawing next on equation (4.103), we have for the average molar mass M_{avg}

$$\begin{aligned} M_{avg} &= (yM)_{CO_2} + (yM)_{H_2O} + (yM)_{O_2} + (yM)_{N_2} \\ &= 0.106 \times 44 + 0.142 \times 18 + 0.01 \times 32 + 0.734 \times 28 \\ M_{avg} &= 28.1 \end{aligned}$$

This value is substituted into equation (4.102) to obtain ρ_{gas}, the density of the hot combustion gas:

$$\rho_{gas} = \frac{pM_{avg}}{RT} = \frac{10^5 \times 28.1}{8.314 \times 2694} = 0.125\ kg/m^3$$

One notes that the gas density is approximately 10 times less than that of the surrounding colder air taken to be 1.17 kg/m^3. The final step of obtaining the lift capacity m_{lift} then follows immediately from equation (4.99). We have

$$\begin{aligned} m_{lift} &= (\rho_{air} - \rho_{gas})V_{balloon} \\ &= (1.17 - 0.125)10^3 \end{aligned}$$

$$\boxed{m_{lift} = 1045\ kg}$$

COMMENTS

- We start by noting that the calculation proceeded over a series of steps that led to the final result. The initial analysis revealed that the first task to be addressed was the calculation of the temperature of the combustion products. To do this we introduced the assumption that heat losses to the surroundings were small. This holds well during the initial launch period, and any losses that do occur can be accommodated in the statement that the calculated value of m_{lift} represents the *maximum* lift capacity of the balloon. We thus establish the upper limit of what the balloon can do. This is in accord with our general philosophy of avoiding precise solutions to complex problems by deriving upper and lower bounds instead. In this particular instance the upper bound comes very close to representing the true value of m_{lift}. We next turned to the task of calculating the gas density using the value of the adiabatic flame temperature calculated earlier. We drew for this purpose on the ideal gas law, which holds well provided the temperature is sufficiently high and the pressure low. Both these conditions are reasonably well satisfied in the present case.

- While the final expression for the lift capacity derives from a force balance that prompted us to place the problem in this particulate chapter, the calculations leading up to the final result required us to use material drawn from the field of physical chemistry. It is not uncommon in modeling to have to use tools from different disciplines. This is one of the many challenges that must be met.
- Is the result obtained reasonable? We answer this question in the affirmative. A balloon typically carries three to four people weighing altogether perhaps 300 kg, while the weight, of the basket, combustion equipment, and balloon fabric might amount to about 500 kg. This still leaves us with a comfortable margin of 250 kg for overcoming air drag during the ascent and for the accommodation of additional equipment. Larger balloons will of course lift a proportionately larger mass, so that larger loads can be easily handled by increasing size. Since the lift capacity varies with the third power of diameter, substantial increases in lift capacity are possible with only modest increases in diameter.
- The combustion gases, which are at their maximum temperature during launch, will ultimately cool during travel. Lift will then decrease, and this can be used as a mechanism to effect a descent of the balloon. Conversely, if it is desired to stay aloft, additional fuel is burned to maintain the temperature of the internal gases.

◆ EXAMPLE 4.9 Work and Energy: Compression of a Gas and Power Output of a Bumblebee

Work in general describes the energy imparted to an object by a force **F** acting on it over a distance $x_2 - x_1$. Only the component acting in the x direction actively affects the energy of the object being moved. We are, in other words, dealing with the projection of the force vector onto the x direction. Since the projection of one vector onto another is given by the scalar product of the two (see Table 2.1), and since the force itself may vary with distance, work can be defined by the following integral of the scalar product of force and distance:

$$W = \int_{x_1}^{x_2} \mathbf{F} \cdot d\mathbf{x} \tag{4.104}$$

This expression leads to a host of different types of work and its equivalent energy, which we have summarized for the convenience of the reader in Table 4.7. The work done in raising a body against gravity by a distance h, for example, results in an increase of its potential energy, and is given by

$$W = E_p = \int_a^h F\cos(F, x)\,dx \tag{4.105a}$$

$$= \int_0^h mg(l)\,dx \tag{4.105b}$$

$$Ep = mgh \tag{4.105c}$$

Table 4.7 Various Types of Energy Resulting from Work Done on an Object

Type of Work/Energy	Expression
General	$W = \int_{x_1}^{x_2} \mathbf{F} \cdot d\mathbf{x}$
	$= \int_{x_1}^{x_2} F \cos(F, x)\, dx$
Potential energy	$E_p = mg\Delta x = mgh$
Translational kinetic energy	$E_K = \frac{1}{2} mv^2$
Rotational kinetic energy	$E_K = \frac{1}{2} I\omega^2$
Extension of a spring	$E_s = \frac{1}{2} kx^2$

The translational kinetic energy resulting from the action of a constant force on an object in the direction of motion can be developed in a similar manner:

$$W = E_K = \int_0^x F \cos(F, x)\, dx = \int_0^x F(1)\, dx \tag{4.106a}$$

$$= \int_0^x ma\, dx = m \int \frac{dv}{dt}\, dx \tag{4.106b}$$

$$= m \int_0^x \frac{dv}{dx}\frac{dx}{dt}\, dx = m \int_0^x \frac{dv}{dx} v\, dx \tag{4.106c}$$

$$= m \int_0^v v\, dv \tag{4.106d}$$

$$E_K = \tfrac{1}{2} mv^2 \tag{4.106e}$$

Note that the counterpart for rotational motion is given by

$$E_K = \tfrac{1}{2} I\omega^2 \tag{4.107}$$

where I = moment of inertia, and ω = angular velocity.

As yet another example of work being transformed into stored energy, consider the process of extending a spring against its restoring force, $F = kx$. Here we are dealing with a force that varies with distance, and we obtain

$$W = E_s = \int_0^x F \cos(F, x)\, dx \tag{4.108a}$$

$$= \int_0^x kx(1)\, dx \tag{4.108b}$$

$$W = \tfrac{1}{2} kx^2 \tag{4.108c}$$

Let us now consider the two cases cited in the title.

Isothermal Compression of an Ideal Gas

Suppose that we wish to compress an ideal gas of initial volume V_1 and at a pressure p_1 to a smaller volume V_2 with an attendant increase in pressure p_1 to p_2.

We start by noting that the travel distance $x_2 - x_1$ of the piston is an inconvenient quantity to deal with because it is rarely known. The force F contained in the work integral is similarly unsuited for direct computation. Rather, we want to be dealing with the known quantities of volume and pressure. These can be easily introduced by multiplying and dividing the work integral by the cross-sectional area of the piston. We obtain

$$W = \int_{x_1}^{x_2} F\,dx = \int_{x_1}^{x_2} \frac{F}{A} A\,dx \tag{4.109a}$$

$$W = \int_{V_1}^{V_2} p\,dV \tag{4.109b}$$

Distance has thus been replaced by volume, and force by pressure. Note that pressure varies with volume, and our first task will therefore be to establish a functional relationship between p and V. This is done by invoking the ideal gas law, which for one mole takes the form

$$pV = RT \tag{4.110}$$

where R = gas constant = 8.314 J/mol K and T = absolute temperature. Substitution into equation (4.109) yields

$$W = RT \int_{V_1}^{V_2} \frac{dV}{V} \tag{4.111}$$

where temperature has been placed outside the integral by assuming isothermal operation. Evaluation of the integral then yields, for one mole of an ideal gas

$$W = RT \ln \frac{V_2}{V_1} = RT \ln \frac{p_1}{p_2} \tag{4.112}$$

Let us examine the magnitude of this term by considering one mole of oxygen (i.e., 32 g) being compressed from an initial pressure of 1 atm to a final pressure of 10 atm at a temperature of 300 K. We obtain

$$|W| = \left| 8.314 \times 300 \ln \frac{1}{10} \right| = 5.74\ \text{kJ/mol}$$

Note that if the compression ratio is raised by another factor of 10, the work required will increase by only a factor of 2. This is a direct consequence of the logarithmic relation established earlier. We obtained a negative quantity (neatly sidestepped by computing the absolute value of W) because work was done *on* the system rather than by the system, which invariably entails a negative quantity.

It is of some interest to compute the *power requirements* for the compression process, since this quantity will determine the size of the motor to be used in driving the compressor. Power is defined as work done per unit time, that is,

$$\text{Power P} = \frac{\text{Work W}}{\text{Time t}} = \frac{\text{Work}}{\text{moles}}\frac{\text{moles}}{\text{time}} \tag{4.113}$$

Usually P is expressed in units of horsepower (hp), thus: 746 W = 746 J/s. Let us assume a production rate of 100 mol oxygen per second, or 3.2 kg O_2/s. We obtain

$$P = 5.74 \times 10^3 \times \frac{100}{746} = 77\,\text{hp}$$

In practice, a motor of at least 100 hp would be chosen to make up for motor and compressor inefficiencies. This is a sizable unit, though not unusual in applications of this magnitude.

The second component of our example is drawn from the field of biomechanics and involves the examination of the work done by a bumblebee.

Power Output of a Bumblebee

As a bumblebee approaches a suitable flower for nectar and begins feeding, it remains airborne above the flower by flapping its wings at the average rate of 100 downward strokes per second! Each wing is 1 cm long and sweeps a total angle of 80° during each stroke (see Fig. 4.15A). The mass of the bumblebee is 0.25 g. The task is to calculate the work done by the bee in maintaining its hovering position above the flower, and the corresponding power output.

We start by noting that force is applied only during the downstroke part of the cycle, while the upstroke provides no force at all (Fig. 4.15B). As a consequence, to yield an *average* force over the *entire* cycle equal to its weight, each downstroke must provide a force equal to *twice* the weight w of the bee. The force is applied at right angles to the arcs of 80° swept during the downstroke, and the corresponding work done is then given

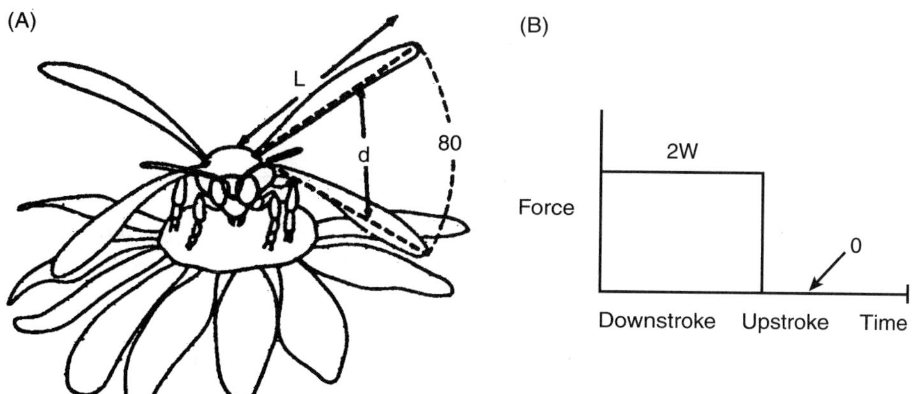

Figure 4.15 (A) Geometry of a bumblebee hovering above a flower. (B) Force applied during downstroke and upstroke.

by the following scalar product integral

$$W = \int \mathbf{F} \cdot \mathbf{ds} = F \times s \tag{4.114}$$

where $F = 2 \times$ weight $= 2$ mg. The arc length s is given in terms of the angle θ swept by $s = 2\pi r\theta/360$, so that

$$W = F \times s = 2\,mg \times 2\pi r \frac{\theta}{360} \tag{4.115}$$

Substituting the pertinent numerical values, we obtain

$$W = 2 \times 0.25 \times 10^{-3} \times 9.81 \times 2\pi \times 10^{-2} \frac{80}{360}$$

$$\boxed{W = 6.9 \times 10^{-5}\ \text{J/stroke}}$$

The corresponding power output is given by

$$P = W \times \frac{\text{strokes}}{\text{s}} = 6.9 \times 10^{-5} \frac{100}{746}$$

$$\boxed{P = 9.3 \times 10^{-6}\ \text{hp}}$$

COMMENTS

- Both these problems involve simple but elegant examples of the application of the work integral. In the first example, force had to be converted to pressure, and distance to volume, to apply the data given for the compression of a gas. The challenge in the second problem was to recognize that the applied force had to equal twice the weight of the bee to maintain the insect in a hovering position. The power required for these two cases, one a large-scale industrial process, the other a delicate biological phenomenon, differed, not too surprisingly, by some eight orders of magnitude. Power output of a different type, that of a car, will be addressed in Example 8.4.

PRACTICE PROBLEMS

4.1 **The suddenly applied static load** Consider a load suspended above a horizontal platform and in contact with it, which is then suddenly released. Show that the resulting stress is exactly twice that caused by the same load at rest. This odd result can be generalized into the following statement.

A *suddenly applied load*, at zero velocity *induces twice the stress* of a gradually applied load of the same magnitude.

4.2 **The neutral axis** Show that the neutral axis always coincides with the centroid of the cross-sectional arc of a beam.

4.3 **Bending of a beam under a lateral load** Derive the deflection δ for the cantilevered beam shown in Fig. 4.3A. (*Hint*: At the clamped end both the deflection and its derivative are zero.)

ANSWER $\delta = -\frac{FL^3}{3EI}$

4.4 **The clamped strut** Show that Euler's formula for a clamped strut is given by

$$F_b = \frac{4\pi EI}{L^2}$$

(*Hint*: The applicable differential equation is in this case

$$EI\frac{d^2y}{dx^2} = -Fy + M_c$$

where M_c is the unknown reactive bending moment of the embedding medium. Three boundary conditions will therefore be required.)

4.5 **The circulating charge in a magnetic field** Figure P4.5A shows a negatively charged particle, $-q$, introduced with velocity $\mathbf{v}$ into a uniform magnetic field $\mathbf{B}$. The magnet straddles the plane of the figure so that $\mathbf{B}$ is at right angles to $\mathbf{v}$ and points *into* the plane. The particle then experiences a sideways force $\mathbf{F}$, much like a stone held by a rope and whirled in a horizontal circle.

Show that the frequency ν of the circular motion is given by

$$\nu = \frac{\omega}{2\pi} = \frac{qB}{2\pi m}$$

(*Hint*: Consult the tabulation on circular motion given by Table 4.1.)

4.6 **The cyclotron** The cyclotron, invented by Ernest Lawrence in 1932, is a device to accelerate positively charged particles such as protons and deuterons (proton + neutron) to high energies for use in atom-smashing experiments. As shown schematically in Fig. P4.5B, essentials of cyclotron are two D-shaped evacuated chambers made of copper sheet and separated by an air gap that contains a source of positively charged particles. The dees are connected to an electric oscillator that imparts a negative potential, first to one and then to the other chamber, at high frequency. This causes the positive particles to be alternately attracted to one or the other of the dees. The dees are also immersed in a magnetic field that points into the plane of the figure, thus imparting a clockwise circular motion to the particles, as shown in Practice Problem 4.5. The two magnetic fields of Fig. P4.5 are in opposite directions because the charges are opposite in sign.

The combined effect of the two fields is to impart a spiral motion of increasing radius to the particles until they reach a negatively charged deflecting plate, which causes them to exit the chamber. The key to the operation of the cyclotron is the

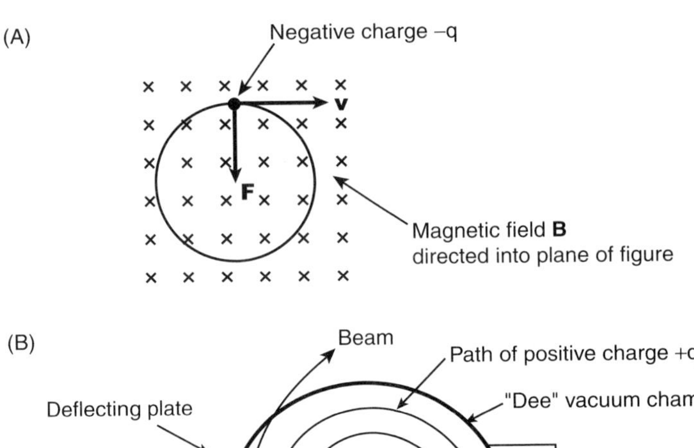

Figure P4.5 Circulating charges in a magnetic field: (A) effect of the Magnetic Field and (B) combined effect of magnetic and oscillating electric fields—the principle of the cyclotron (the magnetic field points into plane of the figure).

resonance condition, which states that if the energy of the circulating ion is to increase, energy must be fed to it at its natural frequency given by the equation in the Practice Problem 4.5. In a similar manner one feeds energy to a swing by pushing it every time it completes an oscillation.

(a) Calculate the magnetic field necessary to accelerate deuterons ($m = 3.3 \times 10^{-27}$ kg, $q = 1.6 \times 10^{-10}$C) to a frequency of 12×10^6.

(b) What is the required cyclotron radius to obtain 500 GeV protons ($1\,eV = 1.6 \times 10^{-19}$ J, $m = 1.67 \times 10^{-27}$ kg) using a magnetic field of 1.5 T (*Hint*: Make a force balance using the relation of Table 4.1 to obtain v, then convert to kinetic energy.)

ANSWERS (a) 1.6 T
(b) 68 m

Note: Since a magnet of this size would have been prohibitively expensive, the operation of the cyclotron was modified. One such modification consists of varying both V and ν in such a way that the orbit radius remains essentially constant. This permits the use of an *annular* magnet in place of the cylindrical configuration of the classical cyclotron, at a great savings in cost. Accelerators based on this principle are known as *synchrotrons*. The lesson here is that successful modeling must include a consideration of cost.

4.7 Wall collision rate

(a) Show that the rate r_C at which a gas undergoes collisions with a container wall is given by

$$\boxed{r_C = (\text{number of collisions/m}^2\ \text{s}) = \frac{N\bar{v}}{4V}} \tag{4.116}$$

(*Hint*: Use part of the procedure used in the derivation of Boyle's law in Example 4.4.)

(b) Calculate the number of collisions hydrogen undergoes with its container walls at $p = 100\,\text{kPa}$, $T = 298\,\text{K}$.

ANSWER (b) 4.35×10^{28} collisions/m^2 s

4.8 Projectile trajectory with air resistance included Derive the differential equations in x and y that describe the path of a projectile fired with a muzzle velocity v_0 and an angle of elevation α. What are the initial conditions for the system? (*Hint*: Establish the component forces for the drag force F_D having a magnitude given by equation (4.59) and opposite in direction to the velocity vector [Fig. 4.11].)

ANSWER

$$\frac{d^2x}{dt^2} = -k\frac{dx}{dt}\left[\left(\frac{dx}{dt}\right)^2 + \left(\frac{dy}{dt}\right)^2\right]^{1/2}$$

$$\frac{d^2y}{dt^2} = -g - k\frac{dy}{dt}\left[\left(\frac{dx}{dt}\right)^2 + \left(\frac{dy}{dt}\right)^2\right]^{1/2}$$

4.9 Projectile fired at a moving target (no drag) Suppose a target moving at a constant velocity v_t (m/s) is level with and at a distance d (m) from a gun at the instant the weapon is fired. If the target moves in a horizontal line directly away from the gun, show that for a given muzzle velocity v_0 the angle of elevation α must satisfy the equation

$$v_0^2 \sin 2\alpha - 2v_t v_0 \sin\alpha - dg = 0$$

4.10 Maximum height attained by a projectile A projectile is fired vertically from the earth's surface with a speed of 10 km/s, just below the escape velocity of 11.2 km/s. What is the maximum height attained by it? (*Hint*: Neglect atmospheric friction.)

4.11 Cruising speed in outer space A rocket is accelerated to a speed of 15 km/s near the earth's surface, which is above the escape velocity of 11.2 km/s. Show that very far from the earth, the cruising speed is 9.98 km/s.

4.12 Maximum performance of a piston pump Manually operated piston pumps can be used to draw water by suction from wells and other water basins. Calculate the maximum height to which water can be raised in this fashion. (*Hint*: Assume a water vapor pressure of 1 kPa.)

ANSWER 10.22 m

4.13 Vertical discharge of a liquid Show that if liquid is discharged vertically upward, an expansion in jet cross-sectional area occurs. Conversely, if the discharge is vertically downward, the jet contracts.

4.14 Discharge from a tank. Minimum constriction diameter Suppose that a constriction is placed within the discharge pipe of the tank shown in Fig. 4.14B. The constriction is smooth and streamlined so that the pipe runs full both upstream and downstream of the narrow passage. The diameter of the constriction is then progressively reduced.

(a) Derive an expression for the minimum diameter that will allow the normal flow rate to be maintained.

(b) What will happen if the constriction is made smaller than the calculated minimum?

ANSWER (a) $(d_C)_{min} = d\left[\frac{g\Delta z}{g\Delta z + (\Delta p)_{max}/\rho}\right]^{1/4}$

4.15 The hot air balloon again: maximizing the flame temperature The use of oxygen instead of air in the combustion of propane has the potential of substantially increasing the adiabatic flame temperature.

(a) Why would this be so present an argument.

(b) Calculate the adiabatic flame temperature that results from the use of a 10% excess of *pure* oxygen.

(c) What is the effect on lift capacity? Discuss.

4.16 Kinetic energy of a charge moving in a uniform electrical field A particle of mass m and charge q is placed at rest in a uniform electrical field set up between two oppositely charged metal plates (Fig. P4.16). It is then released, and the task is to calculate its kinetic energy as a function of the distance y traveled. (*Hint*: Recall that the force exerted by an electric field on a charge q is given by $\mathbf{F} = \mathbf{E}q$.)

ANSWER $E_K = qEy$

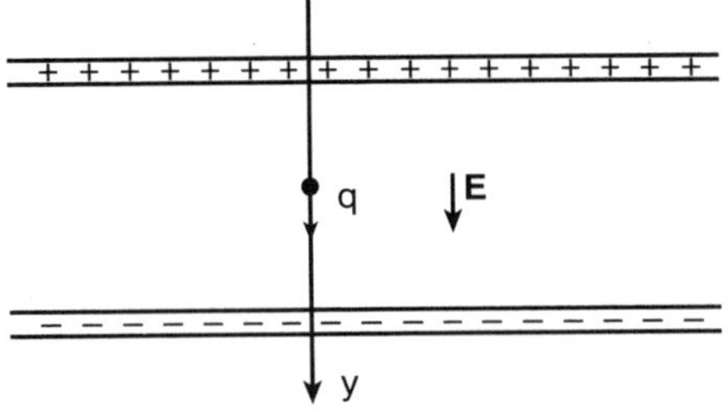

Figure P4.16 Charge moving in a uniform electrical field.

CHAPTER 5

Compartmental Models

In Chapter 1 we introduced the notion of a compartment, which is one of two mainstays of modeling at an elementary level, the other being what we referred to as the one-dimensional pipe. The principal feature of a compartment, or stirred tank as it is often called, is that its contents are assumed to be uniformly distributed so that their properties and the associated state variables vary at most with time, and not at all with spatial distance. In the most common application of this model, mass or energy enters and leaves the compartment, frequently accompanied by chemical reactions, phase changes, or by an exchange of mass and energy with the surroundings. Uniformity may be achieved by thoroughly mixing the contents of the compartment or, more frequently, by conceptually deducing from the physical situation that the state variable is uniformly distributed or very nearly uniformly distributed in space. For example, in the case of a thermocouple response to a change in ambient temperatures, which we considered in Example 1.5, no physical mixing of the contents took place. Instead it was deduced from the small dimensions of the device and its high thermal conductivity that the state variable, here the temperature, would be uniformly distributed in space. The thermocouple could consequently be regarded as a compartment and its response derived from a compartmental model.

Compartmental models are usually set up by performing integral mass and energy balances around the compartment. These can be instantaneous in time, in which case we obtain a first-order differential equation, or cumulative in time, in which case the initial amount present is equated with the amount consumed plus the amount left over at time t. An application of this latter type of balance is found in Chapter 6 (Practice Problem 6.9) and leads to an algebraic equation. Auxiliary relations provide additional algebraic equations so that the complete model is generally composed of a set of simultaneous first-order ODEs combined with a set of algebraic equations. In many cases this reduces to a set of one ODE and one AE.

If both the input to the compartment and its output are steady from the outset, or become steady after a period of time, the transients in the solution will ultimately disappear as $t \to \infty$, and the system will attain a steady state. In the present chapter we shall be dealing with both unsteady- and steady-state models.

Compartmental models occur most commonly in the fields of chemical engineering, environmental engineering, and biomedical engineering. Chemical engineers use compartments to describe and analyze the behavior of tank reactors with inflow and outflow, also known as continuous flow stirred tank reactors, or CSTRs. These reactors may operate under transient conditions, which may occur during start-up or shutdown, or in response to disturbances in input or output; most frequently, however, they are analyzed in the steady state. The information derived from such an analysis includes reactor volume, conversion obtained in a reactor of given size, and, less frequently, the kinetic parameters of the reaction. Chemical engineers, as well as mechanical engineers, also concern themselves with heat transfer operations involving compartments, which are then assumed to have a uniform temperature. These include the heating and cooling of tanks and other similar entities. The thermocouple response to an external disturbance just cited previously (Example 1.5) is another instance of compartmental modeling involving the transfer of energy.

In environmental engineering, the task is usually to track the fate, or course in time, of toxic or other objectionable chemicals, although benign substances such as oxygen may also be monitored. Alternatively, the pertinent concentration variables may be determined experimentally and the data used to extract transport coefficients, kinetic parameters such as the biodegradation rate constants, or what are termed partition coefficients, which relate concentrations in adjacent phases in a state of equilibrium. The phases of concern in environmental problems are three in number: air, water, and soil or sediment. In the case of accidental spills, for example of oil, additional phases may make their appearance. Both transient and steady-state conditions may arise.

In biomedical engineering, the aims of compartmental modeling are similar to those seen in problems involving the environment. One is again interested in obtaining the time history of a substance to which the body has been exposed, but more often than not the substances are benign (e.g., drugs or tracers). The model can be used, for example, to determine how long a constant infusion of a drug takes to reach a useful concentration level, or the size of the dose that should be applied for effective therapy to take place. The parameters needed to make these predictions, such as rate constants, volume of the body compartments, and internal flow rates, are again obtained from appropriate experiments and the fitting to a compartmental model of the data, which are usually concentration histories. Transient conditions are more common here than steady-state conditions.

The problems we have just described are presented in this chapter in a series of illustrative examples applied to chemical, environmental, and biomedical systems. With the exception of some Practice Problems, models involving energy transport are not taken up. Problems of interest in this area, at least at an elementary level, are more likely to be of a steady-state distributed type, and their consideration is consequently relegated to the next chapter.

Example 5.1, which is drawn from the biomedical field, describes the use of tracer data to determine plasma volume and cardiac output in humans or animals. This is followed by three examples dealing with chemical reactors. Example 5.2 presents the model for a traditional CSTR, discusses the important concepts of conversion and residence time, and takes up an interesting optimization problem in which the cost of the reactor is balanced against the benefits of increased yield. Example 5.3 deals with an important subcategory, the bioreactor, in which bacteria, fed with an appropriate nutrient, are converted to beneficial or benign products. The analysis here is entirely a steady-state one and reveals some unexpected results not found in conventional reactors. Departures from ideal, uniform mixing are addressed in Example 5.4, which introduces the concept of residence time distribution (RTD). An outline of the determination of RTDs by appropriate tracer experiments is given.

Examples 5.5 to 5.7 deal with what we term moving boundary problems. These arise when the boundary between two phases undergoes a continuous movement owing to evaporation, condensation, or solidification, or as a result of a chemical reaction, such as the combustion of a liquid or solid fuel particle. Example 5.5 considers a reacting particle and introduces the reader to the notion of the shrinking (or growing) core model and the quasi-steady state. Here the core is viewed as a compartment of uniform composition that shrinks or grows with time, while the external region, consisting of products, grows in size and is assumed to be at steady state at any given instant. This division into two simple and distinct domains brings about an enormous simplification of the problem and is also used in Examples 5.6 and 5.7. The first of these considers a crystallization process that still contains a core compartment but grows with time instead of shrinking. Example 5.7 presents another shrinking core example, this time drawn from the biomedical field. It deals with implants of controlled-release drug delivery systems, which constitute an important and growing phenomenon on the medical scene. An analysis of the model results is made to determine conditions that lead to a constant release rate.

Three environmental problems are taken up in Examples 5.8 to 5.10. They deal, respectively, with the evaporation of a pollutant from a water basin, the penetration of oil from a land-based spill into the ground, and the concentration changes that result in a water basin with stratified layers. The results of Example 5.8 are particularly striking, predicting as they do an enormous drop in pollutant concentration with only a small amount of evaporation.

The final three examples are again drawn from the biomedical field. Example 5.11 deals with a one-compartment model for drug administration and neatly demonstrates the wealth of information to be derived from just one set of experimental data. Example 5.12 strays from the conventional scene of biomedical modeling by addressing the deposition of blood cells from flowing blood onto the vessel wall. It turns out that the system can be treated as a compartment of constant concentration with an adjacent film resistance and an associated concentration gradient. The aim here is to predict the change in surface coverage with time. These results are important in the context of biocompatibility of implant materials. Finally, in Example 5.13 we introduce the contemporary topic of the human immune deficiency virus (HIV), the dynamics of which we examine in some detail. Modeling of the course of the disease has added considerably to our understanding of it and has been used to help devise therapeutic strategies.

We conclude our introduction with a reminder to the reader of the wealth of information and additional insight provided by the Practice Problems. They are worth a look even if no solution is attempted.

◆ EXAMPLE 5.1 Measurement of Plasma Volume and Cardiac Output by the Dye Dilution Method

One standard and historical method of determining the amount of certain body fluids present in the system is the use of a measured amount of dye that is preferentially soluble in the fluid (i.e., is not absorbed by the surrounding tissue). Suppose we wish to find the total volume V of blood plasma in the body. This is done by injecting a measured

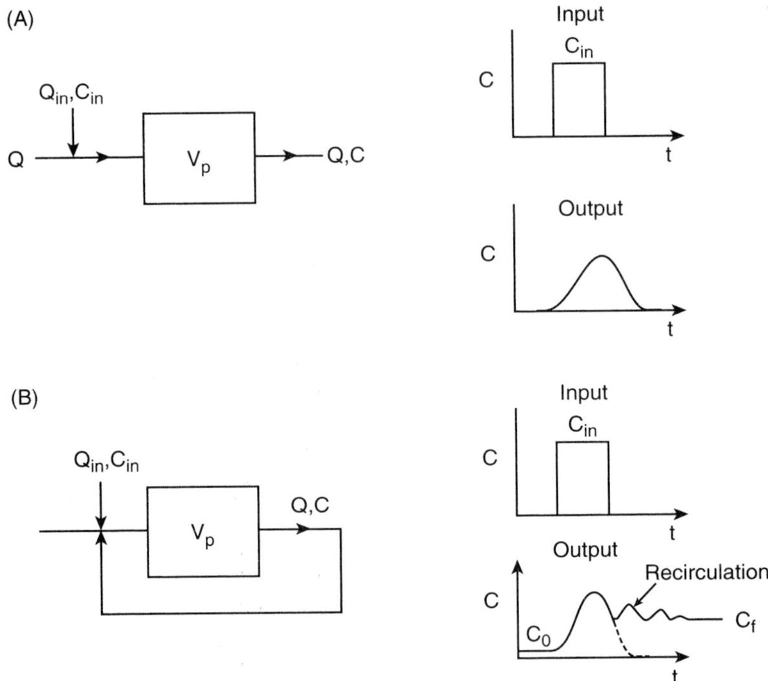

Figure 5.1 Plasma volume V and cardiac output Q by dye dilution: (A) without recirculation and (B) with recirculation.

amount of dye of known concentration into the circulatory system and monitoring its concentration with time at some convenient downstream location (Fig. 5.1). Figure 5.1B shows a typical record of concentration versus time for input and output. The input consists of a square wave of constant concentration, while output concentration exhibits both time lag and a drop in concentration due to the diluting effect of the plasma. The first and principal peak seen in the output is followed by several diminished secondary peaks as the circulating blood adds more dye to the sample region. Eventually the dye concentration subsides to a constant steady-state value C_f, which is recorded and used to compute the unknown plasma volume V_p. The calculation proceeds as follows.

Let us assume that the injection takes place at a flow rate Q_{in}, which may be constant or variable but is generally not known or recorded. This does not pose a problem, however, since the expression containing it can be converted, as we shall shortly see, into the injection volume V_{in}, which is known and can be measured with a great degree of precision. We can then write the following mass balance for the period of injection

$$\begin{array}{ccccc} \text{Rate of dye in} & - & \text{Rate of dye out} & = & \begin{array}{c}\text{Rate of change}\\ \text{of dye content}\end{array} \\ Q_{in}C_{in} & - & 0 & = & V_p \dfrac{dC}{dt} \end{array} \tag{5.1}$$

where C is the average concentration of the compartment and V_p the volume of plasma.

Formal integration over the injection period from 0 to t and assuming zero initial concentration then yields the expression

$$V_p \int_0^{C_f} dC = C_{in} \int_0^t Q\, dt \tag{5.2}$$

where it is quickly recognized that the integral on the right equals the volume of injected solution. We arrive at the relation

$$V_p C_f = V_{in} C_{in} = \text{Mass } m_d \text{ of injected dye}$$

or alternatively

$$\boxed{V_p = \frac{m_d}{C_f}} \tag{5.3}$$

Thus the plasma volume is given simply by the ratio of total mass of dye injected to the final steady-state concentration in the plasma, C_f. Several recirculations are usually necessary to attain that value. Since blood volume "turns over" approximately once every minute, a short time period is sufficient to obtain reliable values of C_f.

Let us now consider the hypothetical case of no recirculation. In Fig. 5.1A, the square input of dye results in an output consisting of a single Gaussian peak. That peak corresponds to the first and principal peak seen in an actual system with recirculation (Fig. 5.1B) and can be reproduced there by extrapolation along the dashed line. It turns out that this simulated case of a system without recirculation can be used to deduce the cardiac output or flow rate produced by the heart muscle. Let us examine the model describing this case.

We consider injection of a dye solution of concentration C_a on the arterial side of the heart, which results in a corresponding output concentration C_v on the venous side. A mass balance applied between the arterial and venous sides results in the expression

$$\begin{array}{ccccc} \text{Rate of dye in} & - & \text{Rate of dye out} & = & \begin{array}{c}\text{Rate of change}\\ \text{of the dye contents}\end{array} \\ QC_a & - & QC_v & = & V_h \dfrac{dC_h}{dt} \end{array} \tag{5.4}$$

where V_h and C_h are the liquid volume and average dye concentration in the heart, respectively. If this volume is assumed to be well mixed, C_h will equal C_v. This assumption, which is commonly used in all simple compartmental models, is not needed here. To show this, we proceed as follows.

Formally integrating equation (5.4), we obtain

$$V_h C_h(t) = Q \int_0^t (C_a - C_v)\, dt \tag{5.5}$$

To eliminate the unknowns V_h and $C_h(t)$, we consider the steady state $t \to \infty$ when all dye has left the system, that is, $C_h(t) = 0$. We obtain

$$0 = Q \int_0^\infty (C_a - C_v)\, dt \tag{5.6}$$

Since C_a need not be monitored, it can be expressed in terms of the known total mass of the dye, m_d. Thus we write

$$m_d = Q \int_0^{\infty} C_a \, dt \tag{5.7}$$

Combining equations (5.6) and (5.7), we arrive at the following expression:

$$Q = \frac{m_d}{\int_0^{\infty} C_v \, dt} \tag{5.8}$$

The cardiac output Q is thus given by the simple ratio of the total mass of injected dye to the integral under the first extrapolated peak. That extrapolation is needed to cancel out the effects of recirculation and can be effected by plotting the data on semilogarithmic paper, extrapolating the resulting straight line and then replotting the data on arithmetic paper. Extrapolation is usually carried to a small finite value C_0, which is induced by preloading the system with a small amount of dye. This yields the dashed curve shown in Fig. 5.1B.

COMMENTS

- Of note in this particular example is the use of a single and simple experiment to extract two relatively inaccessible parameter values, those of the plasma volume V_p and its flow rate Q. In both cases simple unsteady mass balances were used to arrive at the desired result. Unknown quantities were either eliminated by drawing on an equivalent steady-state mass balance $[V_h C_h(t) = 0]$ or by relating them to the known total mass of dye m_d. Note that even though compartments were used, it was not necessary to assume that the concentrations inside them were uniform. We used instead an average and unknown concentration C, which was again either eliminated $[V_h C_h(t) = 0]$ or allowed to come to a constant, measurable steady-state concentration C_f. Thus even though the underlying mass balances were quite elementary, their imaginative use led to the desired solution. The lesson here is that a model need not be complex even if the underlying system or process and the information sought are, at first sight, quite baffling.

  Equations (5.1) and (5.3) are often referred to as "Fick's principle" and are generally attributed to Fick who was among the original, nineteenth-century investigators in the field. It speaks to the genius of these early workers that they were able, with the crude tools available in their day, to determine two of the most important physiological parameters.

◆ EXAMPLE 5.2 The Continuous Stirred Tank Reactor (CSTR): Model and Optimum Size

The continuous stirred tank reactor, or CSTR, is a widely used device for carrying out chemical reactions. A schematic diagram of the unit is shown in Fig. 5.2. Feed of concentration C_{A_0} enters continuously with a volumetric flow rate Q, undergoes a reaction in the well-stirred tank, and leaves with a concentration C_A of reactant along with various products. We assume a system of constant density and constant volume in which the

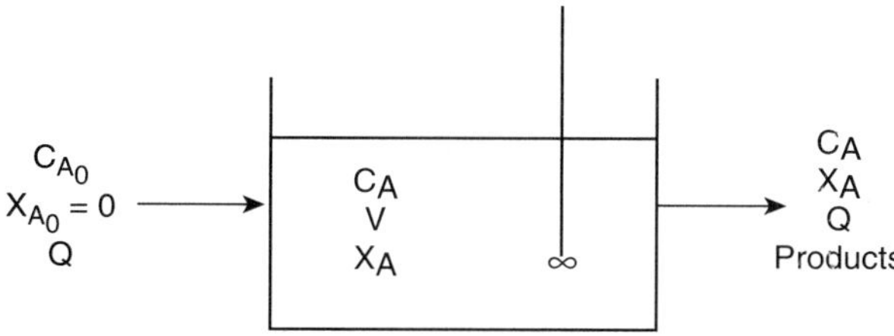

Figure 5.2 The variables in a continuous stirred tank reactor (CSTR).

reaction is of the general form

$$A \rightarrow \text{Products (B, C, etc.)}$$

with a reaction rate r_A in units of moles A reacted per unit time and volume. Polymerization reactions fall into this category.

In setting up the mass balance for the system, we follow the convention of placing the reaction term r_A for the species being consumed in the "out" column, and the corresponding terms for the species produced (r_B, r_C, etc.) under the "in" column. We then obtain two equations:

A Balance

$$\begin{array}{ccc} \text{Rate of A in} & - \text{ Rate of A out} = & \begin{array}{c}\text{Rate of change}\\ \text{of A contents}\end{array} \\ QC_{A_0} & - [QC_A + r_A V] = & V\dfrac{dC_A}{dt} \end{array} \tag{5.9}$$

B Balance

$$\begin{array}{ccc} \text{Rate of B in} & - \text{ Rate of B out} = & \begin{array}{c}\text{Rate of change}\\ \text{of B contents}\end{array} \\ [QC_{B_0} + r_B V] - & QC_B \quad = & V\dfrac{dC_B}{dt} \end{array} \tag{5.10}$$

and similarly for the remaining species.

These expressions are general-purpose equations that describe system behavior under general transient conditions. Such time-dependent behavior arises, as we have seen, during start-up and shutdown of the reactor, or when there are fluctuations in the incoming flow rates and compositions. Of greater practical interest is the important subcategory of steady-state operation. Here we assume that the transients have subsided or that disturbances have been rectified by an appropriate controlling mechanism. This is the normal mode of operation of a reactor. Under these conditions, the time derivatives on the right of the balance equations vanish, and we obtain, for the A balance,

$$Q(C_{A_0} - C_A) - r_A V = 0 \tag{5.11}$$

and similar expressions for the remaining species balances.

We now introduce two concepts designed to put equation (5.11) into a terser and more practical form. The first is the so-called conversion X_A, encountered in earlier in

Example 3.11, which is defined as

$$X_A = \frac{\text{initial moles} - \text{moles left}}{\text{initial moles}} = \frac{C_{A_0} - C_A}{C_{A_0}} \tag{5.12}$$

Thus X_A denotes the fraction of the incoming feed that has been converted to product; X_A is zero when no reaction has taken place and is equal to one when all the feed has been reacted to product. Hence high values of X_A are good, and low values are undesirable.

The second concept is that of residence or holding time τ, in seconds, which is defined as the ratio of reactor volume to volumetric flow rate. Thus

$$\tau = \frac{V}{Q} \tag{5.13}$$

Large values of τ generally lead to high conversions and are thus desirable, while low values of τ produce the opposite result.

With these definitions in hand, equation (5.11) becomes

$$\tau = \frac{V}{Q} = \frac{C_{A0}}{r_{A0}} X_A \tag{5.14}$$

This equation can serve several purposes. It can be used in reactor design to specify the volume V required to obtain a prescribed conversion X_A. It can also be used in performance prediction in the calculation of the conversion X_A, which results in a reactor of given size V operating with a residence time of $\tau = V/Q$. Finally, we can draw on experimental conversion data obtained for different residence times to determine the reaction rate r_A or the kinetic parameters contained in it. Suppose, for example, that we are dealing with a first-order reaction having the rate $r_A = k_r C_A$, where k_r is the reaction rate constant. Using equations (5.12) and (5.14), we obtain

$$\tau = \frac{V}{Q} = \frac{X_A}{k_r(1 - X_A)} \tag{5.15}$$

It follows from this expression that values of k_r can be obtained from the slope of a plot of residence time versus $X_A/(1 - X_A)$.

As a further application of these equations, suppose that 100 mol of R is to be produced hourly from a feed consisting of a saturated solution of A ($C_{A_0} = 0.1$ mol/L) in a continuous stirred tank reactor. The reaction is

$$A \rightarrow R \qquad r_A = k_r C_A = (0.2\,\text{h}^{-1})C_A$$

Cost of the reactant is

$$\$_A = \$0.50/\text{mol A}$$

Cost of the reactor, expressed as a depreciation rate and including installation, auxiliary equipment, instrumentation, overhead, labor, and so on, is

$$\$_E = 0.01/\text{h L}$$

The questions we wish to address are the following:

What reactor size, feed rate, and conversion should be used for optimum operation?

What is the unit cost of R for these conditions if unreacted A is not recycled?

We start by recognizing this as an optimization problem that balances high fractional conversion in a large reactor (high equipment cost) on the one hand against low conversion obtained in a smaller and cheaper reactor. On an hourly basis, the total cost will be

$$\$_{tot} = (\text{Volume of reactor})(\text{Cost/time volume of reactor}) + (\text{Feed rate of reactant})(\text{Unit cost of reactant})$$

or

$$\$_{tot} = V\$_E + F_{A_0}\$_A$$

where F_{A_0} is the feed rate in moles of A fed to the reactor per hour and equals QC_{A_0}, and $\$_E$ and $\$_A$ are the equipment and feed costs, respectively. Equation (5.15) consequently becomes

$$V = \frac{F_{A_0}X_A}{k_r C_{A_0}(1 - X_A)}$$

Noting that the rate of production F_R of R is given by

$$F_R = F_{A_0}X_A = 100\,\text{mol/h}$$

we can eliminate the unknown F_{A_0} and write the total cost expression in terms of one variable alone, X_A. We obtain

$$\$_{tot} = \frac{F_R}{KC_{A_0}(1 - X_A)}\$_E + \frac{F_R}{X_A}\$_A$$

$$= \frac{100}{(0.2)(0.1)(1 - X_A)}0.01 + \frac{100}{X_A}0.5$$

$$\$_{tot} = \frac{50}{1 - X_A} + \frac{50}{X_A}$$

To find the condition for minimum cost, we differentiate this expression and set the result equal to zero. Thus

$$\frac{d}{dt}\$_{tot} = 0 = \frac{50}{(1 - X_A)^2} - \frac{50}{X_A^2}$$

from which we obtain

$$X_A = 0.5$$

Hence the optimum conditions are as follows:

Conversion

$$X_A = 0.5$$

Feed Rate

$$F_{A_0} = \frac{F_R}{X_A} = \frac{100}{0.5} = 200 \text{ mol A/h}$$

or

$$Q = \frac{F_{A_0}}{C_{A_0}} = \frac{200}{0.1} = 2000 \text{ L/h}$$

Reactor Size

$$V = \frac{F_{A_0} X_A}{k_r C_{A_0}(1 - X_A)} = \frac{100}{(0.2)(0.1)(1 - 0.5)} = 10{,}000 \text{ L}$$

Cost of Product (Minimum)

$$\frac{\$_{tot}}{F_R} = \frac{V\$_E + F_{A_0}\$_A}{F_R} = \frac{10{,}000(0.01) + 200(0.5)}{100}$$

$$\boxed{\frac{\$_{tot}}{F_R} = \$2/\text{mol R}}$$

COMMENTS

- This example gives but one small and simple sampling of a much broader subject area, commonly referred to as chemical reaction engineering. The topic addresses, among other things, various reactor types and their advantages, the effect of reaction complexity, catalytic reactions and gas–solid reactions in general, and the departure from ideal, well-mixed conditions in a CSTR. We shall have occasion to address some of these items in subsequent examples (see Examples 5.4, 5.5, and 5.6).

  Optimization problems of the type shown here occur with considerable frequency in chemical reaction engineering. Two examples involving multistep reactions are given in Practice Problems 5.3 and 5.4. The task there will be to determine optimum conditions for maximizing the production of an intermediate species. Example 5.3 provides yet another case of optimization. We note that all these problems assume that the existence of a maximum or minimum at locations other than the obvious end points of the domain has in fact been established. This is not always the clear-cut matter it appears to be and may require some good thinking (one of the aims of the "closure" step alluded to in Chapter 1).

◆ EXAMPLE 5.3 Modeling a Bioreactor: Monod Kinetics and the Optimum Dilution Rate

When a small quantity of cells is placed in a liquid solution of an essential nutrient at an appropriate temperature and pH, the cells will grow. That growth comes about by a complex series of enzymatic reactions that consume the nutrient, also called substrate S, to produce living cell matter. We recall that enzymes are proteins that act as catalysts to convert substrates into some product P, which is often a new protein. The enzymes, being catalysts, emerge from the process unaltered and return to the task of producing more product.

As well as undergoing growth, the cells produce metabolic "wastes," which often turn out to be desirable or valuable substances. Thus, in fermentation processes such as the production of alcoholic beverages, the cells convert carbohydrate substrates into alcohol, while they themselves increase in numbers and size. This is referred to as biomass formation. Here we examine, in the first instance, the pertinent cell growth kinetics, following up with a steady-state analysis of both cell growth and production in a continuous flow stirred tank reactor.

Growth Kinetics

The most commonly used rate expression is that proposed by Monod in 1942, which has the form

$$\mu = \frac{\mu_{max}(S)}{K_s + (S)} \tag{5.16}$$

where (S) is the substrate concentration and μ_{max}, K_S are empirical constants.

The rate of growth μ does not have the conventional rate units of mass/volume time but is instead defined as the rate of growth per unit cell mass, with dimensions of reciprocal time. Thus

$$\mu = \frac{1}{m}\frac{dm}{dt}[t^{-1}] \tag{5.17}$$

where m = mass of cells. We further note that at low substrate concentrations (S), the rate follows first-order kinetics, while at high concentrations the enzyme catalyst becomes essentially saturated with substrate and no longer depends on its concentration. This condition represents the maximum attainable growth rate, $\mu = \mu_{max}$, and is referred to as having zero-order kinetics. A plot of the Monod equation appears in Fig. 5.3.

We note that to use this rate expression in conventional mass balances, we must redimensionalize it by multiplication with the cell concentration C (mass/volume). We then obtain:

$$r(\text{mass of cells/volume time}) = \mu C = \frac{\mu_{max}(S)C}{K_s + (S)} \tag{5.18}$$

This is the expression we use in the reactor model given shortly.

Steady-State Analysis of a Bioreactor

To model a continuous flow stirred bioreactor, we start as usual by composing the pertinent species balances, which are here the cells, the substrate, and the product. For the

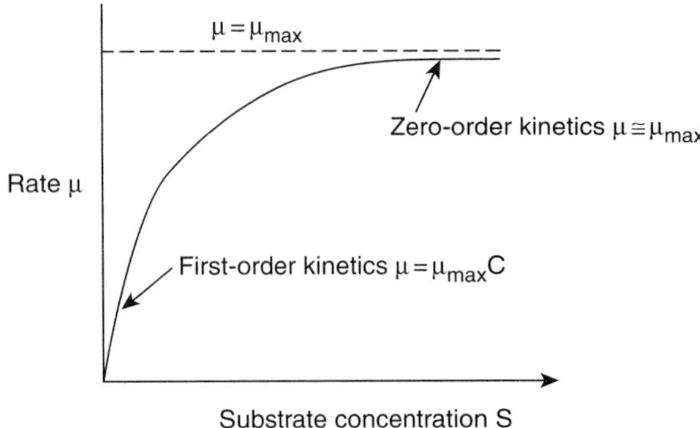

Figure 5.3 Monod kinetics.

cell balance we obtain

$$\begin{aligned} &\text{Rate of cells in} - \text{Rate of cells out} = 0 \\ &[QC_f + \mu CV] \quad - \qquad QC \qquad = 0 \end{aligned} \tag{5.19}$$

where we have placed the cell production term μC under the "in" column and denoted the incoming cell concentration by C_f. This expression can be expanded by using equation (5.18) and rewritten in the form

$$DC_f + \left[\frac{\mu_{max}(S)}{K_s + (S)} - D\right] C = 0 \tag{5.20}$$

where $D = Q/V$ is the *reciprocal* of the residence time τ we have already encountered in connection with the conventional CSTR (see Example 5.2). It is the preferred variable for biochemical engineering calculations and is termed the *dilution rate*. One notes that it equals the number of tank volumes that pass through the vessel per unit time.

A similar steady-state balance for the substrate leads to the expression

$$D[(S)_f - (S)] - \frac{1}{Y_{CS}} \frac{\mu_{max}(S)C}{K_s + (S)} = 0 \tag{5.21}$$

where $Y_{CS} =$ mass of cells produced/mass of substrate consumed. This latter quantity is a characteristic constant for a particular system under consideration. Equations (5.20) and (5.21) are two coupled algebraic equations in the substrate and cell concentrations (S) and C. They are collectively known as the Monod chemostat model.

It often happens that of a reactor initially contains a certain mass of cells and is fed continuously with a cell-free solution of substrate (i.e., nutrient). Setting $C_f = 0$, we obtain from the chemostat equations

$$S = \frac{DK_s}{\mu_{max} - D} \tag{5.22}$$

and

$$C = Y_{CS}[(S)_f - (S)] \tag{5.23}$$

which can be combined into the single expression

$$C = Y_{CS}\left[(S)_f - \frac{DK_s}{\mu_{max} - D}\right] \tag{5.24}$$

One notes from equation (5.24) that the cell concentration C at first declines slowly and then begins to drop off more rapidly as the dilution rate D increases and approaches the value μ_{max}. Simultaneously, the effluent substrate concentration (S) experiences a rise until it reaches a maximum value equal to its feed concentration $(S)_f$. This occurs at a critical dilution rate D_{crit} given by

$$\frac{D_{crit}K_s}{\mu_{max} - D_{crit}} = S_f \tag{5.25}$$

or

$$D_{crit} = \frac{\mu_{max}(S)_f}{K_s + S_f} \tag{5.26}$$

Physically what happens here is that an increase of the dilution rate D causes a rise in the rate of removal of cells that is only partially offset by the rate of cell growth within the reactor. Further increases in D accelerate the divergence of the two rates until a point is reached at D_{crit} where all cells are swept away. That condition, shown graphically in Fig. 5.4, is termed *washout*.

Let us next examine what happens to a second quantity, the product of dilution rate and cell concentration, DC. That quantity has the units of kilograms per cubic meter-second; that is, it represents the cell output per unit reactor volume and time. Since C at first declines only slowly, an increase in D will initially cause DC to increase; as the decrease in C accelerates toward washout, however, the rise in DC will slow and, after passing through a maximum, will decrease at an accelerated pace, ultimately dropping to zero. The peak in DC defines an optimum dilution rate D_{opt}, which maximizes cell output and

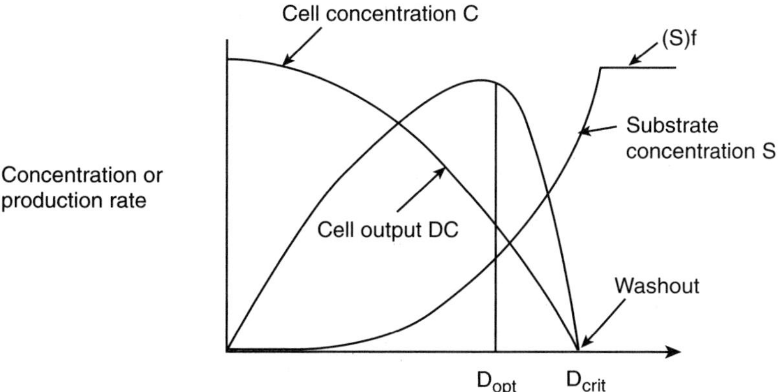

Figure 5.4 Concentration and production of cells in a continuous flow stirred bioreactor.

is thus of considerable practical interest. It is given by the expression

$$D_{opt} = \mu_{max}\left[1 - \sqrt{\frac{K_s}{K_s + (S)_f}}\right] \tag{5.27}$$

the derivation of which is the task of Practice Problem 5.5.

The maximum corresponding to this optimum dilution rate turns out to be a sharp one and is located in close proximity to D_{crit} (see Fig. 5.4). Thus a small increase in dilution rate beyond D_{opt} will cause a rapid drop in both C and DC to zero; that is, washout will occur. Since fluctuations in flow rate can and often do occur, it is advisable to avoid this region of high sensitivity by operating somewhat below the optimum dilution rate. This entails some loss in potential cell production but leads to greater stability and robustness of the system. Identification of such regions is one of the benefits to be derived from our analysis.

Finally we turn to the derivation of the product concentration in the effluent, for which we write the following steady-state mass balance.

$$\begin{aligned} &\text{Rate of production in} - \text{Rate of product out} = 0 \\ &[Q(P)_f + Y_{PC}\mu CV] \quad - \quad Q(P) \quad = 0 \end{aligned}$$

or alternatively

$$C[(P)_s - (P)] + Y_{PC}\mu C = 0 \tag{5.28}$$

where Y_{PC} = mass of product formed per mass of cells, which, like Y_{CS}, is again a characteristic constant for a particular biochemical reaction. We combine this expression with the Monod equation (5.16) and with equations (5.20) and (5.21) to eliminate C and (S) and obtain

$$(P) - (P)_f = \frac{Y_{PC}Y_{CS}}{D}\,\frac{\mu_{max}\left[(S)_f - \dfrac{DK_s}{\mu_{max} - D}\right]}{1 + \dfrac{\mu_{max} - D}{D}} \tag{5.29}$$

One notes from this expression that production shutdown [i.e., $(P) - (P)_f = 0$] occurs at the same critical dilution rate given by equation (5.26), since this value yields a zero numerator in equation (5.29). We do not derive D_{opt} for this case, but it is again located close to D_{crit}, forcing us as before to operate somewhat below optimum dilution rates.

COMMENTS

- This example reveals two features of reactor operation we have not yet encountered. The first is the existence of an optimum residence time (the inverse of the dilution rate D), which is not seen in the operation of conventional reactors involving single reactions. The second novel feature is the occurrence of complete reactor shutdown at a certain critical residence time, which again does not have its counterpart in ordinary reactor operation. These factors, and the use of the unconventional operational parameter D by biotechnologists, require an adjustment in our thinking.

  We had been accustomed to analyzing reactor behavior in direct proportion to the residence time $\tau = V/Q$. The situation here was simple: large values of τ were good, giving high yields; small values of τ were bad [see equation (5.14)]. No maximum occurred between the two extremes of zero and infinite residence time. In a bioreactor, the effect of the inverse quantity, the dilution rate $D = 1/\tau$, is more complex. Here high values of D (i.e., small residence times) can be very good, in fact leading to optimum conversion; but a relatively small increase in D beyond that point leads to complete shutdown of the reactor (i.e., washout). Thus, small residence times can be both very good and very bad, and these conditions are in close proximity to each other. These are all novel features not encountered in conventional reactor engineering and deserve the attention of the analyst.

This example is a particularly fruitful illustration of the benefits to be derived from what we have termed "closure" (see Chapter 1), that is, the analysis of the results obtained from a model. The expression to be scrutinized in this case was equation (5.24), which at first sight appears rather innocuous. We advised the reader in Chapter 1, however, to be particularly alert when dealing with fractions involving differences. This is the case here, and a quick inspection of equation (5.24) shows that it passes, with increasing values of D, from positive values of C through zero to $-\infty$. This provides the first inkling that reactor shutdown may occur at some particular value of D. Further analysis of the quantity DC then reveals that shutdown not only occurs, but is preceded by a maximum, which is associated with an optimum dilution rate D_{opt}. It is precisely this type of "postmortem" analysis and the results that spring from it that are among the most beneficial aspects of modeling.

◆ EXAMPLE 5.4 Nonidealities in a Stirred Tank. Residence Time Distributions from Tracer Experiments

In Examples 5.2 and 5.3, dealing with continuous stirred tank reactors, the tacit assumption was that flow through the tank was "ideal." This implied perfect mixing of the reactor contents, that is, concentrations inside the reactor were everywhere uniform and equal to the concentrations in the exiting stream. Though real reactors never fully follow these flow patterns, a large number of designs approximate these ideals with negligible error. In other cases, however, deviation from ideality can be considerable. This deviation can be caused by channeling or short-circuiting of fluid, by recycling of fluid, or by the creation of stagnant zones. Nonidealities of this type, shown in Fig. 5.5, are to be avoided because they always lower the performance of the unit.

It is evident that elements of a fluid taking different routes though a reactor will require different lengths of time to pass through the vessel. The distribution of these times for the stream of fluid leaving the tank is called the exit age distribution E, or the residence time distribution RTD of the fluid. Here the term "age" for an element of the exit stream refers to the time spent by that element in the vessel. These notions are shown in Fig. 5.6B. We find it convenient to represent the RTD in such a way that the area under the plot of E vs time t is unity, that is,

$$\int_0^\infty E\,dt = 1 \tag{5.30}$$

where E has the units of reciprocal time.

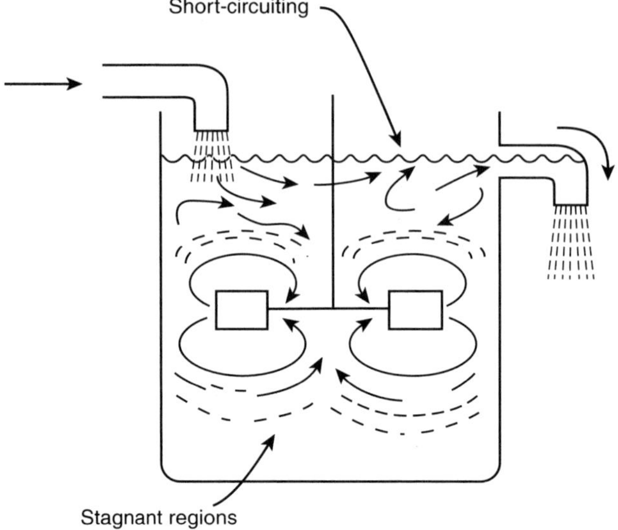

Figure 5.5 Nonidealities in a stirred tank.

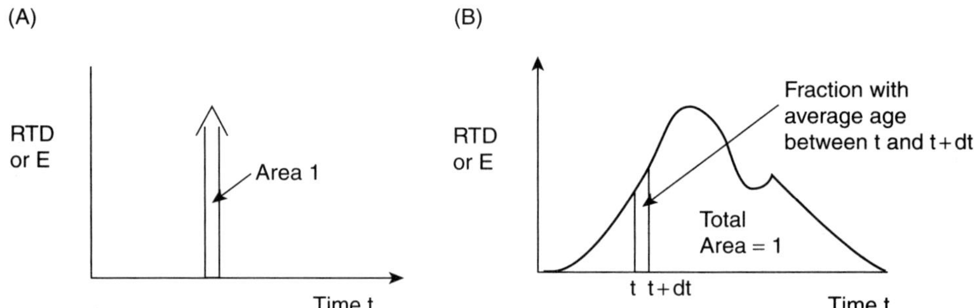

Figure 5.6 (A) tracer input (ideal pulse) and (B) tracer output in a stirred tank.

This procedure is called normalizing the distribution, and Fig. 5.6 shows E in normalized form.

With this representation, the fraction of the exit stream of age between t and t + dt is given by E dt, and the fraction younger than age t is $\int_0^t E\,dt$. A common way of measuring residence time distributions is to introduce a tracer, such as a dye, into the fluid entering the vessel and to record its concentration in the fluid leaving it. A typical tracer output or response, is shown in Fig. 5.6B.

Suppose now that a particular tracer experiment has led to the collection of the data in Table 5.1.

Since the data were collected at finite intervals, a convenient way of interpreting them is to extend each measured concentration over successive halfway marks of the time interval used, as shown in Fig. 5.7. Thus the concentration of 3 g/L obtained at the 5 min mark is assumed to extend over the time interval from 2.5 min to 7.5 min, and so on. The area under the resulting bar plot is seen to very nearly equal the area under the smoothed curve drawn through the experimental points. The distribution pertinent to each interval is given

Table 5.1 Tracer Output from a Stirred Tank

Time t (min)	Tracer Output Concentration (g/L)
0	0
5	3
10	5
15	5
20	4
25	2
30	1
35	0

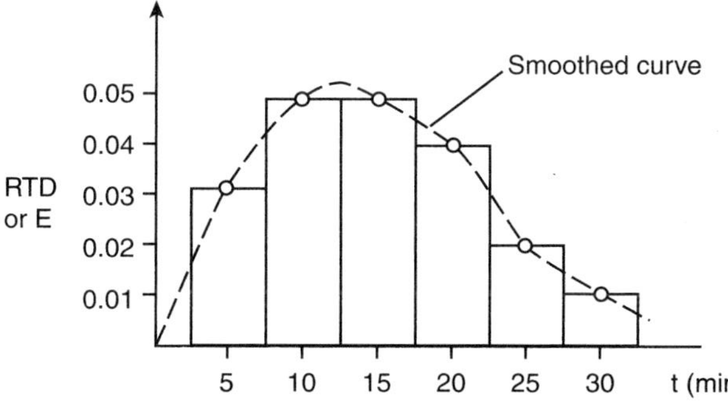

Figure 5.7 Tracer output given in Table 5.1.

by the equation

$$(\mathrm{E})_{\mathrm{interval}} = \frac{(\mathrm{C})_{\mathrm{interval}}}{\sum (\mathrm{C})_{\mathrm{interval}} \Delta t} \tag{5.31}$$

where it is seen that

$$\sum (\mathrm{E})_{\mathrm{interval}} \Delta t = \sum \frac{(\mathrm{C})_{\mathrm{interval}}}{\sum (\mathrm{C})_{\mathrm{interval}}} \Delta t = 1 \tag{5.32}$$

as required. The values of E calculated in this manner are shown in Table 5.2.

Thus, for the first data point for example, we obtain

$$\sum (\mathrm{C})_{\mathrm{interval}} \Delta t = (3 + 5 + 5 + 4 + 2 + 1)5 = 100\,\mathrm{g\,min/L}$$

and hence

$$\mathrm{E} = \frac{(\mathrm{C})_{\mathrm{interval}}}{\sum (\mathrm{C})_{\mathrm{interval}} \Delta t} = \frac{3}{100} = 0.03\,\mathrm{min}^{-1}$$

as shown in Table 5.2.

Let us next examine how these residence time distributions are to be used in the calculation of actual reactor conversions. We start by arguing that the *average* concentration $\overline{C}_A$ of

Table 5.2 Calculated Values of E

t (min)	0	5	10	15	20	25	30
E (min^{-1})	0	0.03	0.05	0.05	0.04	0.02	0.01

reactant concentration in the exit stream, assuming no intermixing of adjacent elements, must be composed of the following product sum

$$\overline{C}_A = \sum_{\substack{\text{all}\\ \text{elements}}} \begin{pmatrix} \text{Concentration} \\ \text{of reactant} \\ \text{of age between} \\ \text{t and t + dt} \end{pmatrix} \begin{pmatrix} \text{Fraction in} \\ \text{exit stream that} \\ \text{is of age between} \\ \text{t and t + dt} \end{pmatrix} \tag{5.33}$$

or, in terms of infinitesimal increments,

$$\overline{C}_A = \int_0^{\aleph} (C_A)_{\text{element}} E \, dt \tag{5.34}$$

Let us now apply this expression to an irreversible first-order reaction. For such a reaction, each element follows the rate law

$$-\frac{d(C_A)_{\text{element}}}{dt} = k_r (C_A)_{\text{element}} \tag{5.35}$$

or in integrated form,

$$(C_A)_{\text{element}} = C_{A_0} e^{-k_r t} \tag{5.36}$$

This formula can now be substituted into equation (5.34) to yield

$$\overline{C}_A = \int_0^{t} C_{A_0} e^{-k_r t} E \, dt \tag{5.37}$$

which is the formula to be used in the calculation of actual conversions. An example of its application appears in Practice Problem 5.6.

COMMENTS

- We see in this example that the effect of the highly complex nonideal flow patterns that arise in actual reactors can be quantified by appropriate experimentation using tracer techniques. This is a major step forward toward the realistic modeling of chemical reactors and has found widespread use both in the laboratory and in industry. No undue hardship is involved in implementing this procedure, since tracer techniques can be applied in a simple and rapid fashion. Dye tracers are usually limited to low temperature ranges. For higher temperatures, radioactive tracers or temperature pulses may be used.

◆ EXAMPLE 5.5 A Moving Boundary Problem: The Shrinking Core Model and the Quasi-Steady State

In a great many physical processes involving two phases, the phase boundary undergoes a continuous movement caused by mass and energy transport, accompanied on occasion by chemical reactions. Processes of this type lead to what are commonly termed moving boundary problems.

Examples of such problems are numerous and important. They arise in the evaporation and condensation of liquids, in freezing and melting phenomena, during crystal growth and dissolution, in the combustion of solid and liquid fuels, and more generally whenever a particle, either solid or liquid, undergoes a chemical reaction. The general aim in modeling these systems is to establish the course in time of the events. One is interested, for example, in obtaining expressions for the rate of crystal growth or dissolution, or in establishing the time needed for a reacting particle to be completely consumed.

The state variables in these processes, such as temperature or concentration, are in principle functions of both distance and time, leading to PDEs that are usually coupled and nonlinear. To reduce the model to a manageable set of algebraic and ordinary differential equations, we introduce two important simplifying concepts: the shrinking (or growing) core model and the attainment of a quasi-steady state.

- The *shrinking (or growing) core* refers to the material contained by the moving front, such as a burning particle (e.g., coal dust), an evaporating droplet, or a growing crystal seed. This core is assumed to have uniform properties and can be treated as an unsteady stirred tank.
- The movement of the front itself is slow enough to allow the transport rates outside the core to attain a *quasi-steady state*. This implies that as the core front slowly recedes or grows, the rate of diffusion or conduction through the external layer quickly and near-instantaneously adjusts itself to a steady-state value given by Fick's law or Fourier's law.
- The processes involved (i.e., phase change, transport, and reaction) are dominated by a single rate-controlling step.

Thus, although both time and distance are retained as variables, distance (expressed through the changing size or mass of the core) becomes a *dependent* variable for the core unsteady balance but is retained as an independent variable for the external, quasi-steady-state balance. Time is an independent variable as well, but it appears only in the core balance. The model is thus composed of unsteady core balances and quasi-steady-state external balances. When both mass and energy changes are involved, two such balances constitute the complete model. When there is a change in mass only, as is the case in our example, single equations suffice. Reaction rates enter the balances as auxiliary relations.

A systematic way of modeling these systems is to start with unsteady mass and energy balances about the core and then consider the quasi-steady-state process outside the core. It is good practice to keep track of the number of dependent variables, which must ultimately be matched by the number of equations.

We illustrate these concepts with the example of a solid spherical particle that undergoes a reaction according to the scheme

$$A(g) + bB(s) \rightarrow \text{Products}$$

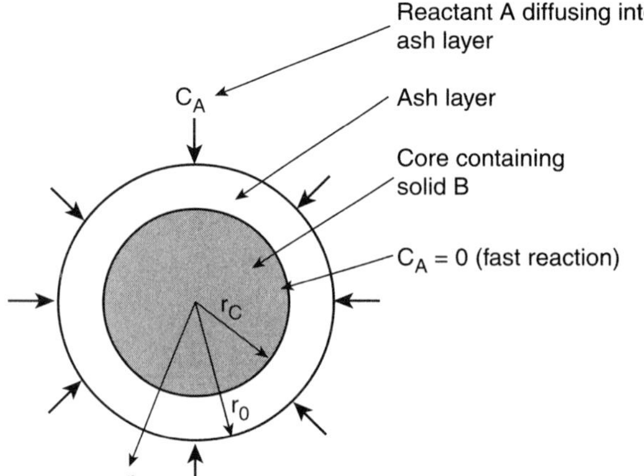

Figure 5.8 A reacting solid particle and its external ash layer.

with rates r_A and r_B, which are related by the stoichiometry of the reaction as follows

$$r_A = \frac{1}{b} r_B \tag{5.38}$$

The rate of reaction is assumed to be fast enough to ensure that all the reactant arriving at the surface of the particle is instantaneously consumed. Hence $(C_A)_{r=r_C} = 0$, as shown in Fig. 5.8. The core front, on the other hand, because of its high density, moves slowly enough to permit the external layer of solid products to be taken to be at steady state. The oxidative degradation of organic aerosols in the atmosphere is an example of such a reaction. The task is to determine the time dependence of particle radius r_C, hence that of its mass.

We start, as recommended, with an unsteady core balance, and write

$$\begin{array}{ccccc} \text{Rate of B in} & - & \text{Rate of B out} & = & \text{Rate of change of B} \\ 0 & - & r_B & = & 4\pi r_C^2 \rho_B \dfrac{dr_C}{dt} \end{array} \tag{5.39}$$

where we have placed the unknown reaction rate r_B in the "out" column and denoted particle density by ρ_B.

This is followed by a quasi-steady-state mass balance in the product layer

$$\begin{array}{ccccc} \text{Rate of A in at r} & - & \text{Rate of A out at } r_C & = & 0 \\ 4\pi r^2 D_{eff} \dfrac{dC_A}{dr} & - & r_A & = & 0 \end{array} \tag{5.40}$$

where D_{eff} is the effective diffusivity of reactant A in the product layer. This last expression can be immediately integrated by separation of variables to yield

$$\frac{4\pi D_{eff}}{r_A} \int_0^{C_{A0}} dC_A = \int_{r_C}^{r_0} \frac{dr}{r^2} \tag{5.41}$$

or

$$\frac{4\pi D_{eff} C_{A_0}}{r_A} - \frac{1}{r_C} - \frac{1}{r_0} \tag{5.42}$$

where $r_0 =$ initial radius of the particle.

We pause at this stage to take a brief inventory. We have, in equations (5.38), (5.39), and (5.42) three equations in the three dependent variables r_A, r_B, and the core radius r_C. The model is consequently complete, and we can proceed with the elimination of r_A and r_B to arrive at the desired relation $r_C = f(t)$. We obtain from the three aforementioned equations

$$-\frac{4\pi D_{eff} b C_{A_0}}{1/r_C - 1/r_0} = 4\pi r_C^2 \rho_B \frac{dr_C}{dt} \tag{5.43}$$

which can again by integrated by separation of variables:

$$-\frac{D_{eff} b C_{A_0}}{\rho_B} \int_0^t dt = \int_{r_0}^{r_C} \left(r_C - \frac{r_C^2}{r_0} \right) dr_C \tag{5.44}$$

The final result is then

$$t = \frac{\rho_B r_0^2}{6bD_{eff} C_{A_0}} \left[1 - 3\left(\frac{r_C}{r_0}\right)^2 + 2\left(\frac{r_C}{r_0}\right)^3 \right] \tag{5.45}$$

COMMENTS

- This example demonstrates how the use of clever simplifying assumptions and some inspired modeling can reduce the complexities of a process to manageable proportions. The assumptions would be violated only if the progress of core consumption were extremely rapid, in which case events in the outer layer become both time and distance dependent, leading to a PDE.

  The time required for the reaction to be complete is obtained from equation (5.45) by setting $r_C = 0$. This leads to

$$t_{tot} = \frac{\rho_b r_0^2}{6bD_{eff} C_{A_0}} \tag{5.46}$$

  where t is seen to vary inversely with diffusivity D_{eff} and the external reactant concentration. This was anticipated on physical grounds because both these factors increase the rate of reactant consumption. Somewhat less expected is the direct dependence on the square of the initial particle radius, r_0^2. It might have been argued that the time necessary for total consumption should vary directly with the mass of the particle, that is, with the radius cubed. As neatly revealed by the model, however, this is not so, for the rate of consumption is dictated by, hence is proportional to, the surface area of the particle at any instant.

  It will also be noted that a knowledge of the reaction rates was not required in this instance. This is a consequence of the assumption that the rate of reaction is very rapid and as a result ceases to affect the overall rate of the process. The rate-determining step here is the speed with which the core front recedes, which is very low given the high density ρ_b of the particle. This, together with the low values of D_{eff} and C_{A_0}, leads to high values of the consumption time, as can be seen from expression (5.46).

◆ EXAMPLE 5.6 More on Moving Boundaries: The Crystallization Process

In this second example of a system with moving boundaries, we draw the reader's attention to some aspects of crystal growth kinetics. The system being considered, shown in Fig. 5.9, is that of a well-stirred crystallization tank. The contents have been cooled to within 1 to 3° below the saturation concentration and are subsequently "seeded" with some crystals of mass m_0. This initiates crystallization, and the task will be to establish the course in time of this process.

In one of the earliest studies of this problem, it assumption was made that crystallization itself is very rapid and that solute molecules arriving at the crystal surface are incorporated into the crystal structure nearly instantaneously. The rate-determining step is that of mass transfer through a boundary layer in which the concentration at the crystal surface is assumed to be that of the equilibrium solubility C_S of the crystallizing substance. The growth of the crystals, however, is sufficiently slow that the boundary layer concentrations are at a quasi-steady state. A mass balance around the crystals then takes the following form:

$$\begin{array}{ccccc} \text{Rate of mass in} & - & \text{Rate of mass out} & = & \begin{array}{c}\text{Rate of change}\\ \text{of crystal growth}\end{array} \\ k_c A_S(C - C_S) & - & 0 & = & \dfrac{dm}{dt} \end{array} \tag{5.47}$$

We note here that both the surface area A and concentration in the liquid C vary with time or indirectly with mass m. For the area, which can be quite irregular, we stipulate, based on the behavior of spherical particles, that it varies with the two-thirds power of volume so that

$$A = \alpha V^{2/3} = \frac{\alpha}{\rho^{2/3}} m^{2/3} = \beta m^{2/3} \tag{5.48}$$

where α is some shape factor.

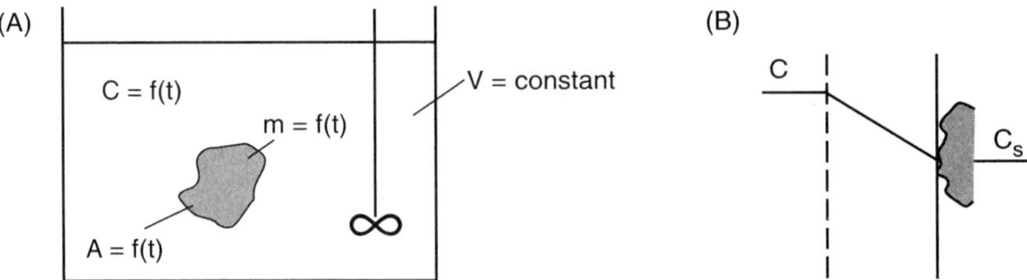

Figure 5.9 Crystal growth: (A) the well-stirred crystallization tank and (B) mass transfer to the crystal surface.

For C, we apply a simple cumulative mass balance, which states

$$\begin{array}{ccccc} \text{Initial concentration} & = & \text{Mass per volume} & + & \text{Concentration} \\ \text{in liquid} & & \text{added to crystals} & & \text{left in liquid} \\ C^0 & = & \dfrac{m - m_0}{V} & + & C \end{array} \tag{5.49}$$

Substituting equations (5.48) and (5.49) into equation (5.47), we obtain

$$k\beta m^{2/3}\left[C^0 - \frac{m - m_0}{V} - C_s\right] = \frac{dm}{dt} \tag{5.50}$$

Formally integrating this expression yields the result

$$A\int_0^t dt = At = \int_{m_0}^{m} \frac{dm}{m^{2/3}(B - m)} \tag{5.51}$$

where $A = k_c\beta/Vm_0^{2/3}$ and $B = V(C - C_S) + m_0$.

Evaluation of the integral is by graphical or numerical means, which we do not address further. We elaborate instead on the special but important case of a constant external concentration C. This condition applies during the early stages of the crystallization process or in systems with a large volume of external solution. Equation (5.51) then becomes

$$A\int_0^t dt = At = \int_{m_0}^{m} \frac{dm}{m^{2/3}(C - C_S)V} = \frac{3}{(C - C_S)V}\left(m^{1/3} - m_0^{1/3}\right) \tag{5.52}$$

The crystal mass m can be expressed in the alternative form

$$m = \gamma N \rho L^3 \tag{5.53}$$

where N is number of crystals, L is a characteristic linear dimension, and γ is a shape factor, equal to $\frac{4}{3}\pi$ for a sphere of radius L. Equation (5.52) then becomes

$$\tfrac{1}{3}A(C - C_S)Vt = m^{1/3} - m_0^{1/3} = (kN\rho)^{1/3}\Delta L \tag{5.54}$$

This expression relates the change in crystal mass m with time to the change in linear crystal dimension ΔL and is confirmed by a considerable body of experimental findings. It is known as the *McCabe ΔL law* and was first proposed in 1929.

COMMENTS

- Considering the complexities of the crystallization process, at least when viewed initially, the model proposed here, which agrees with a considerable body of experimental observations, is one of impressive simplicity and elegance. The complex and generally unknown shape of the crystals was neatly accommodated by the two-thirds power relation (5.48). The mass transfer rate used in equation (5.47) can be applied not only to crystallization, but also in the reverse process of solids dissolution (with a reversal of sign). The model proposed in this example can thus be applied to a host of interesting problems. Note also that it is the linear dimension of the crystal, not its volume, that varies directly with time t. This was not anticipated on physical grounds and shows the power of a model to reveal the unexpected.

EXAMPLE 5.7 Moving Boundaries in Medicine: Controlled-Release Drug Delivery

Administration of medication to a specific part of the body can be done in several ways. In the conventional method, the drug is administered orally or by injection, with the expectation that the blood circulation will convey it to the site where it is required. This procedure, which is widely used, is nevertheless not the most efficient way of delivery because the drug is diluted by passage through the whole body. This carries with it the risk of undesirable side effects in some other parts of the body if the drug concentration at the desired site is to be high enough for optimal effectiveness. One method of overcoming these drawbacks is by implantation of a wafer loaded with the medication near the desired site, with the expectation that the drug will be delivered directly and in the shortest possible time to the affected organ and consumed there, thus avoiding dispersion to other parts of the body. This local delivery strategy has been used to administer the drug pilocaropine directly to a glaucoma patient's eyeball by means of a wafer under the lower eyelid and to deliver progesterone to the uterus for preventing conception.

One type of implant in current use consists of a structurally uniform polymer matrix that is loaded with a saturated solution of the drug, saturation being maintained by drug particles suspended in the solution. The device is analogous to a sponge, with the polymer matrix providing the solid skeleton of the structure and the drug suspension filling the void space.

When such a device is implanted, the drug molecules in solution diffuse across the polymer–body interface and a drug-depleted region of thickness $\delta_D(t)$ forms on the matrix side of the interface (Fig. 5.10). As diffusion continues, $\delta_D(t)$ grows, while the corresponding "hydrodynamic region" outside the device remains at a constant thickness δ_H because of the continuous washing effect of body fluids and constant clearing processes in the body.

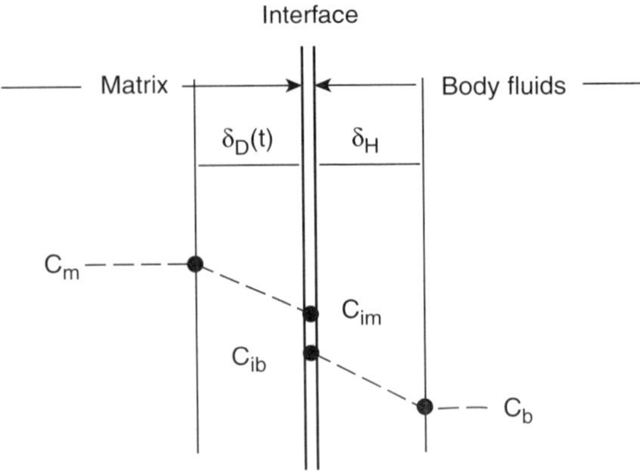

Figure 5.10 Concentration variation in the interface region of a controlled-release device.

The attendant concentrations, schematically plotted in Fig. 5.10 are the matrix-side concentration, C_m, which equals the saturation concentration C_S of the drug and the bulk concentration in the fluid, C_b. It is assumed that the body's metabolic and other processes function at a constant rate, so that the medication is also consumed or cleared at a constant rate, thus maintaining C_b at a constant level. At the interface, the membrane-side and fluid-side concentrations are taken to be at equilibrium and are interrelated through the expression

$$C_{ib} = KC_{im} \tag{5.55}$$

where K is the so-called partition coefficient. We have here again, as in Example 5.5, a case of a shrinking core, the front of which recedes slowly enough to permit us to assume that the external transport rate has attained a quasi-steady state.

We start as usual by composing a core mass balance and follow this up with a mass balance for the external transport layer. We obtain

Core Mass Balance

$$\begin{aligned} \text{Rate of drug in} - \text{Rate of drug out} &= \begin{array}{l}\text{Rate of change}\\ \text{in drug content}\end{array} \\ 0 \qquad - \qquad N \qquad &= C_0 \frac{d(L - \delta_v)}{dt} \end{aligned} \tag{5.56}$$

where C_0 = solid loading (kg/m^3) and L = thickness of the water.

External Mass Balance

$$\begin{aligned} \text{Rate of drug into interface} &= \text{Rate of drug out of interface} \\ N = \frac{D_m}{\delta_D}(C_m - C_{im}) \quad &= \quad \frac{D_b}{\delta_H}(C_{ib} - C_b) \end{aligned} \tag{5.57}$$

where D_m and D_b are the drug diffusivities in the matrix and body fluid, respectively.

We obtain, as a result, three equations, (5.55) to (5.57), in the three dependent variables, C_{im}, C_{ib}, and δ_D. It is the variation of δ_D with time that we wish to establish.

If C_{ib} is eliminated by the interphase equilibrium relation (5.55) we obtain, after some manipulation,

$$N = A\frac{D_m}{\delta_D}(C_m - C_{im}) = A\frac{KC_m - C_b}{\delta_H/D_b + K\delta_D/D_m} \tag{5.58}$$

where δ_H/D_b equals the inverse of the external mass transfer coefficient k_b. Substitution of equation (5.58) into the core mass balance (5.56) yields

$$\frac{KC_m - C_b}{1/k_D + K\delta_D/D_m} = C_0\frac{d\delta_D}{dt} \tag{5.59a}$$

which is the expression now to be integrated. This is done by separation of variables, which leads to the expression

$$\delta_D\left[\delta_D + 2\frac{D_m}{k_b K}\right] = 2D_m\frac{C_m}{C_0}\left(1 - \frac{C_b}{C_m}\right)t \tag{5.59b}$$

This relation gives us the desired time dependence of δ_D, from which the total lifetime of the device can be calculated (see Practice Problem 5.9).

COMMENTS

- We start our observations with some remarks about the expression for the mass transfer rate N, equation (5.58). The numerator of that relation, $KC_m - C_b$, can be viewed as the driving force of the mass transfer process, and the denominator, $1/k_b + K\delta_D/D_m$, as the resistance R to it. In general, whenever mass transfer takes place across two "films," such as those shown in Fig. 5.10, the rate of transport will be given by an expression of the form

$$N = \frac{\text{Driving force}}{\text{Resistance}} = \frac{C^* - C_b}{1/k_b - K/k_m} \tag{5.60}$$

where we use C^* to denote the concentration on the fluid side that would be in equilibrium with the matrix-side concentration. The driving force, then, is made up of the difference of the equilibrium and actual concentrations on the fluid side, while the resistance is composed of the sum of the inverses of the transport coefficients, or "conductivities," of the two films. The underlying concepts that led to equation (5.60) are referred to as the two-film theory, which is used widely to describe mass transfer through two films or resistances in series.

In practical applications of controlled-release implants, it is desirable to have the drug supply proceed at a constant rate, which implies $d\delta_D/dt$ = constant, and from it $\delta_D(t) = at + b$. In other words, for a constant release rate to prevail, the matrix film thickness must vary at most linearly with time. Examination of equation (5.59a) shows that this condition is approached provided $\delta_D \ll 2\ D_m/k_bK$. This can be assured only if matrix diffusivity D_m is very large (high polymer porosity) or the partition coefficient K is very small, meaning that the drug has far more affinity for the matrix than for the body fluids. Under these conditions, $K\delta_D/D_m \ll \delta_H/D_b$; that is, the resistance to mass transport is almost entirely on the fluid side. As a consequence C_b is quite low. Equation (5.58) then simplifies to the expression

$$N = \frac{AKD_bC_m}{\delta_H} \tag{5.61}$$

and we see from this that the release rate N is indeed constant.

The condition that leads to equation (5.61) is said to be "partition controlled," as opposed to "diffusion controlled," which applies whenever there is a substantial resistance on the membrane side. The choice of polymer matrix material to assure a small K or large D_m is one of the tasks of the biomedical engineer. In addition to this condition, the design engineer must satisfy appropriate criteria for biocompatibility and mechanical and chemical stability.

◆ EXAMPLE 5.8 Evaporation of a Pollutant into the Atmosphere

When a pollutant enters a body of water (e.g., by an accidental spill or by intermittent contamination of the incoming flow), part of it may be released to the atmosphere by

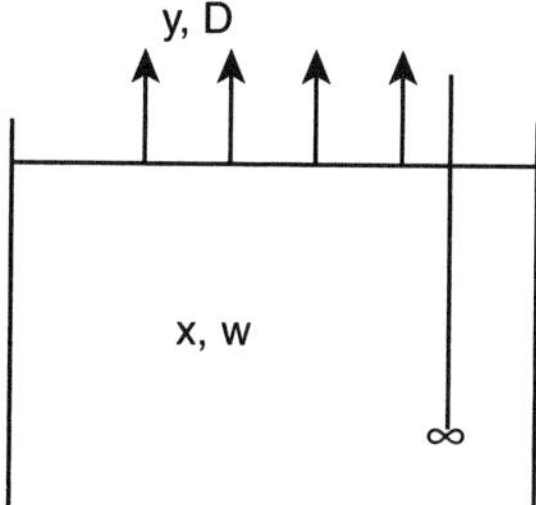

Figure 5.11 Evaporation of a pollutant.

evaporation, while other portions are adsorbed on or react with bottom sediment, or undergo biodegradation through bacterial action.

In this example case (Fig. 5.11), we deal solely with release to the atmosphere and attempt to determine to what extent the pollutant concentration in the water is reduced by evaporative loss. We note that this transfer of objectionable substances is small comfort to the environment as a whole, since the pollutants are merely being transferred from one medium to another. Nevertheless, the process needs to be scrutinized in order to gain an idea of the overall fate of a pollutant.

It is not quite clear at the outset how we should proceed or indeed what exactly it is that we are seeking. Determination of the pollutant concentration with time would be a logical objective, but one senses that this would require experimental data on rates of evaporation, which would be dependent on local temperatures and wind conditions. Let us see whether we can avoid the use of evaporation data by eliminating it from the model. We do not at this stage know how this can be done, or indeed whether it is possible at all. A good strategy in such cases is to start the modeling process and scrutinize the initial results to see whether a worthwhile objective can be attained. The tools we have available are various mass balances, and we start by writing two of them, using mole fractions as concentration units. This is necessary because the physical laws involved here (e.g., ideal gas law, phase equilibrium relations) are based on moles rather than mass. We choose a total mole balance and a pollutant mole balance. (We could also have chosen a water mass balance, but this ultimately turns out to be less convenient.) As usual, the body of water is assumed to be well mixed and of uniform concentration with no in flow or outflow. We write

Total Mole Balance

$$\begin{array}{ccccccc} \text{Rate of moles in} & - & \text{Rate of moles out} & = & \begin{array}{c}\text{Rate of change}\\ \text{of total moles}\end{array} \\ 0 & - & D & = & \dfrac{d}{dt}W \end{array} \tag{5.62}$$

where D = total moles evaporating per unit time and W = moles of water and pollutant in the body of water.

Pollutant Mole Balance

$$\begin{array}{ccccccc} \text{Rate of pollutant in} & - & \text{Rate of pollutant out} & = & \begin{array}{c}\text{Rate of change}\\ \text{of pollutant}\end{array} \\ 0 & - & yD & = & \dfrac{d}{dt}xW \end{array} \tag{5.63}$$

where y and x are the pollutant concentrations in the vapor and aqueous phases, respectively.

Inspection of these two equations shows that they contain four dependent variables: x, y, W, and the evaporation rate D. Hence we need two more equations. One relation can be obtained by assuming that the vapor leaving the liquid is locally at equilibrium with the liquid, which yields $y = f(x)$. We will address the formulation of this relation shortly, but we turn our attention first to the fourth variable, the evaporation rate D. We wish, if possible, to eliminate it entirely. Some thought shows that this can be accomplished by applying one of the tricks reported at the end of Chapter 2. We divide the two equations (5.62) and (5.63), thereby eliminating both D and dt. This is permissible because both expressions are first-order ODEs. We obtain

$$y = x + W\frac{dx}{dW} \tag{5.64}$$

where either x or W can be viewed as the new independent variable.

We now return to the problem of relating the vapor composition y to that of the liquid, x. To do this, we introduce the following relation, which comes from the thermodynamics of phase equilibria. That relation applies to sparsely soluble, low volatility substances and is given by

$$p = \frac{xP^0}{x_s} \tag{5.65}$$

where p = partial pressure of pollutant in the vapor phase, P^0 = vapor pressure of the pure pollutant, and x_s = solubility of the pollutant in water. We do not delve into the derivation of this expression but ask ourselves instead: Does it make physical sense? Equation (5.65) states that p, or the pollutant content in the vapor, increases with the liquid phase concentration x and with the vapor pressure of the pure pollutants. This agrees with physical reasoning. The inverse relation to pollutant solubility is less obvious, but it can be made acceptable on intuitive grounds: the less soluble a substance, the greater its tendency to escape into the adjacent phase. Thermodynamicists have enshrined this formally in the concept of the fugacity function f, some-times termed the escaping tendency of a substance. For sparsely soluble substances, the fugacity equals x/x_s and enters the relation (5.65) as such.

To make use of equation (5.65), we must convert the partial pressure p of the pollutant, to its mole fraction y. From the ideal gas law, we know that at a given temperature, pressure is proportional to the number of moles, so that

For the Pollutant

$$n \sim p \tag{5.66a}$$

For the Total Moles

$$n_{tot} \propto p + P^0_{H_2O} \tag{5.66b}$$

where we have used the additivity of partial pressures (Dalton's law) to express the total number of moles. Note that since pollutant concentration is very low because of its assumed low concentration, the partial pressure of water in the vapor phase will equal its pure component vapor pressure $P^0_{H_2O}$. By dividing the proportionalities that have the same constant V/RT, we obtain

$$y = \frac{n}{n_{tot}} = \frac{p}{p + P^0_{H_2O}} \approx \frac{p}{P^0_{H_2O}} \tag{5.67}$$

Pollutant partial pressure p in the denominator can be neglected because the water vapor pressure $P^0_{H_2O}$ is orders of magnitude higher than the partial pressures of the common pollutants considered here, which include heavy oils, mercury, pesticides, polychlorinated biphenyls (PCBs), and the like. Equation (5.64) then becomes

$$\frac{xP^0}{x_S P^0_{H_2O}} = x + W\frac{dx}{dW} \tag{5.68}$$

which is easily integrated by separation of variables to yield

$$\left[\frac{P^0}{x_S P^0_{H_2O}} - 1\right] \ln\frac{W}{W_0} = \ln\frac{x}{x_0} \tag{5.69}$$

Let us examine the predictions of this equation for a particular pollutant, mercury. The relevant data are as follows:

Mercury solubility (25 °C): 3×10^{-2} mg/L = 2.7×10^{-9} mole fraction

Mercury vapor pressure (25 °C): 0.173 Pa

Water vapor pressure (25 °C): 3.17×10^3 Pa

A list of solubilities of various potential pollutants appears in Table 5.3.

We wish to consider the concentration change that results when a small percentage of the water basin, say 0.01%, evaporates. Substitution of these values into equation (5.69)

Table 5.3 Solubility for Various Compounds at 25 °C

Compound	Solubility (mg/L)	Vapor Pressure (mmHg)
Hydrocarbons		
n-Octane	0.66	14.1
Benzene	17780	95.2
Toluene	515	28.4
Biphenyl	7.48	0.057
Pesticides		
DDT	0.0012	10^{-7}
Lindane	7.3	9.4×10^{-6}
Aldrin	0.2	6×10^{-6}
PCBs		
Aroclor 1248	0.054	4.94×10^{-4}
Aroclor 1260	0.0027	4.05×10^{-5}

yields

$$\left[\frac{0.173}{2.7 \times 10^{-3}(3170)} - 1\right] \ln(1 - 10^{-4}) = \ln \frac{x}{x_0}$$

from which we obtain

$$\frac{x}{x_0} = 0.133$$

In other words, some 87% of the initial mercury content has passed into the atmosphere. This is a remarkably large drop in concentration, considering that only one ten-thousandth of the contents of the water pool has evaporated.

COMMENTS

- We start by noting that our initial goal of monitoring the variation of concentration with time was not attained because no data on evaporation rates were immediately available. We were nevertheless able, by clever maneuvering, to obtain information on pollutant depletion with the fraction of water evaporated. These highly useful results dramatize the extent of release of pollutants to the atmosphere from bodies of water. The unpalatable fact emerges that these objectionable substances will not be confined to a particular contaminated water basin but can travel long distances through the atmosphere, carried by wind and other air currents. Part of the reason for the contamination of arctic regions is to be found in this effect.

  In the process of deriving the model for this example, we had to confront the reader with certain relations drawn from physical chemistry and thermodynamics without providing a detailed derivation. Such shortcuts are often necessary in modeling, and one has to accept equations reported in the literature without quite knowing, or understanding, the manner in which they were derived. Our advice to the reader is that such acceptance should not be automatic; hence an examination on the functional relations expressed in the equations is necessary to ensure that they make physical sense. Confidence in the validity of the model is thus enhanced, and one can proceed with the solution with a greater degree of assurance.

◆ EXAMPLE 5.9 Ground Penetration from an Oil Spill

Accidental oil spills can occur both in oceans (from oil tankers) and on land (from tank trucks and following train derailments). One question that needs to be answered in land-based oil spills is the depth to which the oil will penetrate in a given time, since this will determine the strategy for cleanup operations. This is the problem to be addressed here.

We assume that initially an oil slick of depth h_0 is deposited, and begins to seep into the ground, with distance z being measured from the surface of the ground (see Fig. 5.12). Our modeling strategy will be to compose appropriate mass balances, and it is clear even at the outset that this will require the use of an auxiliary relation that expresses the velocity of flow in the porous soil. To express this velocity in terms of the system variables, we draw

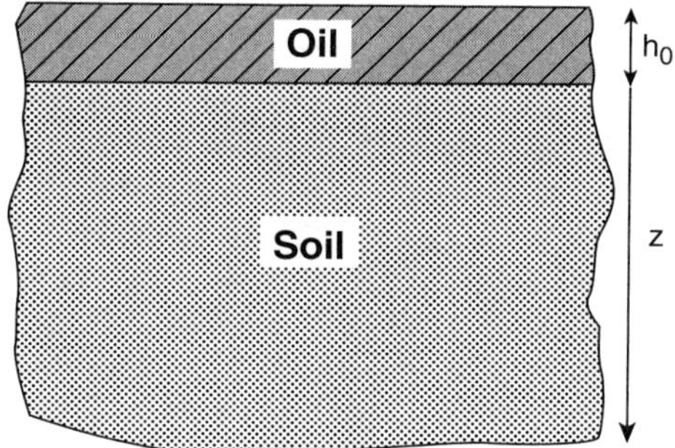

Figure 5.12 Ground penetration from an oil spill.

on D'Arcy's law for flow in porous media, which has the form

$$v = \frac{K}{\mu}\frac{\Delta p}{L} \tag{5.70}$$

where K is the so-called permeability of the soil, μ is the viscosity of the oil, L is the depth of the porous medium, and Δp represents a driving force for the process, which is an applied pressure difference. For the oil spill in question, the driving force is the weight per unit area of the oil column, that is, $\Delta p = \rho g(z + h)$, where h is the variable thickness of the oil slick covering the top of the soil. We also recognize that L in equation (5.70), being the depth of the porous medium in contact with the seeping oil, must equal the penetration depth z. Equation (5.70) then becomes

$$v = \frac{K}{\mu}\rho g\frac{z+h}{z} \tag{5.71}$$

Having just gone through a rather convoluted argument to arrive at this expression, let us see whether it makes physical sense. We start by observing that v is proportional to K/μ; that is, the more permeable the soil and the less viscous the oil, the higher the flow rate. This agrees with physical reasoning. Moreover, v varies directly with the column weight, $\rho g(z + h)$, which again makes physical sense. Since, however, the resistance to flow increases with the depth to which the soil is filled with oil, v will also vary inversely with z. Thus a dual dependence on penetration depth z appears in both numerator and denominator of equation (5.71). This may at first sight seem surprising, but it follows logically from our process of physical reasoning.

Let us now construct an appropriate model. We start with a mass balance around the soil and write

$$\begin{array}{ccccc} \text{Rate of oil in} & - & \text{Rate of oil out} & = & \begin{array}{c}\text{Rate of change}\\ \text{of oil content}\end{array} \\ \rho v A & - & 0 & = & \varepsilon\rho A\dfrac{dz}{dt} \end{array} \tag{5.72}$$

where ε = porosity of the soil.

This gives us two equations, D'Arcy's law (5.70) and equation (5.72), but still leaves us with three dependent variables v, h, and z. A third relation is consequently required. We have the choice of making a second unsteady oil balance, this time over the oil layer, or performing a cumulative oil balance for the entire system. The latter is algebraic and avoids a second integration, which would be needed for the unsteady balance. We obtain

$$\begin{array}{ccccc} \text{Initial amount of oil in the layer} & = & \text{Oil left in the layer} & + & \text{Oil contained in the soil} \\ h_0\rho A & = & h\rho A & + & z\rho\varepsilon A \end{array} \tag{5.73}$$

and from it

$$h = (h_0 - \varepsilon z) \tag{5.74}$$

The model is now complete [equations (5.70), (5.72), and (5.74)]. Eliminating h and v from these three equations, we obtain

$$\frac{dz}{dt} = \frac{K\rho g}{\varepsilon\mu}\left[\frac{h_0}{z} + (1 - \varepsilon)\right] \tag{5.75}$$

Integrating between $z = 0$ and $z = L$ by separation of variables and using the integration formula

$$\int \frac{x\,dx}{a + bx} = \frac{1}{b^2}[a + bx - a\ln(a + bx)] \tag{5.76}$$

we obtain

$$t = \frac{\varepsilon\mu}{K\rho g}\left\{\frac{L}{1 - \varepsilon} - \frac{h_0}{(1 - \varepsilon)^2}\ln\left[1 + \frac{L(1 - \varepsilon)}{h_0}\right]\right\} \tag{5.77}$$

Let us obtain a sense of the time scale involved by working out a specific numerical example. Suppose it is desired to calculate the time it takes for the spilled oil to penetrate 4 cm into the ground, given the following data:

Initial height of oil level: $h_0 = 6$ cm
Porosity of ground: $\varepsilon = 0.7\ \text{cm}^3/\text{cm}^3$
Permeability of ground: $K = 10^{-6}\ \text{cm}^2$
Oil viscosity: $\mu = 0.02$ Pa s
Oil density: $\rho = 800\ \text{kg/m}^3$

We obtain

$$t = \frac{0.7 \times 0.02}{10^{-10} 800 \times 9.81}\left\{\frac{0.04}{1 - 0.7} - \frac{0.06}{(1 - 0.7)^2}\ln\left[1 + \frac{0.04(1 - 0.7)}{0.06}\right]\right\}$$

$$t = 210\,\text{s}$$

Thus the penetration into the ground is quite fast, of the order of minutes. This is partly because a relatively high permeability of $K = 10^{-6}$ cm was chosen. For dense soils, K values of the order of $10^{-8}\ \text{cm}^2$ are more common, and ε is of the order 0.1. The time scale then becomes one of hours, rather than seconds or minutes.

◆ EXAMPLE 5.10 Concentration Variations in Stratified Layers

Lakes and similar water basins often exhibit departures from well-mixed conditions. We investigate such nonideal behavior in Example 5.4 and used residence time distributions to quantify it in some detail. This approach took account of channeling, stagnant zones, and other nonideal phenomena. The model we wish to examine here is built on a simpler premise, which does however reflect conditions that arise in real situations. It assumes that an unmixed layer of thickness Δh is sandwiched between two well-mixed compartments of height h_1 and h_2, respectively (see Fig. 5.13). Chamber 1 adjoining the bottom of the basin has an initial concentration C_1^0 while that of chamber 2 is C_2^0. There is no inflow or outflow or chemical reaction. The concentration gradient across the unmixed layer, termed a *pycnocline*, can be assumed to be linear. Such pycnoclines, which roughly translate into density gradients, have been observed in a natural context and arise primarily as a result of heat flow from the bottom and the top, which causes mixing due to natural convection in the upper and lower portions of the basin, leaving an intermediate unstirred layer.

We start the model with a solute balance on the lower chamber, using Fick's law for the diffusional term:

Rate of solute in − Rate of solute out = Rate of change of contents

$$0 \quad - \quad \frac{DA(C_1 - C_2)}{\Delta h} \quad = \quad h_1 A \frac{dC_1}{dt} \tag{5.78}$$

Since there are two dependent variables C_1 and C_2, a second equation is needed, which is most conveniently obtained from a cumulative balance. The latter is algebraic and avoids the need to solve two simultaneous differential equations. We have

Total mass = Mass in the chambers + Mass in the pycnocline

$$m_T \quad = \quad (h_1 AC_1 + h_2 AC_2) \quad + \quad (AC_1 + AC_2)\frac{\Delta h}{2} \tag{5.79}$$

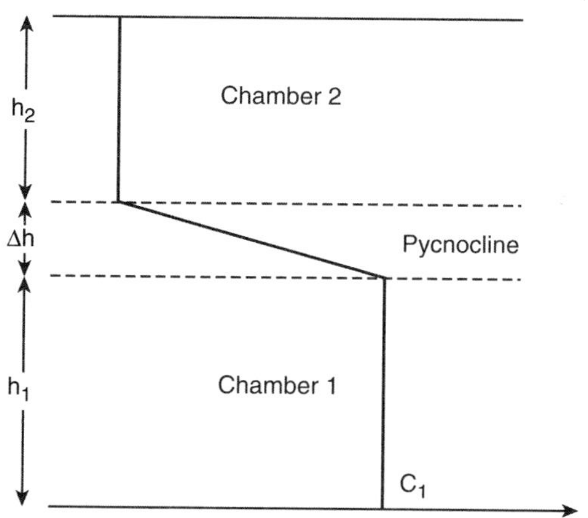

Figure 5.13 Transport in stratified layers.

from which we obtain the relation

$$C_2 = \frac{m_T/A}{h_2 + \Delta h/2} - \frac{h_1 + \Delta h/2}{h_2 + \Delta h/2} C_1 \tag{5.80}$$

Substitution of equation (5.80) into (5.78) yields the ODE

$$\frac{dC_1}{dt} = -\frac{D}{h_1 \Delta h} \frac{h_1 + h_2 + \Delta h}{h_2 + \Delta h/2} \left(C_1 - \frac{m_T/A}{h_1 + h_2 + \Delta h} \right) \tag{5.81}$$

which can be integrated by separation of variables. Solving the result for C_1 we obtain

$$C_1 = \frac{m_T/A}{h_1 + h_2 + \Delta h} \left(1 - \exp\left[-\frac{Dt}{h_1 \Delta h} \frac{h_1 + h_2 + \Delta h}{h_2 + \Delta h/2} \right] + C_1^0 \exp\left[-\frac{Dt}{h_1 \Delta h} \frac{h_1 + h_2 + \Delta h}{h_2 + \Delta h/2} \right] \right) \tag{5.82}$$

The corresponding expression for the second chamber comes from the cumulative balance, equation (5.79), and is given by

$$C_2 = \frac{m_T/A - C_1(h_1 + \Delta h/2)}{h_2 + \Delta h/2} \tag{5.83}$$

When the initial concentration in the upper chamber is zero, $C_2^0 = 0$, equation (5.83) reduces to the expression

$$C_2 = C_1^0 \frac{h_1 + \Delta h/2}{h_1 + h_2 + \Delta h} \left(1 - \exp\left[-\frac{Dt}{h_1 \Delta h} \frac{h_1 + h_2 + \Delta h}{h_2 + \Delta h/2} \right] \right) \tag{5.84}$$

Some results from this equation have been calculated and are shown in Fig. 5.14. They are for $h_2 = 60\,\text{m}$, $h_1 = 25\,\text{m}$, a pycnocline thickness $\Delta h = 10\,\text{m}$, and two diffusivities of 0.05 and 0.005 cm^2/s, which span the range reported for pycnoclines in stratified lakes. The curves show that the concentrations in the two layers would equalize in 10 to 40 years.

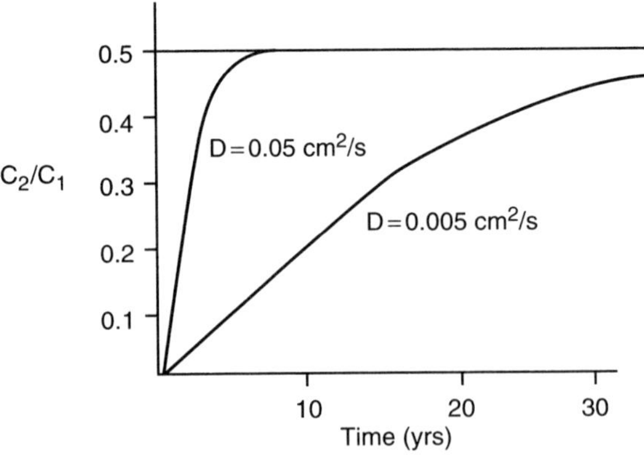

Figure 5.14 Concentration variations in the system of stratified layers shown in Fig. 2.13.

There is a fourfold change in this span, which is considerable. But we are interested in the general time frame of the event, which is found to be of the order of 10 years. Thus we have succeeded in bracketing the solution, which is all that could have been done, given the uncertainty in the diffusivity. This is yet another case in which even a good modeler must be content with an approximate answer or a range of answers.

◆ EXAMPLE 5.11 One-Compartment Pharmacokinetics

Compartmental models are used extensively in biomedical engineering, particularly in the context of tracer or drug administration and the uptake of toxic substances in various body compartments. These compartments encompass body fluids such as plasma and intercellular fluids, as well as body tissues (fat, muscle, bone). By following the time course of the concentration of an administered substance and fitting the experimental data to a compartmental model, one can extract a host of important parameters, including local flow rates, compartment volumes, rate constants, and transport or partition coefficients. A limited example of parameter estimation from tracer output was considered in Example 5.1. This example uses the measured time course of an administered drug, often referred to as pharmacokinetics, to extract a broader range of physiological parameters by means of a one-compartment model. In spite of their often limited validity, such one-compartment models remain, the most popular device for the prediction of body functions or for the extraction of parameters from the response to external inputs. Three specific situations arise.

1. *Injection or short-term exposure to a substance*: Blood volume and cardiac output in the adult human are approximately 5 L and 5 L/min, respectively. An instantaneous or short-term input, denoted a "dose" in Fig. 5.15A is thus "turned over," or distributed

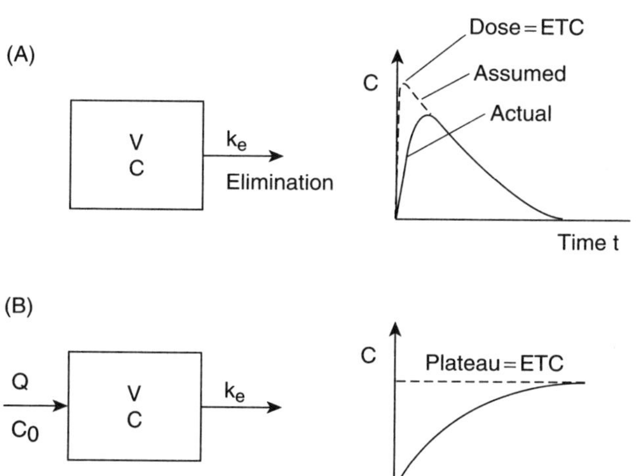

Figure 5.15 One-compartment models and the time course of drug concentration: (A) clearance following injection and (B) steady infusion.

within the bloodstream, approximately once every minute. This is the reason for the almost instantaneous effect of many drug injections, and it allows us to assume that the distribution of drugs within the plasma approaches uniformity shortly after injection. By contrast, however, the total volume of body fluids is approximately 50 L, so that turnover there occurs only once every 10 minutes.

2. *Clearance*: In the process of clearance, the body slowly metabolizes, excretes, or otherwise disposes of an injected substance (Fig. 5.15A). Here it is assumed that the bloodstream, or some other compartment, including the totality of body fluids, has reached a near-uniform level, that exposure has ceased, and that the concentration starts to decline in accordance with a compartmental model and an elimination rate law, usually taken to be linear in the concentration of the substance.
3. *Infusion*: In this situation, there is a continuous influx and elimination of the drug or other substance, with the concentration ultimately rising to a constant plateau, referred to as the "effective therapeutic concentration," or ETC (Fig. 5.15B). When that plateau is reached, influx and elimination are exactly in balance, and the system is at steady state. Note that the ETC is also obtained by extrapolation of the clearance curve (Fig. 5.15A), to time $t = 0$, for at that point the influx and elimination rates are, albeit only briefly, in exact balance.

We now consider a specific example. In a well-known classic study a 392 mg dose of the stimulant aminophylline was administered to a patient. Blood samples taken over the next 10 h yielded the following data:

t (h)	1	2	5	6	8	10
C (mg/L)	8.53	7.28	4.52	3.85	2.81	2.04

The following questions are to be addressed:

What is the ETC?

What is the apparent rate constant of elimination k_e?

What is the apparent volume of distribution V?

What is the apparent flow rate of elimination Q?

For a required ETC of 10 mg/L, what infusion rate should be used and what is the time necessary to attain 95% of ETC?

We start with a leadoff model describing the clearing process. After cessation of the injection, we have

$$\begin{array}{ccccc} \text{Rate of drug in} & - & \text{Rate of drug out} & = & \begin{array}{c}\text{Rate of change}\\ \text{in drug content}\end{array} \\ 0 & - & k_e C V & = & V\dfrac{dC}{dt} \end{array} \tag{5.85}$$

where V = volume of body fluids.

One obtains by integration by separation of variables

$$\ln\frac{C}{C_0} = -k_e t \tag{5.86a}$$

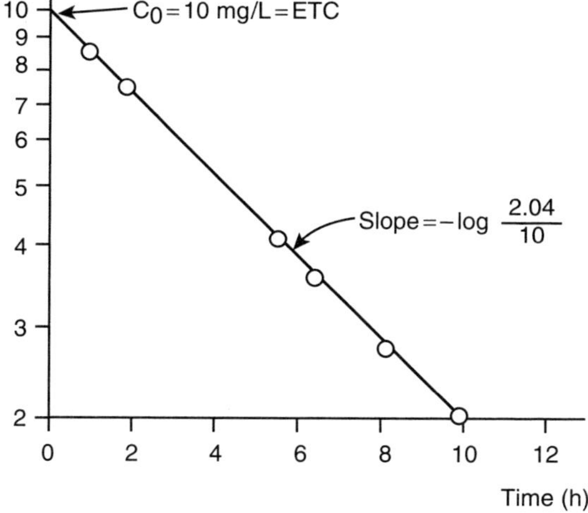

Figure 5.16 Experimental data for the clearance of the drug aminophylline from a patient.

or

$$\frac{\log C}{C_0} = \frac{k_e}{2.303} t \tag{5.86b}$$

The semilog plot of the given data in accordance with equation (5.86b) shown in Fig. 5.16 yields the following data.

Effective Therapeutic Concentration (ETC)

The ETC is obtained by extrapolation of the semilog plot 5.16 to $t = 0$ and yields the approximate value of ETC = 10 mg/L.

Apparent Elimination Rate Constant k_e

The value of this parameter is obtained from the slope of Fig. 5.16. Using the value of $C = 2.04$ mg/L obtained after 10 h, we have for that slope

$$-\frac{k_e}{2.303}\Delta t = \frac{\log C}{C_0} \tag{5.87a}$$

$$-\frac{k_e}{2.303}10 = \frac{\log 2.04}{10} = -0.69 \tag{5.87b}$$

Hence

$$k_e = \frac{0.69 \times 2.303}{10} = 0.150\,\mathrm{h}^{-1} \tag{5.87c}$$

Apparent Distribution Volume V

We obtain this from the relation

$$V = \frac{\text{Dose (mg)}}{\text{ETC(mg/L)}} = \frac{392}{10} = 39.2\,\mathrm{L} \tag{5.88}$$

Note that this value is not too far from the value of approximately 50 L for the total volume of body fluids (blood and intracellular fluids).

Apparent Flow Rate of Elimination Q

The "out" term k_eCV in equation (5.85) can be written in the equivalent form QC, so that $Q = k_eV$. We obtain

$$Q = k_eV = (0.159\,\text{h}^{-1})(39.2\,\text{L}) = 6.23\,\text{L h} \tag{5.89}$$

This term expresses the rate of passage from the intracellular fluid into tissue, where it metabolizes and is ultimately excreted. Since this process is diffusive rather than a convective one, we must regard Q as a fictitious but equivalent flow rate.

Required Infusion Rate

This quantity is obtained directly from the steady-state condition:

$$\begin{aligned} &\text{Rate of infusion} - \text{Rate of elimination} = 0 \\ &\quad Q_iC_i \quad - \quad k_eC_{ETC}V \quad = 0 \end{aligned} \tag{5.90a}$$

$$Q_iC_i = k_eC_{ETC}V = 0.159 \times 10 \times 39.2 = 62.3\,\text{mg/h} \tag{5.90b}$$

Time to 95% of ETC

Here we invoke the mass balance applicable to infusion (Fig. 5.15B). We obtain

$$\begin{aligned} &\text{Rate of drug in} - \text{Rate of drug out} = \begin{array}{c}\text{Rate of change}\\ \text{of drug content}\end{array} \\ &\quad Q_iC_i \quad - \quad k_eCV \quad = \quad V\frac{dC}{dt} \end{aligned} \tag{5.91}$$

Integrating by separation of variables, we obtain

$$\frac{1}{k_e}\ln\frac{Q_iC_i/V}{Q_iC_i/V - k_e0.95\,C_{ETC}} = t \tag{5.92a}$$

or

$$t = \frac{1}{0.159}\ln\frac{62.3/39.2}{62.3/39.2 - 0.159 \times 0.95 \times 10} \tag{5.92b}$$

$$t = 18.9\,\text{h} \tag{5.92c}$$

COMMENTS

- One notes here the wealth of information obtained from just a single set of experimental data and a single mass balance, equation (5.85). Not only were we able to derive a rate constant and physiological volume and transport rate, but these parameters could then be used in a different model, that of infusion, to derive required drug infusion rates and the approximate time necessary to attain a given physiological concentration level, the ETC. Single-compartment models, though capable of providing much useful information, are often refined into two-compartment or multicompartment models encompassing several body organs and fluids. Single sets of data are then no longer sufficient, and one must resort to more sophisticated experimentation and analysis. We note in addition that in biomedical modeling, body processes such as excretion are expressed as either first-order reactions (kCV) or in terms of equivalent apparent flow rates QC. This duality may lead to some confusion, but it has become a standard feature in biomedical modeling and must therefore be accommodated.

◆ EXAMPLE 5.12 Deposition of Platelets from Flowing Blood

When blood comes into contact with a site of injury or with an artificial implant, the body reacts by covering the site with platelets, which ultimately form a blood clot. We address here the problem of determining the course in time of the platelet deposition in terms of the fractional surface area covered, S, as a function of time. The process has been sketched in Fig. 5.17.

We assume that the deposition rate is determined by a film resistance near the wall and that the rate of adhesion r_A is proportional to the local concentration C_w of platelets and the fraction of surface area that remains uncovered, that is,

$$r_a = k_w C_w (1 - S) \tag{5.93}$$

Because the blood has a high concentration of platelets, this level can be assumed to remain constant at C_b^0 during the deposition process. We are thus dealing with the equivalent of a stirred tank, with the rate of deposition representing an "out" term. For a unit area of wall surface, we can write

$$\begin{matrix} \text{Rate of platelets} \\ \text{arriving at the wall} \end{matrix} - \begin{matrix} \text{Rate of platelets} \\ \text{leaving the wall} \end{matrix} = \begin{matrix} \text{Rate of platelet} \\ \text{accumulation} \end{matrix}$$
$$k_f(C_b^0 - C_w) \quad - \quad 0 \quad = \quad \rho_p \frac{dS}{dt} \tag{5.94}$$

where k_f is the film mass transfer coefficient (m/s) and ρ_p is the surface density of the deposited platelets (kg/m^2). Since we have two dependent variables, C_w and S, a second equation is required to complete the model. This is obtained by equating the rate of arrival

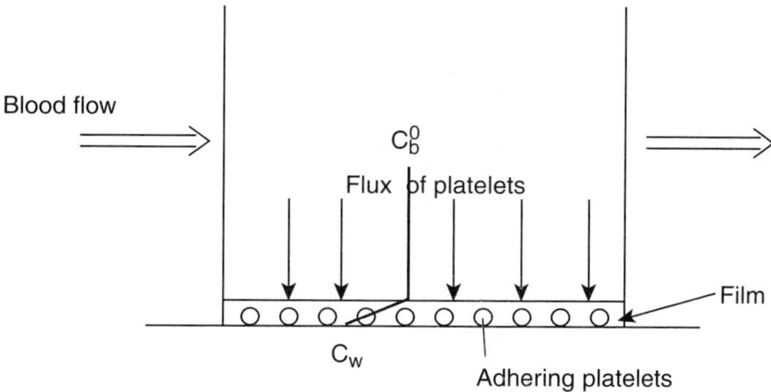

Figure 5.17 Deposition of platelets from flowing blood.

with the rate of adhesion, that is,

$$\begin{aligned} \text{Rate of arrival} &= \text{Rate of adhesion} \\ k_f(C_b^0 - C_w) &= k_w C_w (1 - S) \end{aligned} \tag{5.95}$$

which yields

$$C_w = \frac{C_b^0}{1 + (k_w/k_f)(1 - S)} \tag{5.96}$$

Substituting C_w into equation (5.94) we obtain

$$\frac{C_b^0 k_w (1 - S)}{1 + (k_w k_f)(1 - S)} = \rho_p \frac{dS}{dt} \tag{5.97}$$

which can be integrated by separation of variables to yield the final result

$$\ln(1 - S) - \frac{k_w}{k_f} S + \frac{k_w C_b^0}{\rho_p} t = 0 \tag{5.98}$$

COMMENTS

- Equation (5.98) shows that at low coverages [i.e., with $\ln(1 - S) \approx \ln 1 = 0$], the fractional surface area covered increases linearly with time, followed by a slow exponential approach to full coverage $S = 1$, which is in principle attained as $t \to \infty$. In practice, deposition will also take place on adhering platelets, which complicates the model considerably. However, this happens mainly when the surface coverage has attained substantial values. The model we have presented can thus be viewed as a valid representation of the events at low and intermediate values of S. We note that it can be used in a quite general way to describe deposition of particles form a flowing medium onto a confining wall. This happens, for example, in the fouling of tubes exposed to fluids carrying finely suspended solids in which gravitational forces plays a minor role. The deposition will then be solely dictated by the adhesion rate constant k_w and the mass transfer coefficient k_f.

  We emphasize that while blood clot formation on an injured site is desirable, the deposition of blood cells at the site of an artificial implant must be avoided at all costs. It is the task of the biomedical engineer to develop materials that interact with blood cells as little as possible. There has been considerable progress in the design of such "biocompatible" materials, and the topic remains one of intense ongoing research.

◆ EXAMPLE 5.13 Dynamics of the Human Immunodeficiency Virus (HIV)

Infection by the human immunodeficiency virus, regarded as a rare event before 1970, has grown to epidemic proportions, becoming a dreaded addition to the long list of potentially fatal diseases. It has also elicited an unparalleled research effort to understand the

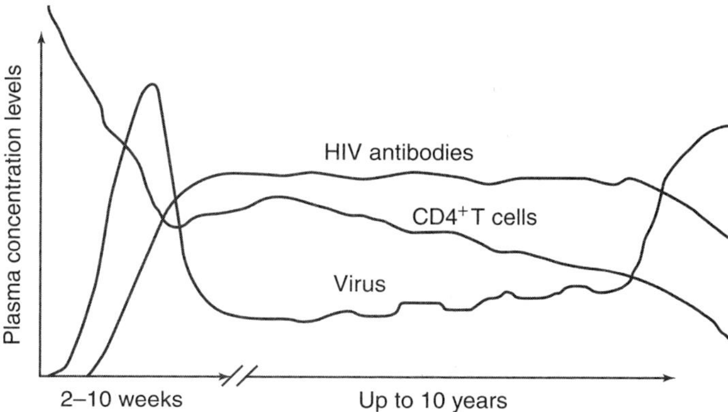

Figure 5.18 The course of HIV infection in a typical patient.

mechanism of its progress and to develop drugs aimed at slowing the viral infection and possibly eradicating it altogether.

Mathematical modeling, which at one time was essentially ignored by the experimental community, has since the mid-1990s become an important tool, and almost all major experimental groups now collaborate with a theoretician. This example provides a glimpse of the type of modeling that has been applied and suggests some of the conclusions that may be drawn from it. It will reveal that simple compartmental models are capable of describing and quantifying some important aspects of the disease. Although we will not attempt to show this here, these models have also been extended to assist in developing strategies for drug therapy.

We start by providing a brief description of the disease and its progress. The major target of HIV infection is a class of white blood cells known as $CD4^{+}T$ cells, or T cells. These cells secrete growth and differentiation factors that are required by other cell populations in the immune system. The course of the disorder is illustrated in Fig. 5.18. Immediately after infection, the amount of virus detected in the blood rises dramatically, accompanied by the appearance of flu-like symptoms. After a few weeks to months, these symptoms disappear and the virus concentration subsides to a lower level. An immune response to the virus has occurred, and antibodies against the virus can be detected in the blood. Persons in whom such antibodies are found are declared to be HIV positive.

After the primary infection the viral concentration remains relatively constant while the T-cell level slowly declines (see Fig. 5.18). During this period, which may last as long as 10 years, the infected person has no disease symptoms. Ultimately, however, the drop in T-cell levels accelerates and the viral concentrations start to climb again. The T-cell count in a healthy person is of the order of 1000 cells/mm^3. If in these latter stages the cell count drops below 200 cells/mm^3, the patient is classified as having AIDS (Acquired immune deficiency syndrome).

Early models of HIV dynamics developed in the 1990s were confined to single compartments and time-invariant parameters. They are nevertheless quite revealing, and it is these one-compartment models we shall be addressing here. Later models and those currently being developed, which use multicompartments as well as time-varying parameters, are expected to provide further insight into the course of HIV infections and the therapies being applied.

We start by considering three dependent variables, the viral concentration C_v and the concentrations of infected and uninfected T cells, C_i and C_u, respectively. Infected cells can themselves produce additional viruses, and we express this fact by means of the following first-order rate law

$$\text{Rate of virus production } r_v = \frac{N}{\tau} C_i \tag{5.99}$$

where N is the number of viruses, or "virions," produced by an infected cell during its lifetime τ.

The rate of production of infected cells, on the other hand, is assumed to follow a second-order rate law, since it depends on the concentrations of both infected and uninfected T cells. We have

$$\text{Rate of infected cell production } r_i = k_r C_v C_u \tag{5.100}$$

where k_p is a second-order rate constant.

Elimination or "clearance" of virus and infected T cells is assumed to follow the usual first-order rate law in accordance with what was established in Example 5.11.

We have

$$\text{Elimination rate for virus } r_{ev} = k_{ev} C_v \tag{5.101}$$

$$\text{Elimination rate for infected T} - \text{cells } r_{ei} = k_{ei} C_i \tag{5.102}$$

where we note that k_{ei} is the inverse of the lifetime τ of the infected cells.

We are now in a position to perform integral balances over the compartment for each of these two species. We obtain

$$\begin{aligned} \text{Rate of virus in} - \text{Rate of virus out} &= \frac{d}{dt}\text{contents} \\ N k_{ei} C_i V \quad - \quad k_{ev} C_v V &= V \frac{d}{dt} C_v \end{aligned} \tag{5.103a}$$

where V = volume of the compartment and

$$\begin{aligned} \text{Rate of infected cells in} - \text{Rate of infected cells out} &= \frac{d}{dt}\text{contents} \\ k_r C_v C_u V \quad - \quad k_{ei} C_i V &= V \frac{dC_i}{dt} \end{aligned} \tag{5.103b}$$

Since we have three dependent variables, a third mass balance on uninfected T cells is required. This complication is often avoided by setting $C_u = c_u^0 =$ constant, a condition that applies reasonably well during the pre-AIDS period (see Fig. 5.18). The model is consequently reduced to the following simultaneous first-order linear ODEs:

$$\frac{dC_v}{dt} = -k_{ev} C_v + N k_{ei} C_i \tag{5.104a}$$

$$\frac{dC_i}{dt} = -k_{ei} C_i + k_r C_v C_u^0 \tag{5.104b}$$

Although a full solution of this system can be easily obtained by Laplace transformation, it is more instructive to combine the equations into a single second-order ODE and examine the resulting characteristic roots. We do this by solving equation (5.104a) for C_i and substituting the result into equation (5.104b). This gives

$$\frac{d^2C_v}{dt^2} + (k_{ei} + k_{ev})\frac{dC_v}{dt} + k_{ei}\left(k_{ev} - Nk_rC_u^0\right)C_v = 0 \tag{5.105}$$

where as noted before, k_{ei} and k_{ev} are elimination constants and k_r is the production rate constant. The characteristic equation for this expression is given by the quadratic equation

$$D^2 + (k_{ei} + k_{ev})D + k_{ei}\left(k_{ev} - Nk_rC_u^0\right) = 0 \tag{5.106}$$

with roots

$$D_{1,2} = \frac{(k_{ei} + k_{ev})}{2} \pm \frac{1}{2}\left[(k_{ei} + k_{ev})^2 - 4k_{ei}\left(k_{ev} - Nk_rC_u^0\right)\right]^{1/2} \tag{5.107}$$

These roots are the exponents that enter the solution of equation (5.105) given in Table 2.9. Positive roots result in an unbounded increase in concentration, while negative roots lead to a decline and ultimate eradication of the virus. Inspection of equation (5.106) shows that one positive and one negative real root are obtained when $k_{ev} < Nk_rC_u^0$, while the condition $k_{ev} > Nk_rC_u^0$ results in two negative real roots. Thus the virus can either proliferate or become extinct, depending on the relative magnitude of these parameters. The special case $k_{ev} = Nk_rC_u^0$ yields $D_1 < 0$, $D_2 = 0$ and consequently results in the ultimate attainment of a constant virus concentration. To facilitate the interpretation of these findings, we have sketched the concentration histories for the three cases in Fig. 5.19. They are composed of the sum of the two exponential terms, which incorporate the two characteristic roots $D_{1,2}$.

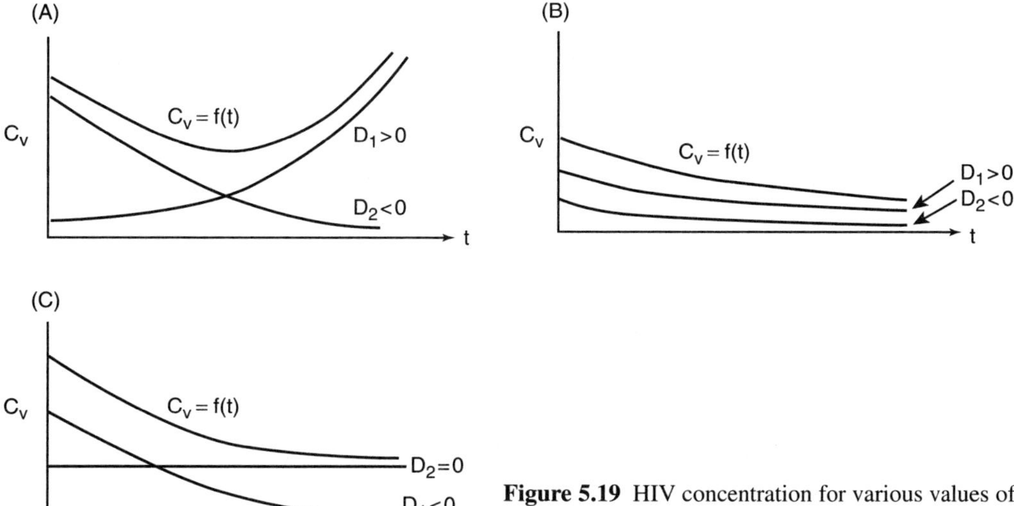

Figure 5.19 HIV concentration for various values of the characteristic roots $D_{1,2}$.

COMMENTS

- Although the model is confined to the pretreatment and pre-AIDS period of a patient's history, and only two variables were considered, the results are nevertheless revealing. First and foremost, they predict that AIDS is not an inevitable consequence of exposure to HIV. The virus concentration can drop to zero even without intervention (Fig. 5.18B), as demonstrated by experimental findings. For example, when health care workers are stuck by needles contaminated with blood from AIDS patients, the frequency at which such people become HIV positive is very low, perhaps 1 in 200 such incidents. However, the initial virus concentration must be small, for other experiments have shown that if the virus load gets high enough to be reliably measured, it is extremely unlikely that the virus will be cleared spontaneously. We have indicated this by holding virus concentrations low in Fig. 5.19B but allowing higher virus loads in Fig. 5.19A.

  The finding for the case $k_{ev} = Nk_rC_u^0$ is noteworthy because it allows for the stable maintenance of virus concentration C_v, and by extension of C_i, at finite positive values (Fig. 5.19C). Furthermore, depending on the parameters characteristic of a particular virus/host combination, those positive values can vary from one patient to another. This, too, has been observed experimentally.

  Finally, the very slow onset of AIDS can be explained by setting D_1 in Fig. 5.19A very low and close to zero. The rising curve will then be an extremely shallow one, and the initial decline of C_v will be overtaken by the growing component only at an extremely low rate. The result is that C_v remains flat, or very nearly so, for extended periods of time.

  As we had noted, the model we have presented does not consider the effect of medical intervention. This topic is taken up in Practice Problem 5.13.

PRACTICE PROBLEMS

5.1 **Blood uptake by the heart muscle** The problem here, in contrast to that dealt with in Example 5.1 is to determine not the flow rate through the heart chamber (ventricle) itself, but the rate at which blood from that chamber is taken up by the heart muscle (myocardium). An early method for determining this rate consisted of perfusing the heart with a hydrogen-saturated solution of saline and blood. Hydrogen, being highly mobile, is quickly and uniformly distributed through the myocardium. Clearance of the muscle with H_2-free blood, during which the hydrogen remains almost uniformly distributed in the muscle at any given instant, yielded the following results:

H_2 concentration	100	50	25	10	2.5
Time (min)	0	1	1.8	2.9	4.65

Assuming a myocardium volume of 100 mL, which is that of a medium-sized dog, calculate the rate of blood flow through the myocardium.

ANSWER 79.5 ml/min

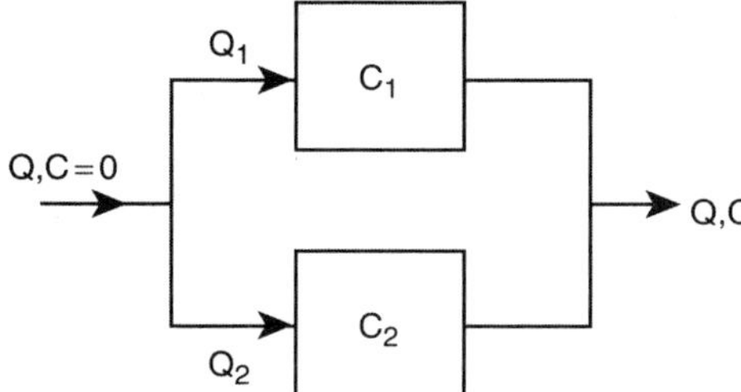

Figure P5.2 Two-compartment model in parallel.

5.2 **A two-compartment model in parallel** Consider the two compartments in parallel shown in Fig. P5.2. Each is initially loaded with an active substance at the level $C_1(0)$ and $C_2(0)$. The system is cleared by perfusing it with clear fluid. Show that the average concentration of the two mixed effluent streams is given by

$$C = C_1(0)f_1e^{-k_1t} + C_2(0)f_2e^{-k_2t}$$

and identify the parameters.

5.3 **Maximizing conversion of an intermediate in a batch reactor** Consider the series reaction

$$A \xrightarrow{k_1} R \xrightarrow{k_2} S$$

taking place in a batch reactor (no in- or outflow) in which the intermediate R is the desired product whose yield is to be maximized. We recognize that the concentration of R does in fact pass through a maximum, since it will initially rise owing to its production from reactant A, but the rate at which this occurs will slow down as degradation to the undesirable product S sets in. The concentration C_R consequently peaks at a particular point in time, after which it declines at a diminishing rate until it reaches the final concentration of zero. Show that the optimum reaction time that maximizes the yield of R is given by

$$t_{opt} = \frac{\ln(k_2/k_1)}{k_2 - k_1}$$

(*Hint*: Eliminate C_A and solve the resulting ODE.)

5.4 **Maximizing conversion of an intermediate in a CSTR** Rework Practice Problem 5.3 for the case of a continuous stirred tank reactor and show that the optimum residence time in the reactor is given by

$$\tau_{opt} = (k_1k_2)^{-1/2}$$

5.5 **The optimum dilution rate of a bioreactor** Derive the expression for optimum dilution rate D_{opt} given by equation (5.27).

5.6 **Conversion in a nonideal reactor** Using the residence time distribution given in Example 5.4, compute the exit concentration $\overline{C}_A$ for a first-order reaction ($k_r = 0.307\ \text{min}^{-1}$). Assume a constant density liquid system.

5.7 **Freeze-drying of food** A slab of frozen meat, assumed to be rectangular and with an initial moisture content of m_0 kg, is to be freeze-dried in a controlled experiment. The aim of the experiment is to derive relevant transport coefficients from measured freeze-drying rate data. To this end, the meat is heated with an electric heater and surface temperature T_g is measured by means of thermocouples. Sublimation of the ice takes place in a vacuum chamber, and water loss is monitored by means of a spring balance. A sketch of the configuration appears in Fig. P5.7. As sublimation progresses, the core ice front, assumed to be at a constant temperature T_i, recedes into the interior, exposing an ice-free matrix that grows thicker with time. Heat conduction through this matrix is assumed to be at a quasi-steady state, so that a linear temperature gradient prevails at any given instant.

We are asked to show that the time dependence of the ice removal is given by the expression

$$\frac{t}{f} = af + b$$

where f = fraction of ice evaporated. (*Hint*: Make an unsteady mass and energy balance around the core and a steady-state energy balance at the ice–matrix boundary with $1/U = 1/h + z/k$, where U = overall heat transfer coefficient, h = gas film heat transfer coefficient, and k = thermal conductivity. Note that the effective thermal conductivity decreases with increasing distance z.)

5.8 **Melting of a solid in its own liquid** A solid of mass m_s and temperature T_s is to be melted by placing it in a well-stirred tank containing its molten liquid form of mass m_L and temperature T_L.

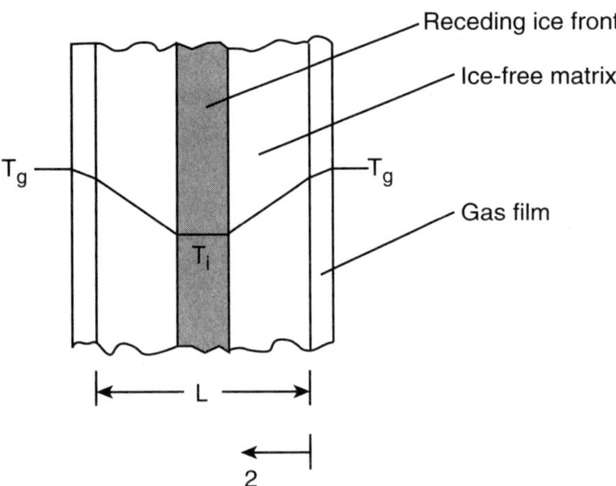

Figure P5.7 Freeze-drying of food.

(a) Formulate a model whose solution will describe the time course of the melting process.

(b) Derive an expression for the total mass of solid melted. Note that not all the solid necessarily melts.

Heat losses to the surroundings are neglected. No solution of the model is required.

5.9 **Useful life time of a drug delivery wafer** Calculate the lifetime of a 0.5 mm thick wafer having a drug partition coefficient of 10^{-3}. Assume a constant release rate; and use the following data:

$D_b = 10^{-5}\,cm^2/s$

$\delta = 100\,\mu m$

$C_0/C_m = 100$

ANSWER $t = 5 \times 10^6\,s$

5.10 **Time dependence of pollutant concentration during evaporation into the atmosphere** Refine the model solution of Example 5.8 to obtain the time dependence of pollutant concentration x, assuming that the evaporation rate D (mol/s) is known.

5.11 **Evaporation from water with suspended solids** The water supply to a well-mixed basin is contaminated with a chemical that enters with a concentration in the water phase of C_{w0} and leaves with a concentration of C_w. Both streams also carry suspended matter that retains the chemical in equilibrium with the water phase. The adsorbed phase concentration C_a is related to the aqueous concentration by the Henry constant (or partition coefficient) H, and the concentration of solids is constant at C_s kg/m^3. While in the basin, the water releases the chemical to the atmosphere by evaporation, simultaneously receiving additional chemical from the suspended solids. Both processes are transport controlled, and an overall mass transfer controlled and an overall mass transfer coefficient Ka can be applied in each case, where a = surface area per unit volume fluid.

Derive an expression for the steady-state concentration C_w in the effluent.

ANSWER
$$\frac{C_w}{C_{w0}} = \left[1 + \frac{\tau_{wa}}{1} + HC_s^{-1} - \tau_{sw}\frac{1 - HC_s^{-1}}{1 + HC_s^{-1}}\right]^{-1}$$

5.12 **More pharmacokinetics** A male patient who has undergone major surgery requires a slow, intravenous infusion of aminocaproic acid to control hemorrhage. Infusion, at a rate of 1 g/h is planned, and the drug is known to have a half-life of 3.9 h; that is, 3.9 h after infusion has ceased, the drug concentration will have been reduced to half its initial level. If the drug concentration is effective only above 0.048 g/L, how long does it take to reach this level (C_{50}) after the start of infusion? Assume a whole-body fluid volume of 50 L.

ANSWER 3.15 h

5.13 More on HIV infection

(a) Use the eigenvalue method described in Chapter 2, Section 2.3.5, to derive the characteristic equation (5.106).

(b) Suppose an antiviral drug, which has an effectiveness E, $1 \geq E > 0$, is administered. Derive the characteristic equation for this case, and show that its roots are both real and negative when $E = 1$.

CHAPTER 6

One-Dimensional Distributed Systems

In the chapter on compartmental models, we dealt with systems or processes in which the state variables varied at most with time, but not at all with spatial distance. We now reverse the situation and examine time-invariant systems; that is, steady-state systems in which the state variables vary in one direction. We refer to these as one-dimensional distributed systems.

Physical systems that lead to such one-dimensional distributions include flow processes in which the state variables (e.g., temperature, concentration, pressure) vary in the direction of flow, diffusion, or conduction, usually taking place in a circular or spherical geometry, and strictly static processes in which the distributions are the result of static considerations. We examined two such static cases earlier (Example 4.2 and Practice Problem 4.2).

To model flow processes, we introduced the reader, to the concept of the one-dimensional pipe (Chapter 1) in which the principal variations took place in the direction of flow, and any radial or lateral variations were lumped at the conduit wall by means of appropriate rate equations and resistance (Figs. 1.3 and 1.4). An early example of the application of the one-dimensional pipe was given in example 1.7, Release of a Substance into a Flowing Fluid. In this chapter we will apply the concept to model a countercurrent heat exchanger (Example 6.6), to derive oxygen profiles in a polluted river (Example 6.10), and to compute the heat released by an electrically heated wire (Example 6.11). Diffusional processes make use of a modified version of the one-dimensional pipe in which transport into and out of an incremental element is by diffusion or conduction, rather than by convection or bulk flow. Such examples are taken up in Examples 6.4, 6.5, and 6.7 to 6.9. The remaining examples do not make use of the one-dimensional pipe model but draw instead on the general modeling principles and guidelines laid down in Chapter 2. We now turn to a more detailed examination of these examples.

In Example 6.1 we consider the variations in temperature and pressure with atmospheric height. The principles we apply here are those of fluid statics and physical chemistry (ideal gas law). A noteworthy feature is that the temperature distributions must be obtained experimentally and are incorporated into the fluid static model to arrive at the pressure distributions.

Examples 6.2 and 6.3 consider systems in flow, in essence deriving relations between flow rate Q and the pressure required to produce that flow rate. The tools used here are force and mass balances. Example 6.2 deals mainly with liquid flow, leading to the Poiseuille equation, while Example 6.3 examines the behavior of flowing gases and introduces the notion of a limiting flow rate.

Examples 6.4 to 6.6 deal with heat transfer problems. In Examples 6.4 and 6.5 we consider the temperature variations and associated heat fluxes that arise in heat conduction through various geometries with and without heat generation. Heat transfer in flowing systems is examined in Example 6.6 in the context of the countercurrent heat exchanger. The primary information again comes in the form of temperature distributions, but we also show how the model equations can be adroitly manipulated to arrive at compact design equations for immediate engineering use.

Examples 6.7, 6.8, and 6.9 consider three completely dissimilar processes that nevertheless lead to identical second-order ODEs. The primary information consists, as usual, of concentration, temperature, and pressure distributions, but these data are quickly set aside and used merely as stepping-stones to arrive at performance indices for engineering use. In the case of the catalyst pellet (Example 6.7), that index is the so-called effectiveness factor E, which is a measure of the extent to which diffusional retardation affects the performance of the catalyst. The heat exchanger fin (Example 6.8) has a similar index associated with it: the fin efficiency E, which measures the extent to which the conductive resistance within the fin diminishes its potential for heat transfer. In Example 6.9, finally, we consider the process of polymer sheet extrusion and define a uniformity index E, which measures variations in sheet thickness due to a drop in the extrusion pressure. In all three cases, $E = 1$ denotes ideal performance, while values below 1 indicate a reduction in efficiency.

Example 6.10 draws our attention to concentration profiles that arise in an environmental context. The focus here is on the distribution of oxygen in a river, which needs to be determined in connection with efforts to use biodegradation to cleanse waterways of pollutants. The model is based on two simple mass balances, which lead to the Streeter–Phelps equation.

In Examples 6.11 and 6.12, we take up two topics of an electrical nature. Example 6.11 examines the heat produced by an electrically heated wire. The tools here are an energy balance and certain elementary notions drawn from the field of electricity. Example 6.12 is an application of the theory of electrostatics. To derive the potential distribution due to a charged disk, we must apply Coulomb's law, equation (1.9), and the concept of potential as work done to move a charge through an electrical field.

To conclude the chapter, Example 6.13 deals with the manufacture of silicon crystals. We address the problem of calculating the velocity at which the solidifying crystal should be drawn from its melt. Some difficulties are encountered, which are finally resolved, perhaps not to the satisfaction of everyone.

Again, the Practice Problems at the end of the chapter offer additional insight into the subject material.

◆ EXAMPLE 6.1 The Hypsometric Formulae

Hypsometric formulae are expressions that relate atmospheric temperature and pressure to atmospheric altitude. To obtain these formulae, one starts with an expression that relates the height h of a static fluid column to the pressure p exerted by it. This relation is given by the equation

$$p = \rho g h \tag{6.1a}$$

or in differential form

$$dp = -\rho g \, dh \tag{6.1b}$$

where ρ = density of the fluid.

Equation (6.1) is known as the *fundamental equation of fluid statics*. It finds its way into all problems involving the calculation of forces exerted on submerged bodies. This includes the derivation of the forces acting on a dam or on the hull of a deep-sea exploration module. Archimides' law, which states that the buoyant force acting on a submerged body equals the weight of the displaced fluid, is a direct consequence of the fundamental equation of fluid statics.

To obtain an expression for pressure as a function of atmospheric altitude, $p = f(z)$, one must establish a relation between density ρ of the air and the prevailing pressure at a particular altitude. A first temptation is to substitute the isothermal density of an ideal gas into equation (6.1b). that is, we set

$$\rho = \frac{pM}{RT_0} \tag{6.2}$$

where M is its molar mass.

This is clearly in violation of established facts, since the temperature of the atmosphere is known to vary with altitude. Early work on this problem then turned to what is termed the adiabatic p–ρ relation, which is based on the premise that the air does not exchange heat with its surroundings during its expansion to a lower pressure. The relevant expression is given by

$$\left(\frac{p}{\rho}\right)^{\gamma} = \text{constant} \tag{6.3}$$

where γ is a dimensionless constant related to the specific heat of the gas. Its use will be demonstrated in Practice Problem 6.1, but we note here that the results obtained from the integrated expression were found to deviate from measured values. These measurements, taken over a period of years and at different locations, revealed that while pressure generally declined smoothly and exponentially with altitude, as one would expect, the recorded temperatures behaved less predictably. For example, in the troposphere, which extends to an altitude of 11 km, a simple linear relation applies:

$$T\,(\text{K}) = T_0 - Bz \tag{6.4}$$

where $T_0 = 288.16$ K or 15 °C, whereupon a reversal occurs that causes it to rise to −3 °C at an altitude of approximately 48 km. It remains at that temperature to 52 km, after which it resumes a rapid linear decline.

Evidently, to obtain the proper pressure variation prevailing in the atmosphere, equation (6.1b) must be integrated in segments, using the measured temperature variations in equation (6.2). We do this for the troposphere by substituting the relations (6.2) and (6.4) into equation (6.1b). We obtain

$$dp = -\rho g\, dz = -\frac{pM}{RT} g\, dz \tag{6.5a}$$

or

$$dp = -\frac{pMg}{R(T_0 - Bz)} dz \tag{6.5b}$$

Integration by separation of variables yields

$$p = p_0 \left(1 - \frac{Bz}{T_0}\right)^{Mg/RB} \tag{6.6}$$

where $Mg/RB = 5.26$ for air and p_0 is the pressure at ground level. This is the hypsometric formula for pressure, with the corresponding temperature coming from equation (6.4).

Let us, as an example, calculate the conditions that prevail at an altitude of 3000 m. From equation (6.4) we obtain

$$T = 288.16 - (6.5 \times 10^{-3})(3000) = 268.66\ \text{K}$$

or approximately −5 °C. By using equation (6.6) for the corresponding pressure, we can write

$$p = 101.3 \left[1 - \frac{(6.5 \times 10^{-3})(3000)}{288.16}\right]^{5.26} = 70.1\ \text{kPa}$$

Thus there is a 30% drop in pressure over that prevailing at ground level. When we go to the upper limit of the troposphere (i.e., 11 km), we obtain $p = 22.6$ kPa. This fivefold drop in pressure over ground level values is a much more substantial decrease.

COMMENTS

- In this first example of a simple one-dimensional distributed model, we have introduced some elementary notions of fluid statics. The results were not based on model predictions alone but had to be supplemented by experimental observations, expressed by equation (6.4). This twinning of experiment with theory is frequently resorted to whenever a part or some parts of a process are too complex to yield to a theoretical treatment. This is the case with atmospheric temperature, which will evidently depend in a complex manner on heat emitted from the sun and the earth, as well as on conduction through the air itself. Atmospheric air currents add to the difficulties, and it is therefore preferable, in this particular instance, to rely on measured temperature profiles.

◆ EXAMPLE 6.2 Poiseuille's Equation for Laminar Flow in a Pipe

We make a transition here from fluid statics, which we just applied to derive the hypsometric formulae, to fluid dynamics, which refers to the study of flowing fluids. We do this by taking up the case of laminar flow in a cylindrical pipe.

Laminar flow is one of two major flow regimes distinguished in fluid dynamics, the other being turbulent flow. The latter is characterized by swirling eddies that extend over the entire cross section of the pipe (Fig. 6.1). This results in a smoothing out of local velocities, which undergo only minor fluctuations about an average velocity v_{avg}. At the pipe wall it is customary to assume that the velocity is zero, since the fluid "sticks" to the wall and undergoes no motion relative to the bounding surface. This is referred to as the *no-slip condition* and applies to all conventional fluids including water and air. The drop from v_{avg} to $v = 0$ occurs in a region very close to the wall, which is referred to as a boundary layer. This layer also constitutes the major resistance to heat and mass transfer, and it is across this near–stagnant film that the flux of mass and energy must take place (see Figs. 1.3 and 1.4). As we have seen this fact has led to the formulation of transport driving forces ΔC and ΔT, which extend from a value in the bulk fluid that is assumed to be constant and uniform to the value of C or T prevailing at the wall.

Let us next examine what happens in laminar flow. The no-slip condition applies here as well, and it is in the bulk fluid that the major divergence from turbulent flow occurs. Here undisturbed layers of fluid slide past each other, driven by a pressure gradient and resisted by the "friction" or shear stress between adjacent layers (Fig. 6.2A). This results in a departure from the flat velocity profile, which we saw in turbulent flow. We now have a smooth parabolic profile, undisturbed by fluctuations, which rises from zero at the wall to a maximum at the centerline (Fig. 6.2B).

The aim in the present example is to derive, in the first instance, an expression for the velocity profile $v = f(r)$ by performing a force balance over a finite element of fluid of axial extent Δz and radial length r. This yields a relation between the driving pressure drop Δp and the opposing shear stress τ. The latter is not of direct interest, and we eliminate it by drawing on Newton's viscosity law [see equation (1.13), Table 1.2] as an auxiliary relation. This introduces the desired variable, velocity v, and leads to a differential equation in v as a function of r. This is precisely where we want to be, since a simple integration will now lead us to the desired results, the radial velocity distribution $v = f(r)$. We then show how this expression can be further manipulated to yield expressions of more direct usefulness for engineering purposes. Let us examine these steps in more detail.

We start with the relevant force balance over the element sketched in Fig. 6.2A and obtain at steady state

$$\sum \text{Forces} = 0 \tag{6.7}$$

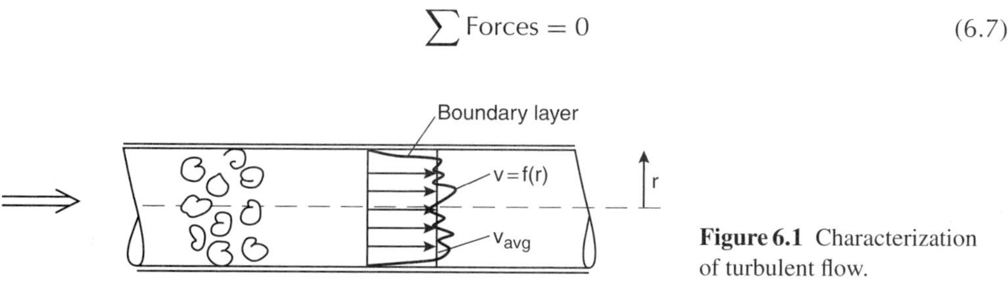

Figure 6.1 Characterization of turbulent flow.

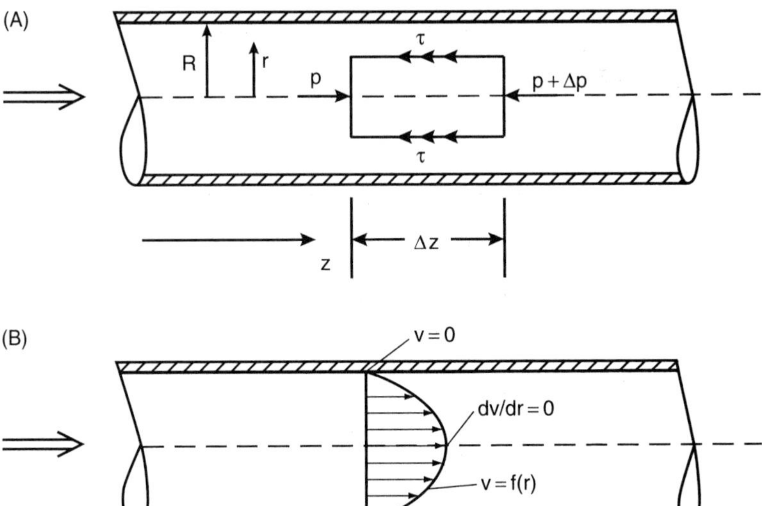

Figure 6.2 Laminar flow in a pipe: (A) force balance and (B) parabolic velocity profile.

and therefore

$$\tau = \frac{\Delta p}{2\Delta z} r \tag{6.8}$$

which shows that shear stress varies linearly with radial distance r, from zero at the centerline to a maximum at the pipe wall. Introducing Newton's viscosity law gives

$$-\mu = \frac{dv}{dr} = \frac{\Delta p}{2\Delta z} r \tag{6.9}$$

which immediately integrates by separation of variables to the expression

$$v = -\frac{\Delta p}{4\mu \Delta z} r^2 + C \tag{6.10}$$

The integration constant C is evaluated from the no-slip condition [i.e., $v(R) = 0$] so that $C = \Delta p R^2/4\mu L$. There results for the velocity distribution

$$v(r) = \frac{\Delta p R^2}{4\mu \Delta z}\left[1 - \left(\frac{r}{R}\right)^2\right] \tag{6.11}$$

This expression shows that the velocity profile is parabolic and varies from zero at the wall to a maximum of $\Delta p R^2/4\mu \Delta z$ at the centerline. This is interesting information but does not address the questions an engineer is most often called upon to answer. Those questions are:

- Given the length Δz and radius R of a pipe, what pressure must we apply to achieve a prescribed volumetric flow rate Q (m^3/s) or mass flow rate F (kg/s)?

- Conversely, if pressure drop Δp is prescribed, what flow rate will result in a pipe of given dimensions?
- Finally, suppose the required pressure drop is excessive and we wish to reduce it by choosing a pipe of different dimension. The question then is: By how much should we change R or Δz to achieve a desired reduction in the pressure drop?

The principal task, then, is to convert the velocity distribution (6.11) into the corresponding volumetric flow rate Q. We do this in two steps: we obtain the average velocity v_{avg} by integrating v(r) over the cross section and then divide by the cross sectional area. We obtain in this case

$$v_{avg} = \frac{\Delta p R^2}{4\mu \Delta z} \frac{\int_0^R \left[1 - \left(\frac{r}{R}\right)^2\right] 2\pi\, dr}{\pi R^2} \tag{6.12a}$$

$$= \frac{\Delta p}{2\mu \Delta z} = \left[\frac{r^2}{2}\bigg|_0^R - \frac{r^4}{4R^2}\bigg|_0^R\right] \tag{6.12b}$$

or

$$v_{avg} = \frac{R^2}{8\mu} \frac{\Delta p}{\Delta z} \tag{6.12c}$$

and for the volumetric flow rate

$$Q = v_{avg} \pi R^2 = \frac{\pi R^4}{8\mu} \frac{\Delta p}{\Delta z} = \frac{\pi d^4}{128\mu} \frac{\Delta p}{\Delta z} \tag{6.13}$$

This is the celebrated Poiseuille or Hagen–Poiseuille equation for viscous flow through a cylindrical tube.

Mass flow rate F follows immediately by multiplication of Q by fluid density ρ, that is,

$$F = \rho Q = \frac{\pi \rho R^4}{8\mu} \frac{\Delta p}{\Delta z} = \frac{\pi \rho d^4}{128\mu} \frac{\Delta p}{\Delta z} \tag{6.14}$$

Let us demonstrate the use of these equations with a numerical example. Suppose a viscous oil, whose flow is in the laminar regime, is to be pumped through a 10 cm diameter horizontal pipe over a distance of 15 km at a rate of 10^{-3} m^3/s. Viscosity of the oil is 1.0 Pa s. Note that the units of viscosity follow from Newton's viscosity law.

We obtain from equation (6.14)

$$\Delta p = \frac{8\mu \Delta z Q}{\pi R^4} \tag{6.15}$$

$$\Delta p = \frac{8 \times 1.0 \times 15 \times 10^3 \times 10^{-3}}{\pi (0.05)^4}$$

$$\Delta p = 6.1 \text{ MPa}$$

which corresponds to approximately 60 atm. This is considered to be excessive, and a reduction through use of a larger diameter pipe seems desirable. Suppose we double the

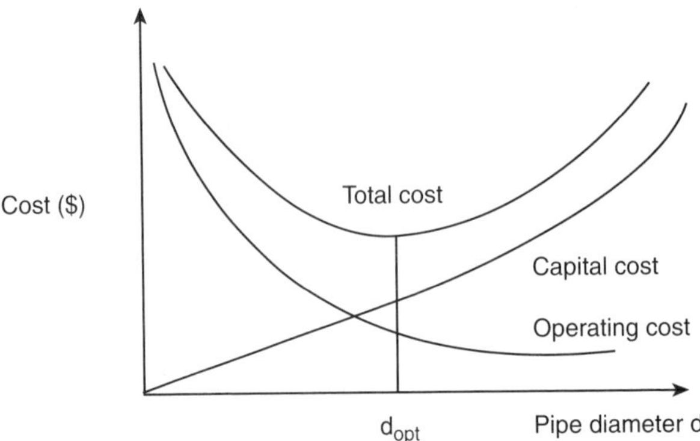

Figure 6.3 Optimum pipe diameter.

pipe diameter to 20 cm. Then Δp becomes $6.1/(2^4) = 0.38$ MPa or 380 kPa, corresponding to about 4 atm. This is a much more reasonable, value and allows us to avoid the use of heavy gauge steel pipe, which would be required at the higher pressure.

COMMENTS

- A number of noteworthy features emerge from this treatment. First and foremost is the strong variation of pressure and flow rate with tubular radius. The fourth-power dependence on radius, which is both startling and unexpected, gives the designer considerable leeway in adjusting Q or Δp through relatively modest changes in pipe diameter. The linear dependence of pressure drop on viscosity is more unexceptional and was to be expected. Finally we had arbitrarily assumed a direct variation of pressure drop with distance. This follows from simple physical reasoning and enabled us to use a fluid element of finite length Δz to make our force balance.

  The discerning reader may have deduced that behind the simple façade of Poiseuille's equation, there lurks an important optimization problem. This comes about as follows. The operating costs, which are almost entirely dictated by pressure drop, vary approximately linearly with Δp, hence [see equation (6.9)] inversely with R^4 or d^4. These costs consequently decline sharply as pipe diameter is increased (see Fig. 6.3). The capital costs, on the other hand, consisting largely of the cost of the pipe, increase almost linearly with the diameter of the pipe, since the volume of material used in its manufacture is approximately proportional to surface area $\pi d\Delta z$. Larger pipe diameters deviate somewhat from this linear relation because they require a somewhat larger wall thickness for structural reasons, and their costs of fabrication and installation generally increase as well. Adding these two costs results in a total cost curve that passes through a minimum, yielding an optimum pipe diameter d_{opt} as shown in Fig. 6.3 We note that the recognition that such an optimum exists relies on what we termed "closure" (see Chapter 1), that is, a careful analysis of the results that emerge from the modeling process.

◆ EXAMPLE 6.3 Compressible Laminar Flow in a Horizontal Pipe

In deriving the Poiseuille equation, we tacitly assumed that velocity remains constant in the direction of flow. This condition holds only for the flow of liquids, or in gas flows where the pressure drop is limited to a few percent of total pressure. Airflow in large ventilation ducts falls in this category. When pressure drop is more substantial, the gas will expand significantly in the direction of flow, and with such expansions will come a decrease in density. This in turn induces an increase in velocity, which is required to maintain a constant mass flow rate F, given by F (kg/s) = $v\rho A$. These changes are all nonlinear in the direction of flow and are a direct consequence of the fact that gases are compressible and respond to a reduction in pressure, by expanding. We term this type of flow *compressible flow*. To accommodate these changes, two things need to be done. First, the difference formulation $\Delta p/\Delta z$, which we used in deriving the Poiseuille equation, must be replaced by a *differential* version. That is, we write for the frictional pressure drop of equation (6.12c)

$$dp = \frac{8\mu}{R^2} v \, dz = \frac{32\mu}{d^2} v \, dz \tag{6.16}$$

Second, we need to take into account the kinetic energy changes that were brought about by the increase in velocity and clearly imply the presence of an accelerating force. That force will have to be incorporated in the overall force balance. We accomplish this in a slightly roundabout way by returning to the Bernoulli equation (Example 4.7), which relates pressure drop to velocity changes in *frictionless* flow, and adding to it a term that expresses the purely frictional pressure loss drawn from equation (6.15). We obtain

$$\frac{dp}{\rho} + v \, dv + \frac{32\mu}{\rho d^2} v \, dz = 0 \tag{6.17}$$

The task now becomes one of integrating this expression, noting that ρ, v, and p all vary with distance z. Evidently, ρ has to be expressed in terms of p and v to make integration possible. We draw for this purpose on the ideal gas law, which relates density to pressure via the relation $\rho = pM/RT$, where M is the molar mass, and utilize the fact that the mass velocity (i.e., the mass flow rate per unit cross-sectional area G)is constant and is given by

$$G \text{ (kg/s m}^2) = \rho v = \text{constant} \tag{6.18}$$

so that

$$\rho = \frac{G}{v}$$

Note that in these expressions v is the average velocity v_{avg}, where we have for convenience omitted the subscript avg. Substituting these expressions into equation (6.11), we obtain, in the first instance

$$\frac{RT}{M}\frac{dp}{p} + v \, dv + \frac{32\mu}{Gd^2} v^2 \, dz = 0 \tag{6.19}$$

Noting from equation (6.18) that $v^2 = G^2/(pM/RT)^2$, we divide equation (6.19) by that expression and obtain

$$\frac{Mp \, dp}{G^2RT} + \frac{dv}{v} + \frac{32\mu}{Gd^2} dz = 0 \tag{6.20}$$

Straightforward integration then leads to the result

$$\frac{1}{2}\frac{M}{G^2RT}(p_2^2 - p_1^2) + \ln\frac{v_2}{v_1} + \frac{32\mu}{Gd^2}L = 0 \tag{6.21}$$

where L = total length of the pipe.

Recognizing from the ideal gas law and relation (6.18) that $v_2/v_2 = p_1/p_2$, we can cast equation (6.21) into the alternative form

$$\frac{1}{2}\frac{M}{G^2RT}p_1^2\left[\left(\frac{p_2}{p_1}\right)^2 - 1\right] - \ln\frac{p_2}{p_1} + \frac{32\mu}{Gd^2}L = 0 \tag{6.22}$$

Given the inlet pressure p_1 to the system, one can calculate from this expression the outlet pressure p_2, hence the pressure drop required to produce a given mass flow rate per unit pipe cross-sectional area G. This is easily done numerically. Conversely, G for a given pressure ratio p_2/p_1 can be calculated by solving a quadratic equation in G.

COMMENTS

- One notes here the combination of the differential forms of the Poiseuille and Bernoulli equations to arrive at a description of compressible flow, usually taken to be gas flow, with substantial changes in pressure and significant frictional pressure losses. By substituting different values for L into equation (6.22) and solving for p_2, we can derive the pressure distribution $p = f(z)$ for the entire pipe length.
- The corresponding velocity profile v(z) can be obtained from the relation $v = G/\rho = RTG/pM$. We note that v does not increase indefinitely but rather attains an upper value that equals the velocity of sound in the gas. Further increases in either pressure or pressure drop will not increase velocity, hence will not increase flow rate. The underlying theory and derivation require somewhat extensive treatment and are not taken up here. The important fact to retain is that for a pipe of given dimensions, the gas flow rate cannot be pushed to arbitrarily high values by simply increasing the driving pressure drop. This is an important factor in the transmission of gases over long distances, such as natural gas pipe lines. Since the maximum flow rate is associated with a downstream pressure minimum, the remedy lies in recompressing the gas at intervals along the pipeline, to prevent this minimum from being reached. This is done in all long-distance transmission lines, where the pertinent compressor facilities are typically stationed at intervals of 100 km.
- Laminar compressible flow is encountered less frequently than its turbulent counterpart. For the latter case, a similar expression applies:

$$\frac{1}{2}\frac{M}{G^2RT}(p_2^2 - p_1^2) + \ln\frac{p_1}{p_2} + 2f\frac{L}{d} = 0 \tag{6.23}$$

where f is an empirical friction coefficient. The last term of this equation differs from the earlier laminar case [equation (6.22)], where $(32\mu/Gd^2)L$ represents the frictional pressure loss.

◆ EXAMPLE 6.4 Conduction of Heat Through Various Geometries

We consider in this example the steady heat conduction through solids with different geometries. The simplest of these is a slab, a solid bounded by two parallel planes. Conduction in cylinders and spheres usually is unsteady, requiring a partial differential equation to describe the process. However, in the special case of *hollow* cylinders and spheres, a steady heat flux can be set up after an initial transient period has subsided. We consider these processes for different boundary conditions, particularly those of types I and III (see Chapter 2, Table 2.7), and derive expressions both for the temperature profiles and the steady flux of heat q. Note that we have confined ourselves to conduction through *solids* and omitted consideration of liquid and gases. This is because the latter two cases entail the possibility of additional heat flow due to *free* or *natural convection*.

Conduction Through Media Bounded by Parallel Planes

We start by considering the simplest of several possible cases, that of a slab composed of a single material (Fig. 6.4A), and we assume boundary conditions of type I at the surfaces, that is,

$$T(x_1) = T_1$$
$$T(x_2) = T_2 \tag{6.24}$$

We introduce Fourier's law $q' = -k(dT/dx)$, where $q' = q/A$ is the constant heat flux. Simple integration by separation of variables then yields

$$\int_{x_1}^{x_2} q'\,dx = -k \int_{T_1}^{T_2} dT \tag{6.25a}$$

$$T_1 - T_2 = \frac{q'(x_2 - x_1)}{k} \tag{6.25b}$$

or

$$q' = \frac{T_1 - T_2}{(x_2 - x_1)/k} = \frac{T_1 - T_2}{R} \tag{6.25c}$$

When written in the form of equation (6.25c), the heat flux q' is seen to be proportional to a driving temperature difference $T_1 - T_2$ and to vary inversely with $(x_2 - x_1)/k$, which can

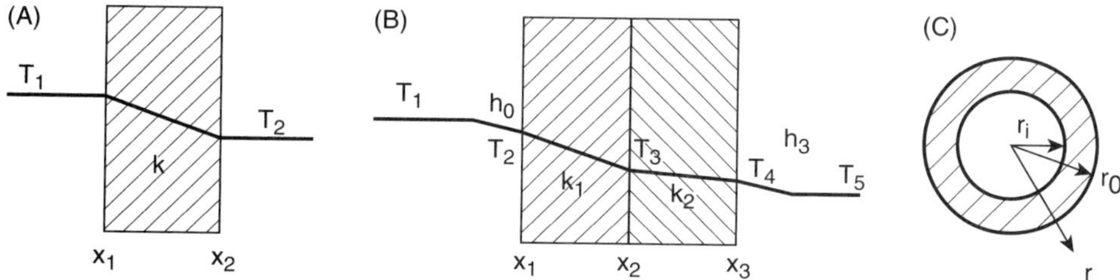

Figure 6.4 Conduction through various geometries: (A) slab, (B) composite slab with film resistance, and (C) hollow cylinder.

Table 6.1 Driving Forces and Resistance in Various Physical Laws

Process	Flux Equation	Driving Force	Resistance
Conductivity of heat	$qA = q' = k\frac{\Delta T}{\Delta x}$	ΔT	$\frac{\Delta x}{k}$
Diffusion of mass	$\frac{N}{A} = D\frac{\Delta C}{\delta} = D\frac{\Delta C}{\Delta x}$	ΔC	$\frac{\delta}{D} = \frac{\Delta x}{D}$
Flow through porous media	$v = \frac{K}{\mu}\frac{\Delta p}{\Delta x}$	Δp	$\frac{\mu \Delta x}{K}$
Flow of electricity	$i = \frac{V}{R}$	V	R

be considered to be a resistance R to the conduction process. This is precisely analogous to Ohm's law for the steady flow of electric current: the flux q' corresponds to the electric current, the fall in temperature to the voltage drop, and the term $(x_2 - x_1)/k$ to the resistance of the conductor. We have also seen similar concepts applied to the case of mass diffusion through a stagnant film, where the driving force was a concentration difference ΔC, and the resistance was composed of the ratio of film thickness to diffusivity δ/D (see Example 5.7). Yet another example is D'Arcy's law, which has a driving force Δp and a resistance $\mu \Delta z/K$ (see Example 5.9). What all these processes have in common, then, is the flow of a physical entity (heat, mass, electric current) that is proportional to a driving force or driving potential (ΔT, ΔC, Δp, V) and varies inversely with a resistance. We have summarized these features for the convenience of the reader in Table 6.1.

The linear temperature profile that arises here can be derived again by integrating Fourier's law, but this time over the limits x_1 to x. We obtain

$$\int_{x_1}^{x} q\,dx = \int_{x_1}^{x} k\frac{T_1 - T_2}{x_2 - x_1}\,dx = -k\int_{T_1}^{T} dT \tag{6.26}$$

from which we obtain the result

$$T_1 - T = \frac{(T_1 - T_2)(x - x_1)}{x_2 - x_1} \tag{6.27}$$

Note that the thermal conductivity k does not appear in this expression.

We next turn to the composite slab shown in Fig. 6.4B. There are four distinct regions, composed of the two external film resistances and the two materials of the slab. We apply equation (6.25c) to each region in turn and obtain

$$T_1 - T_2 = \frac{q'}{h_0} \tag{6.28a}$$

$$T_2 - T_3 = \frac{q'(x_2 - x_1)}{k_1} \tag{6.28b}$$

$$T_3 - T_4 = \frac{q'(x_3 - x_2)}{k_2} \tag{6.28c}$$

$$T_4 - T_5 = \frac{q'}{h_3} \tag{6.28d}$$

Summing these equations and solving for q' yields

$$q' = \frac{T_1 - T_5}{\dfrac{1}{h_0} + \dfrac{x_2 - x_1}{k_1} + \dfrac{x_3 - x_2}{k_2} + \dfrac{1}{h_3}} = \frac{\Delta T}{\sum R} \tag{6.29}$$

where it is seen that the driving force is now composed of the overall temperature difference $T_1 - T_5$ and the resistance equals the *sum* of the individual resistances offered by each segment.

Finally, we turn to the problem of accommodating variations of thermal conductivity with temperature. For this case we write

$$q' = -k(T)\frac{dT}{dx} \tag{6.30a}$$

from which we obtain by formal integration

$$q'(x_2 - x_1) = -\int_{T_1}^{T_2} k\,dT \tag{6.30b}$$

or equivalently

$$q' = \frac{T_1 - T_2}{x_2 - x}k_{avg} \tag{6.30c}$$

where

$$k_{avg} = -\frac{1}{T_1 - T_2}\int_{T_1}^{T_2} k\,dT \tag{6.30d}$$

Conduction in a Hollow Cylinder

We start as before with Fourier's law, which is applied to some arbitrary point in the cylinder wall and takes the form

$$q = -k2\pi rL\frac{dT}{dr} \tag{6.31}$$

Note that we are by necessity dealing with heat flow q (J/s) rather than heat flux q' (J/s m^2), since the latter has now become a variable. We integrate by separation of variables between inner and outer radius (Fig. 6.4C) assuming an absence of film resistance. Thus

$$q\int_{r_i}^{r_o} \frac{dr}{r} = -2\pi kL\int_{T_i}^{T_o} dT \tag{6.32a}$$

or

$$q = 2\pi kL\frac{(T_i - T_o)}{\ln r_o/r_i} \tag{6.32b}$$

This is the equation for the steady heat flow (not flux) through a cylindrical wall. To obtain the corresponding temperature we integrate Fourier's equation in similar fashion, but this time set the upper limits equal to the radius variable r and its temperature T. We obtain

$$q\int_{r_i}^{r}\frac{dr}{r} = -2\pi kL\int_{T_i}^{T} dT \tag{6.33}$$

which becomes, after substitution of equation (6.32b) for q,

$$T(r) = T_i - \frac{T_i - T_o}{\ln r_o/r_i}\ln\frac{r}{r_i} \tag{6.34}$$

Note again the absence of thermal conductivity in this expression: the temperature profile will be the same for *all* materials, given fixed boundary conditions.

Let us now consider the same case with internal and external film resistances. This situation can again by simplified by using the concept of additivity of resistances expressed in equation (6.29). For a film resistance, we have

$$q = h2\pi rL(\Delta T) \tag{6.35a}$$

with an equivalent resistance R_f of

$$R_f = \frac{1}{h2\pi rL} \tag{6.35b}$$

For the cylinder wall, the resistance, obtained from equation (6.32b) is given by

$$R_c = \frac{\ln r_o/r_i}{2\pi kL} \tag{6.36}$$

Consequently we obtain for the heat flow

$$q = \frac{\text{Driving force}}{\sum \text{Resistances}}$$

or

$$q = \frac{T_i - T_o}{\dfrac{1}{2\pi r_i L h_i} + \dfrac{\ln r_o/r_i}{2\pi kL} + \dfrac{1}{2\pi r_o L h_o}} \tag{6.37}$$

This is the equation for heat flow through a cylindrical wall with internal and external film resistances. It can be used, for example, to calculate the heat loss from a steam pipe to the atmosphere. Since steam pipes are usually insulated, the pipe wall will be a composite one, which can again be accommodated by the additivity of resistances concept.

An interesting optimization problem arises in this case, which will be taken up in Practice Problem 6.3.

Conduction in a Hollow Sphere

Here Fourier's equation, applied to a spherical surface, is given by

$$q = -k4\pi r^2 \frac{dT}{dr} \tag{6.38}$$

which upon integration by separation of variables leads to

$$q \int_{r_i}^{r_o} \frac{dr}{r^2} = -k4\pi \int_{T_i}^{T_o} dT \tag{6.39a}$$

and

$$q\left(\frac{1}{r_i} - \frac{1}{r_o}\right) = k4\pi(T_i - T_o) \tag{6.39b}$$

or

$$q = \frac{4\pi k(T_i - T_o)}{1/r_i - 1/r_o} \tag{6.39c}$$

The corresponding temperature profile, obtained by integrating to the upper limits (r, T) and substitution for q using equation (6.34) is then given by

$$T(r) = T_i - \frac{1/r_i - 1/r}{1/r_i - 1/r_o}(T_i - T_o) \tag{6.40}$$

This case of conduction in a hollow sphere is relatively rare but can arise in principle when the interior surface is maintained at a constant temperature by means of a thermal or nuclear reaction. Ultimately, when the fuel nears exhaustion, the conduction process enters a transient period and finally comes to a complete halt.

◆ EXAMPLE 6.5 Conduction in Systems with Heat Sources

In Example 6.5 we considered purely conductive processes without internal heat generation. Now we address the case of heat production superposed on heat conduction, and we examine the resulting temperature distribution and heat fluxes. Such internal heat production can be brought about by embedded electrical heating elements or by chemical or nuclear reactions. We assume that these heat sources are *uniformly* distributed in the conducting matrix and that heat is produced at the constant volumetric rate S (J/m^3 s). The geometries to be considered are the slab of thickness L and the sphere and the infinitely long cylinder, both of radius R. The surfaces are in each case held at a constant temperature T_s by appropriate cooling; that is, we are dealing with a boundary condition of type I. Note that we need not in this instance confine ourselves to hollow spheres and cylinders, since throughout the solid medium the heat sources provide a steady supply of heat that is removed by conduction at the same rate at which it is produced. Let us now examine each geometry in turn.

The Slab

An energy balance performed over a difference element Δx of the slab leads to the following expressions:

$$\text{Rate of energy in} - \text{Rate of energy out} = 0$$

$$\left[\begin{matrix} -kA \left.\dfrac{dT}{dx}\right|_x \\ +SA\,\Delta x \end{matrix}\right] - \left[kA \left.\dfrac{dT}{dx}\right|_{x+\Delta x}\right] = 0 \tag{6.41a}$$

Upon dividing by $A\Delta x$ and letting $\Delta x \to 0$ we obtain:

$$-k\frac{d}{dx}\left(\frac{dT}{dx}\right) = S \tag{6.41b}$$

This is the differential equation that describes the temperature distribution in a slab with constant internal heat generation S. A first integration by separation of variables gives

$$-k\frac{dT}{dx} = Sx + C_1 \tag{6.42}$$

The integration constant C_1 is most conveniently evaluated at the midplane $x = L/2$, where for symmetry reasons the temperature gradient dT/dx must be zero. We obtain

$$0 = \frac{SL}{2} + C_1 \tag{6.43a}$$

or

$$C_1 = -\frac{SL}{2} \tag{6.43b}$$

and

$$-k\frac{dT}{dx} = S\left(x - \frac{L}{2}\right) \tag{6.43c}$$

It follows that the heat flux leaving the surface $x = L$ is given by

$$q' = -k\frac{dT}{dx} = \frac{SL}{2} \tag{6.44}$$

or twice that amount from both sides.

A second integration is next performed to obtain the full temperature distribution. We obtain

$$-kT = \frac{Sx^2}{2} - \frac{SLx}{2} + C_2 \tag{6.45}$$

where C_2 is obtained from the boundary condition $T(0) = T_s$ (i.e., $C_2 = -kT_s$). The full temperature profile is consequently given by

$$T - T_s = \frac{SL^2}{2k}\left[\frac{x}{L} - \left(\frac{x}{L}\right)^2\right] \tag{6.46}$$

That is, temperature is seen to have a parabolic distribution.

The Sphere

Here the energy balance has the form

$$\text{Rate of energy in} - \text{Rate of energy out} = 0$$

$$\begin{bmatrix} -k4\pi r^2 \left.\dfrac{dT}{dr}\right|_r \\ +S4\pi r^2 \Delta r \end{bmatrix} - \left[-k4\pi r^2 \left.\frac{dT}{dr}\right|_{r+\Delta r}\right] = 0 \tag{6.47}$$

Dividing by $4\pi\Delta r$ and letting $\Delta r \to 0$ we obtain

$$-k\frac{d}{dr}\left(r^2\frac{dT}{dr}\right) = Sr^2 \tag{6.48}$$

A first integration leads to

$$-kr^2\frac{dT}{dr} = \frac{Sr^3}{3} + C_1 \tag{6.49}$$

where $C_1 = 0$ by virtue of the fact that the gradient equals zero at $r = 0$. The flux q at the surface is consequently given by

$$q' = -k\left.\frac{dT}{dr}\right|_R = \frac{SR}{3} = \frac{Sd}{6} \tag{6.50}$$

This is three times less than the one-sided flux from a slab.

A second integration is next performed using as a starting point

$$-k\frac{dT}{dr} = \frac{Sr}{3} \tag{6.51}$$

Hence

$$-kT = \frac{Sr^2}{6} + C_2 \tag{6.52}$$

Since $T(R) = T_s$, the integration constant C_2 becomes $C_2 = -kT_s - SR^2/6$, and the resulting profile is given by

$$T - T_s = \frac{SR^2}{6k}\left[1 - \left(\frac{r}{R}\right)^2\right] \tag{6.53}$$

which is again parabolic in form.

The Cylinder

We had in our preamble assumed an infinitely long cylinder to eliminate conduction through the ends. The same result is obtained by taking the ends to be insulated. In either case, conduction is confined to the radial direction and we obtain

$$\text{Rate of energy in} - \text{Rate of energy out} = 0$$

$$\begin{bmatrix} -k2\pi rL \left.\dfrac{dT}{dr}\right|_r \\ +S2\pi rL\Delta r \end{bmatrix} - \left[-k2\pi rL \left.\frac{dT}{dr}\right|_{r+\Delta r} \right] = 0 \tag{6.54}$$

Dividing by $2\pi L\Delta r$ and letting $\Delta r \to 0$ yields

$$-k\left(\frac{d}{dr}r\frac{dT}{dr}\right) = Sr \tag{6.55a}$$

which upon integration gives

$$-kr\frac{dT}{dr} = \frac{Sr^2}{2} + C_1 \tag{6.55b}$$

From the symmetry condition $(dT/dr)_{r=0} = 0$, we obtain as before $C_1 = 0$ and the flux at R then becomes

$$q' = -k \left.\frac{dT}{dr}\right|_R = \frac{SR}{2} = \frac{Sd}{4} \tag{6.56}$$

This is again less than the flux obtained in a slab, this time by a factor of 2.

The final step consists of determining the temperature profile by integrating equation (6.55b). Thus

$$-kT = \frac{Sr^2}{4} + C_2 \tag{6.57}$$

where $C_2 = -KT_s - SR^2/4$ from the surface boundary condition $T(R) = T_s$. The profile is then given by

$$T - T_s = \frac{SR^2}{4k}\left[1 - \left(\frac{r}{R^2}\right)\right] \tag{6.58}$$

which is parabolic, like the sphere and the slab.

COMMENTS

- We have here, by means of some simple models, established some important parameters for systems with internal heat generation. We considered three geometries that shared some common features. All three led to second-order ODEs that could be integrated by repeated application of separation of variables. This demonstrates again the power of that method. Second-order ODEs were anticipated because the auxiliary relation used, Fourier's equation, already contained a derivative. As pointed out in Chapter 2, systems of this type invariably lead to higher order ODEs. All three geometries yielded parabolic temperature distributions, with a zero gradient at the center. This was again expected on physical grounds. They differed, however, in the magnitude of the heat flux, the slab yielding the highest value. This is an important revelation, which can be used to advantage in the design of such systems.

Finally, we draw the reader's attention to Practice Problem 6.4 (A Biomedical Problem: The Krogh Cylinder). In that example nutrients contained in the bloodstream pass into the surrounding tissue, where they are metabolized (i.e., consumed at a constant rate). The tissue is given the geometry of a cylindrical shell, the outer surface of which is assumed to be impermeable ($dC/dr = 0$). At the inner surface the concentration is assumed to be constant. We thus have a boundary condition of type II at one end and type I at the other (see Table 2.7). The task is to derive the concentration distribution of nutrient in the tissue.

At first sight this problem appears to be quite different, at least physically, from what was considered in the present example. Some thought, however, reveals that there are a number of important similarities between the two cases.

Both the cylinder decribed by the equations (6.54) to (6.58) and the Krogh cylinder use a cylindrical geometry.

Transport is described by a rate equation that contains a derivative (Fick's law), as was the case in the present example (Fourier's law).

Both cases exhibit a constant internal production rate that is positive in heat generation, and negative in nutrient consumption.

We can consequently expect that the resulting ODE will be of the form given by equation (6.55a), with the source term now taking a negative sign. This is indeed the case. Although the final solution will have a somewhat altered form owing to the differences in boundary conditions, the important fact remains that the two constitutive ODEs are similar in form and differ only in the sign of one term. The solution methods are therefore expected to be identical. This leads us to suggest a search in related disciplines for ways of solving a particular problem at hand. This requires a good sense of what is happening outside one's own field but does avoid a good deal of unnecessary work. Furthermore, unusual features discovered in one system may translate directly into similar behavior in the other. Such guidelines obtained from external sources thus carry important advantages.

◆ EXAMPLE 6.6 The Countercurrent Heat Exchanger

In Examples 6.4 and 6.5 we concerned ourselves with heat transfer in stationary media in which transport occurred by a conductive mechanism. In the present example we consider in some detail the transfer of heat to and from flowing fluids, using what is termed a shell-and-tube heat exchanger as an illustration. In this device, a limited application of which is considered in Chapter 1 (Practice Problem 1.5), two fluids at different temperatures and separated by a tubular wall are contacted to effect a transfer of heat from one fluid to the other without mixing the two (Fig. 6.5A). The flow can be in the same direction, in which case we speak of cocurrent flow, or in opposite directions, which is termed countercurrent flow. It is the latter we address here, the primary task being the derivation of the fluid temperature profiles.

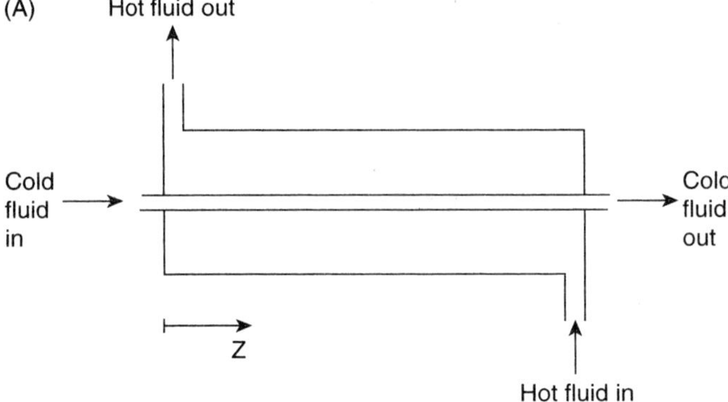

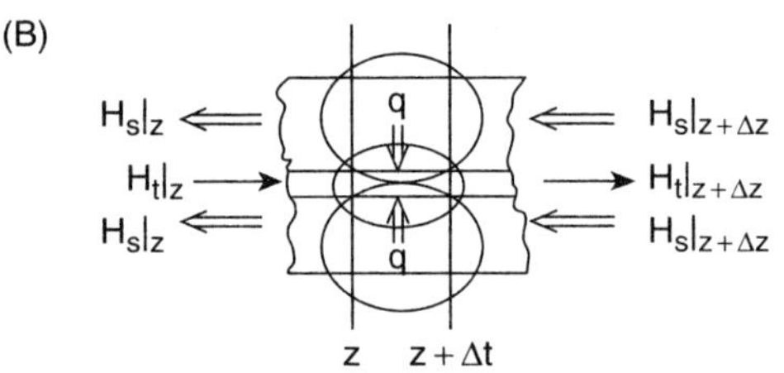

Figure 6.5 The counter-current shell-and-tube heat exchanger: (A) schematic diagram and (B) difference element with envelopes for the energy balances.

We start by noting that since two state variables are involved, the shell-side and the tube-side temperatures, at least two equations will be required. We can obtain these in a variety of ways by making integral balances over different parts of the system or by making differential balances over one or both phases. Nine such balances can be made, but we recognize that the integral balances cannot be the sole model equations because they do not contain distance z as the required independent variable. This variable can be brought in only by making two differential balances, one over each of the two phases (Fig. 6.5B). As we shall, see these two balances not only yield the most general solution in the form of temperature profiles, but also, by various combinations and manipulations, can be reduced to yield solutions obtained from integral balances. We assume that the heating medium is located in the shell, and obtain the following:

Tube-Side Balance

$$\begin{aligned} &\text{Rate of energy in} - \text{Rate of energy out} = 0 \\ &\left(H_t|_z + q_{avg}\right) \quad - \quad \left(H_t|_{z+\Delta g}\right) \quad = 0 \end{aligned} \tag{6.59a}$$

or, replacing the enthalpy flow rate H_t(J/s) by flow rate F × specific enthalpy $\overline{H}_t$ (J/kg),

$$F_t \Delta \overline{H}_t - q_{avg} = 0 \tag{6.59b}$$

We now express $\overline{H}$ in terms of temperature and specific heat and introduce as well the auxiliary relation for convective heat transfer, listed in Table 1.3, equation (1.18). Since several resistances in series are involved, we use an overall heat transfer coefficient U, which is the inverse of the sum of the two film resistances. That is, we have

$$\frac{1}{U} = \frac{1}{h_i} + \frac{1}{h_o} \tag{6.59c}$$

where h_i and h_o are the internal (i.e., tube-side) and h_o the outside (i.e., shell-side) heat transferred coefficient. Equation (6.59b) then becomes

$$F_t C_{pt} \Delta T_t - U\pi\, d\Delta z(T_s - T_t) = 0 \tag{6.59d}$$

Dividing by Δz and letting $\Delta z \to 0$ yields

$$F_t C_{pt} \frac{dT_t}{dz} - U\pi\, d(T_s - T_t) = 0 \tag{6.59e}$$

where the subscripts t and s refer to the tube- and shell-side variables, respectively.

Then, using the same procedure as before, we obtain

Shell-Side Balance

$$F_s C_{ps} \frac{dT_s}{dz} - U\pi\, d(T_s - T_t) = 0 \tag{6.60}$$

which has *exactly* the same form as the tube-side balance. This may at first seem surprising, but no changes in sign occur because both flow and heat transfer change direction.

We now proceed to manipulate these differential balances to arrive at various results of interest.

1. Subtracting the two balances leads to cancellation of the heat transfer terms as well as dz and yields

$$F_s C_{ps}\, dT_s - F_t C_{pt}\, dT_t = 0 \tag{6.61a}$$

which upon integration over *part* of the exchanger yields

$$F_s C_{ps}(T_s - T_{s\,out}) = F_t C_{pt}(T_t - T_{t\,in}) \tag{6.61b}$$

and over the *entire* exchanger

$$q_{tot} - F_s C_{ps}(T_{s\,in} - T_{s\,out}) = F_t C_{pt}(T_{t\,out} - T_{t\,in}) \tag{6.61c}$$

It is immediately seen that the same results could have been obtained by algebraic integral energy balances taken over both phases. Since z disappeared, they tell us nothing about the *profiles* or the size (length) of heat exchanger required for a particular application. As we shall see, however, these results can be used to supplement other model equations.

2. In a second manipulation, we first divide each differential balance by FC_p and *then* subtract. This is a clever device to reduce T_s and T_t to a *single* variable ($Ts - T_t$). We obtain

$$\frac{d(T_s - T_t)}{dz} = U\pi\, d\left[\frac{1}{F_s C_{ps}} - \frac{1}{F_t C_{pt}}\right] \tag{6.62a}$$

which when integrated by separation of variables over the entire exchanger length L yields

$$\ln\frac{(T_s - T_t)\,|_L}{(T_s - T_t)\,|_0} = U\pi\, dL\left[\frac{1}{F_sC_{ps}} - \frac{1}{F_tC_{pt}}\right] \tag{6.62b}$$

with $\pi dL = A =$ total heat transfer area.

The equation does not provide an individual temperature profile but, by replacing L with z (integration over part of the exchanger), we can obtain the profile of the temperature *differences*. Equation (6.62b) can be transformed into a more convenient form by using equation (6.61c) replace the FC_p terms with the corresponding temperature differences; after rearrangement, we have

$$q_{tot} = UA\frac{(T_s - T_t)|_L - (T_s - T_t)|_0}{\ln\dfrac{(T_s - T_t)|_L}{(T_s - T_t)|_0}} \tag{6.63}$$

The fraction in this expression is commonly known as the log mean temperature difference (LMTD), so that equation (6.63) becomes in abbreviated form

$$q_{tot} = UA(\text{LMTD}) \tag{6.64}$$

This is a frequently used design equation. It is a design equation because by knowing the inlet temperatures $T_{t\,in}$ and $T_{s\,in}$, and *specifying* the desired outlet temperature $T_{t\,out}$, one can calculate the fourth outlet temperature $T_{s\,out}$ from the integral balance (6.61c). Both q_{tot} and LMTD can then be evaluated, thus allowing the design area A to be obtained. It is assumed that U has been estimated a priori by means of empirical film coefficient correlations. The equation cannot be used in problems where only the inlet temperatures are known and the third temperature remains unspecified. This occurs, for example, if for an existing heat exchanger (A known) one wishes to calculate the effect of a change in inlet conditions on the outlet temperatures. Similarly, one cannot predict the effect of changes in heat exchanger length or area on outlet conditions.

3. Problems not amenable to solution by equation (6.64) must be handled by solving the differential balances of equations (6.59) and (6.60) to obtain full profiles. We note that these equations are coupled so that a simultaneous solution seems to be indicated. This can, however, be avoided by the use of a clever "trick," demonstrated in Chapter 2 [equation (2.90)]. We solve equation (6.59e) *algebraically* for T_s and substitute the result into equation (6.60). The penalty we pay is that the result is a *second-order* ODE, given by

$$T_t'' - \frac{\alpha - \beta}{\alpha\beta}T_t' = 0 \tag{6.65}$$

where $\alpha = F_tC_{pt}/U\pi d$ and $\beta = F_sC_{ps}/U\pi d$.

Equation (6.65) can be solved either by the p-substitution or D-operator method given in Chapter 2. We choose the latter and obtain

$$\left(D_2 - \frac{\alpha - \beta}{\alpha\beta}D\right)T_t = 0 \tag{6.66}$$

and from Table 2.9 the solution

$$T_t = A + B\exp\left(\frac{\alpha - \beta}{\alpha\beta} z\right) \tag{6.67}$$

From the boundary condition for the tube inlet we obtain

$$(T_t)_{in} = A + B \tag{6.68}$$

To use the second boundary condition $(T_s)_{in}$, which is given at the other end of the exchanger, we must return to the differential balance [equations (6.59) or (6.60)] and write

$$\alpha T_t''(L) - T_s(L) + T_t(L) = 0 \tag{6.69}$$

where tube-side quantities are obtained from the solution obtained earlier in equation (6.67). There results

$$B\frac{\alpha - \beta}{\alpha\beta}\exp\left(\frac{\alpha - \beta}{\alpha\beta} L\right) + A + B\exp\left(\frac{\alpha - \beta}{\alpha\beta} L\right) - T_{s\,in} = 0 \tag{6.70}$$

After solving equations (6.68) and (6.70) for the integration constants A and B and noting from the definitions for α and β that

$$\frac{\alpha - \beta}{\beta} = U\pi d\left[\frac{1}{F_s C_{ps}} - \frac{1}{F_t C_{pt}}\right] \tag{6.71}$$

we obtain the final result

$$\frac{T_t - T_{t\,in}}{T_{s\,in} - T_{t\,in}} = \frac{F_s C_{ps}}{F_t C_{pt}}\left[\frac{\exp U\pi d\left(\dfrac{1}{F_s C_{ps}} - \dfrac{1}{F_t C_{pt}}\right) z - 1}{\exp U\pi d\left(\dfrac{1}{F_s C_{ps}} - \dfrac{1}{F_t C_{pt}}\right) L - \dfrac{F_s C_{ps}}{F_t C_{pt}}}\right] \tag{6.72}$$

A similar expression applies to the shell-side profile $T_s = f(z)$.

COMMENTS

- The two profile equations that result from our procedure are fairly formidable expressions, but they carry the advantage of having much greater versatility than the LMTD equation (6.64). Thus to calculate the outlet temperature $T_{t\,out}$ for an existing heat exchanger of area πdL and taking flow rates F_s and F_t, one simply sets $z = L$ and solves for T_t. The equation can also be used to obtain a value for U by *measuring* the tube-side outlet temperature $(T_{t\,out})$ and setting it equal to T_t, with L being the length of the experimental heat exchanger. Design problems can be accommodated by setting T_t equal to the desired or specified temperature and z equal to L, and solving for length L or heat exchanger area $U\pi dL = A$. To make these equations easier to use, plots of them have been constructed in terms of convenient nondimensional quantities that encompass flow parameters FC_p, transport resistance, area, and temperature. Such plots are found in all standard texts on heat transfer.

◆ EXAMPLE 6.7 Diffusion and Reaction in a Catalyst Pellet: The Effectiveness Factor

This example is the first in a series of three having the following features in common:

All three give rise to second-order ODEs that are the result of conductive or diffusive processes.

All three are solved by the D-operator method to yield temperature and concentration profiles.

In all three cases this primary information is converted into useful engineering indices by differentiation or integration.

The present example deals with a catalyst pellet exposed to a reactant with a constant concentration C_A^0. This reactant diffuses into the pellet in accordance with Fick's law and undergoes a first-order reaction with the rate $r_A = k_r C_A$. The pellet comes in the shape of a flat slab whose edges are assumed to be impermeable. In other words, diffusion takes place only into the two major exposed surfaces, both of which we assume to be square. A diagram of the pellet appears in Fig. 6.6A. A mass balance with respect to the reacting species A over the difference element shown in Fig. 6.6B leads to the following result.

$$\text{Rate of A in} - \text{Rate of A out} = 0$$

$$-DA\frac{dC_A}{dx}\bigg|_x - \left[\begin{array}{c} DA\dfrac{dC_A}{dx}\bigg|_{x+\Delta x} \\ +k_r C_A A\Delta x \end{array}\right] = 0 \tag{6.73}$$

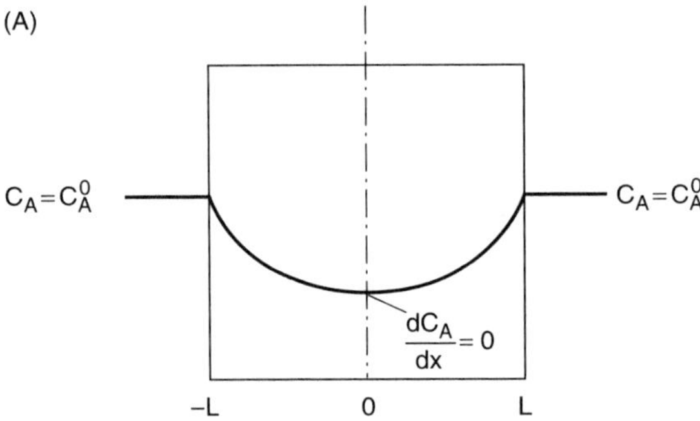

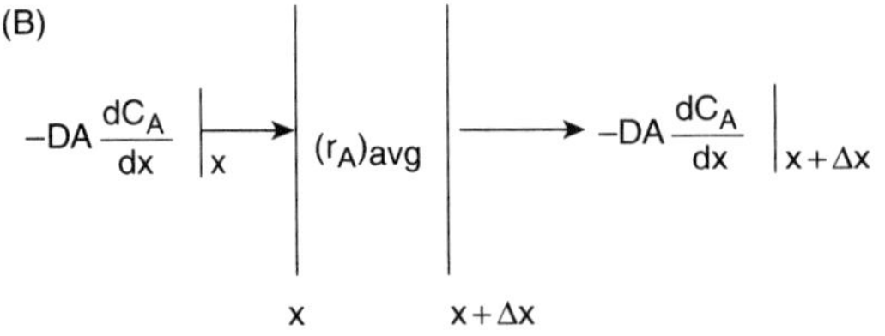

Figure 6.6 Diffusion and reaction in a catalyst pellet: (A) concentration profile and (B) the difference element.

Division by $A\Delta x$ and going to the limit $\Delta x \to 0$ yields the linear ODE

$$\frac{d^2C_A}{dx^2} - \alpha^2 C_A = 0 \tag{6.74}$$

where $\alpha^2 = k_r/D$

Applying the D-operator method (see Chapter 2), we obtain

$$(D^2 - \alpha^2)C_A = 0 \tag{6.75a}$$

with

$$D_{1,2} = \pm\alpha \tag{6.75b}$$

The solution is consequently given by (see Table 2.9):

$$C_A = A\cosh\alpha x + B\sinh\alpha x \tag{6.76}$$

Noting that

$$\sinh 0 = 0, \quad \cosh 0 = 1, \quad \frac{d}{dx}\sinh x = \cosh x, \quad \frac{d}{dx}\cosh x = \sinh x \tag{6.77}$$

we obtain from the symmetry condition $(dC_A/dx)_{x=0} = 0$

$$B = 0 \tag{6.78}$$

and from the condition $C_A(L) = C_A^0$

$$A = \frac{C_A^0}{\cosh\alpha L} \tag{6.79}$$

so that the profile becomes

$$\frac{C_A}{C_A^0} = \frac{\cosh\alpha x}{\cosh\alpha L} \tag{6.80}$$

To translate this into a quantity of engineering usefulness, we make use of the effectiveness factor E, which is defined as

$$E = \frac{\text{Reaction rate with diffusion}}{\text{Reaction rate without diffusion}} = \frac{\int r\,dV}{rV} \tag{6.81}$$

where V = volume of the catalyst pellet.

We will discuss the benefits of this concept in our comments and proceed directly to an evaluation of E. We obtain

$$E = \frac{2k_r\int_0^L C_A A\,dx}{2k_r C_A^0 LA} \tag{6.82a}$$

or

$$E = \frac{\int_0^L \cosh \alpha x \, dx}{L \cosh \alpha L} = \frac{\sinh \alpha L}{\alpha L \cosh \alpha L} \tag{6.82b}$$

Since the ratio of the hyperbolic sine to the hyperbolic cosine equals the hyperbolic tangent, we can also write

$$E = \frac{\tanh \alpha L}{\alpha L} \tag{6.83}$$

where $\alpha = (k_r/D)^{1/2}$.

Equation (6.83) tends toward 1 as $\alpha L \to 0$ and conversely goes to zero for large values of αL. In other words, for the small values of αL that correspond to small pellet size L and high diffusivities, the rates with and without diffusional effects are essentially equal, and the reaction proceeds at maximum efficiency. When α is large (i.e., the reaction proceeds at a high rate) or diffusion is very slow, the reactant concentration in the interior of the pellet will drop to low values and with it there will be a reduction in reaction rate.

Thus, under these conditions, the pellet will have a low efficiency, that is, a low value of E.

COMMENTS

- By introducing the concept of effectiveness, we have achieved two goals. First, we have established a criterion for the efficiency of the pellet, which varies between the limits 0 and 1, or 0 and 100%. High efficiencies result when the pellet is small, diffusion is fast, and the reaction rate is slow relative to the diffusion rate. Under these conditions, the interior reactant concentration will essentially equal that prevailing at the surface, and conversion to product will be at its maximum. Conversely, when the pellet dimension is large and diffusion slow in comparison to the reaction rate, reactant concentration in the interior of the pellet will drop below the surface concentration and there will consequently be a reduction in reaction rate.

  A second goal we have achieved is to eliminate distance within the pellet as a variable. Had it been necessary to incorporate it in a model for a catalytic reactor, a PDE would have resulted, since axial distance in the reactor would have been an additional independent variable to the lateral distance within the pellet. By introducing the concept of effectiveness E, we can now write

$$\text{Actual rate in the pellet } r_p = E \times \text{Intrinsic rate in the pellet } r'_p \tag{6.84}$$

  where r'_p is the rate that prevails in the absence of diffusional effects. That rate can in principle be obtained from experimental conversion measurements on finely powdered catalysts. Hence, a catalytic reactor can be modeled simply by expressing the reaction rate, which appears in the mass balance, by means of equation (6.84). For a first-order intrinsic reaction rate, we write $r'_p = k_r C_A$, and similarly for more complex reactions. Distance within the pellet is eliminated as a variable, and concentration changes occur only in the direction of flow. The model is consequently a one-dimensional one, leading to an ODE in axial distance z.

◆ EXAMPLE 6.8 The Heat Exchanger Fin

Finned tubular heat exchangers are used extensively to enhance heat transfer areas. The fins are usually attached externally to the exchanger tubes and are typically mounted in a longitudinal rectangular or circular radial mode. Transport of heat is, in the first instance, from the external (hot) fluid to the fin by convection, radially by conduction through the fin to the fin base and tubular wall, from which the heat is transferred by further convective transport to the internal (and colder) tubular fluid.

The question that arises in the design and analysis of these devices is the extent to which the heat transfer area is enhanced. If conduction within the fin were infinitely fast, the effective transfer area would be simply the sum of the fin and tubular areas: $A_{eff} = A_{fin} + A_{tube}$, and temperature throughout the fin would be equal that of the tubular wall. This corresponds to a catalyst effectiveness factor $E = 1$, which arises when there is no diffusional resistance and concentration throughout the pellet is equal to the external concentration. Because of the finite conduction rate, however, a temperature gradient develops within the fin, reducing the local convective heat transfer driving force. To assess this effect, one must first derive the temperature profile within the fin and then compute the total heat transfer rate from the external fluid by integration of the local rates over the entire fin. This is the actual rate, which can then be compared with the ideal rate that would prevail if the entire fin were at the tube wall temperature, that is, conduction were infinitely fast. From the ratio of the two, then, springs the concept of the fin efficiency E, which is completely analogous to the catalyst pellet effectiveness factor and is a measure of the degree to which one is able to approach the ideal case. For example, a value of $E = 0.8$, or 80%, means that the effective heat transfer area now becomes $A_{eff} = 0.8\,A_{fin} + A_{tube}$. Thus E can be incorporated in simple fashion into heat exchanger models or their solutions as a mere correction factor.

The fin we consider here is a rectangular longitudinal one that extends over the entire external length of the tube (Fig. 6.7). To avoid the complication of axial temperature variations, we limit consideration to an infinitesimally small length dz, along which the temperature variations can be neglected.

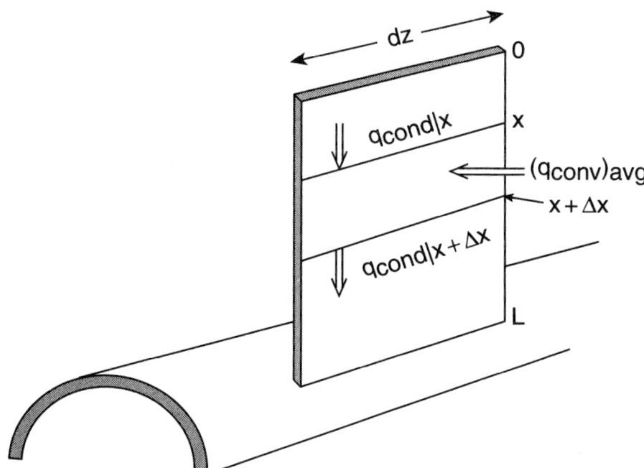

Figure 6.7 Difference element for a longitudinal heat exchanger fin.

In the model fin shown in Fig. 6.7, convective transport enters the element over Δx, while conductive heat enters and leaves at positions x and $x + \Delta x$. The energy balance over this increment then becomes

$$\text{Rate of energy in} \quad - \quad \text{Rate of energy out} \quad = 0$$

$$\begin{bmatrix} 2h_f \Delta x\, dz (T_h - T_f) \\ -kW\, dz \left.\dfrac{dT_v}{dx}\right|_x \end{bmatrix} - \left[-kW\, dz \left.\dfrac{dT_f}{dx}\right|_{x+\Delta x} \right] = 0 \tag{6.85}$$

where W = thickness of the fin.

Dividing by $kW\Delta x\, dz$ and letting $\Delta x \to 0$, we obtain

$$\frac{d^2}{dx^2}(T_h - T_f) - \frac{2h_f}{kW}(T_h - T_f) = 0 \tag{6.86}$$

This equation is a second-order linear ODE with constant coefficients, identical in form to equation (6.74) describing diffusion and reaction in a catalyst pellet. It is solved by the D-operator method as was done in Example 6.7 and yields

$$T_h - T_f = C_1 \sinh mx + C_2 \cosh mx \tag{6.87}$$

with $m^2 = 2h_f/kW$.

The two boundary conditions we need require some thought. Formulation at one end, $x = L$, is straightforward: fin temperature equals the tubular wall temperature, $T_{fL} = T_t$. At the other end, the formulation is more uncertain. It is best to argue that because of the small thickness of the wedge, heat transfer from the hot fluid at this point is negligible, so that $(dT_f/dx)_{x=0} = 0$.

From the second boundary condition we obtain

$$C_1 = 0 \tag{6.88a}$$

and from the first BC

$$C_2 = \frac{T_h - T_t}{2\, mL} \tag{6.88b}$$

so that the solution (6.87) now becomes

$$\frac{T_h - T_f}{T_k - T_t} = \frac{\cosh mx}{\cosh mL} \tag{6.89}$$

We can now proceed to evaluate the fin efficiency E, which we defined earlier as

$$E = \frac{\text{Actual heat transferred } q_a}{\text{Ideal heat transferred } q_i} \tag{6.90}$$

We note that q_a equals the heat leaving the fin at the base, so that we can write

$$q_a = -kW\, dz \left.\frac{d}{dx} T_f\right|_{x=L} = kW\, dz \left.\frac{d}{dx}(T_h - T_f)\right|_{x=L} \tag{6.91a}$$

Introducing equation (6.89) for $(T_h - T_f)$ yields

$$q_a = kW\,dz\frac{T_h - T_f}{\cosh mL}\frac{d}{dx}\cosh mx\bigg|_{x=L} \tag{6.91b}$$

or equivalently

$$q_a = kW\,dz(T_h - T_f)m\frac{\sinh mL}{\cosh mL} \tag{6.91c}$$

from which there results

$$q_a = kW\,dz(T_h - T_f)m\tanh mL \tag{6.91d}$$

Introducing this relation into the fin efficiency, equation (6.90), we obtain

$$E = \frac{kW\,dz(T_h - T_f)}{h_f 2L\,dz(T_h - T_f)}m\tanh mL \tag{6.92a}$$

or, since $2h_f/kW = mz$,

$$E = \frac{\tanh mL}{mL} \tag{6.92b}$$

which is identical in form to the expression (6.83) obtained for the catalyst pellet.

COMMENTS

- We have here an example of how two quite dissimilar physical processes, diffusion and reaction in a catalyst pellet and heat transfer in a heat exchanger fin, lead to models and solutions that are identical in form. This is not an uncommon feature of modeling and will be encountered again in Practice Problem 6.5 (Analogy Between Countercurrent Gas Absorption and Heat Exchange). A second point of note is that in both cases the primary information obtained, here the temperature profile in the fin, was converted to a more practical entity, the fin efficiency E. In the case of the catalyst pellet, that conversion was achieved by *integration* of the concentration profile, while in the present example, differentiation was used to achieve that result.

◆ EXAMPLE 6.9 Polymer Sheet Extrusion: The Uniformity Index

In this example we consider the performance of a simple device used in the extrusion of polymer sheets. The molten polymer is forced with an inlet pressure p_0 into a pipe with a lateral slit or lip extending outward from the pipe wall (Fig. 6.8A). The polymer flows into the pipe axially and exits radially through the extruder lip. A problem that arises in these devices is that the pressure driving the melt through the slit diminishes in the axial direction, causing nonuniformity in the thickness of the extruded sheet. A model is required to relate sheet thickness to the system parameters and to axial distance z so that

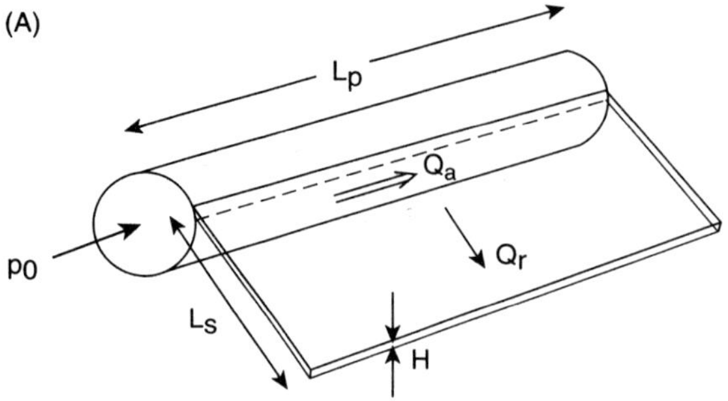

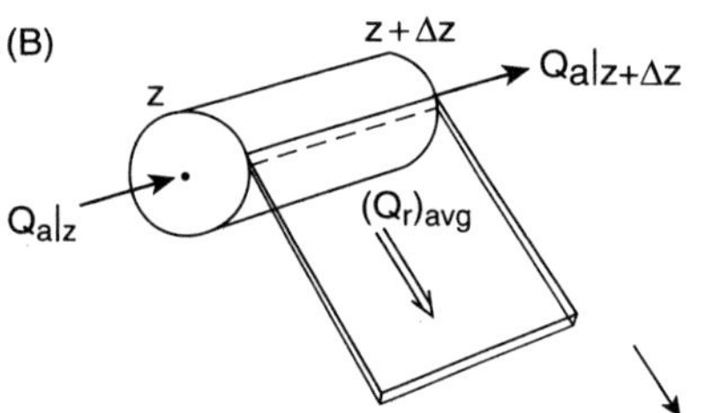

Figure 6.8 (A) Schematic diagram of a polymer sheet extruder. (B) Difference element for the mass balance.

these can be properly manipulated to ensure high uniformity. This is done by considering the ratio of radial flow at the inlet, $Q_r(0)$, and corresponding flow $Q_r(L_p)$ at the pipe end, which is sealed off. The ratio of these two values is known as the uniformity index $E = Q_r(L_p)/Q_r(0)$. Thus, for complete uniformity, $E = 1$, and for nonuniform sheets, $E < 1$. In analyzing the system, we shall assume that flow is Newtonian, so that standard flow rate–pressure drop relations may be applied. These are drawn from Example 6.2 for the axial flow (Poiseuille's equation) and from Practice Problem 6.2 for the radial flow.

We start by considering a mass balance over the difference element shown in Fig. 6.8B. Axial flow Q_a enters and leaves at the position z and $z + \Delta z$, while at the same time there is radial outflow $(Q_r)_{avg}$ over the distance Δz. From Example 6.2 we have for axial flow, suitably transformed to the conditions and symbols used here,

$$Q_a = -\frac{\pi R^4}{8\mu}\frac{dp}{dz} \tag{6.93a}$$

Drawing on the solution to Practice Problem 6.2 for the radial flow, we obtain

$$(Q_r)_{avg} = \frac{H^3 \Delta z}{12\mu}\frac{(\Delta P_r)_{avg}}{L_s} \tag{6.93b}$$

Note that for the axial flow, the pressure derivative rather than $\Delta p/L_p$ is used, since p varies nonlinearly with z because of outflow through the lip. For radial flow, the variation is linear, and we can write $dp/dr = \Delta p_r/L_s = (p - p_{ext})/L_s$. The mass balance for the increment then becomes

$$\text{Rate of flow in} - \text{Rate of flow out} = 0$$

$$[Q_a|_z] - \left[Q_a|_{z+\Delta z} + (Q_r)_{avg}\right] = 0 \tag{6.94a}$$

and after introduction of the auxiliary relations (6.93a) and (6.93b), we write

$$\frac{\pi R^4}{8\mu}\Delta\left(\frac{dp}{dz}\right) - \frac{H^3\Delta z}{12\mu}\frac{(\Delta p_r)_{avg}}{L_s} = 0 \tag{6.94b}$$

Dividing by Δz and letting $\Delta z \to 0$ yields the second-order ODE

$$\frac{d^2p}{dz^2} - m^2 p = 0 \tag{6.94c}$$

where $m^2 = \frac{2}{3}(H^3/\pi R^4 L_s)$ and p_{ext} has been set equal to zero. Solution of this equation by the D-operator method yields the same form of solution obtained in the preceding two examples:

$$p = C_1 \sinh mx + C_2 \cosh mx \tag{6.94d}$$

Two boundary conditions are required, which are as follows:

BC 1. $\left.\frac{dp}{dz}\right|_{z=L_p} = 0$ (No flow at sealed end)

BC 2. $p(0) = p_0$ (Inlet pressure $> p_{ext} = 0$)

We obtain from BC 1:

$$C_1 = -C_2 \tanh(m\,Lp) \tag{6.95}$$

and from BC 2

$$C_2 = p^0$$

since, as noted earlier, $\sinh(0) = 0$ and $\cosh(0) = 1$. The solution then becomes

$$\frac{p}{p^0} = \cosh mx - \sinh(mx)\tanh mx \tag{6.96}$$

From this expression we can immediately derive the uniformity index E, which for our purposes is given by

$$E = \frac{Q_r(L_p)}{Q_r(0)} = \frac{p(L_p)}{p^0} \tag{6.97a}$$

Drawing on equation (6.96), we obtain

$$E = \cosh(mL_p) - \sinh(mL_p)\tanh(mL_p) \tag{6.97b}$$

or equivalently

$$E = \frac{\cosh^2(mL_p) - \sinh^2(mL_p)}{\cosh(mL_p)} \tag{6.97c}$$

Noting (see standard mathematical handbooks) that $\cosh^2 x - \sinh^2 x = 1$ and $1/\cosh x = \operatorname{sech} x$, we obtain

$$E = \operatorname{sech}(mL_p) \tag{6.97d}$$

We note that the argument mL_p is usually much less than one. This enables us to expand equation (6.97b) in terms of power series, which are published in mathematical handbooks and can be truncated after the first term. The result is

$$E = 1 - \frac{(mL_p)^2}{2} = 1 - \frac{H^3 L_p}{3\pi L_s R^4} \tag{6.98}$$

To obtain a sense of parameter sensitivity, suppose that for a given configuration, E was found to be 0.95 (i.e., $H^3 L_P/3\pi L_s R^4 = 0.05$). It is now proposed to double the sheet thickness H. How will this affect sheet uniformity? We find $H^3_{new}/H^3_{old} = 8$, which translates into a new index value of $E_{new} = 0.60$. This is an unacceptable value, but it can easily be rectified by taking advantage of the strong fourth-power dependence on pipe radius. By reducing R by a mere 10%, $E_{new} = 0.60(1.1)^4 = 0.88$, which is close to being acceptable.

COMMENTS

- The first impression one gains from the formulation of the problem is that it called for a PDE model. Velocities vary in a complex way radially, axially, and in the angular direction. In addition, the geometry is discontinuous, leading to discontinuous boundary conditions, which increase the complexity of the problem. The principal tool in sidestepping these difficulties was the tacit assumption that the opening width of the lip is small in comparison to the circumference of the pipe and, consequently, the normal parabolic velocity profile remains essentially undisturbed. This enables us to use Poiseuille's equation as a close approximation of the axial flow. This is a reasonable simplification in view of the small thickness of normal polymer sheets. Its consequences, however, are quite considerable, since we can now lump the radial flow into the axial mass balance as a "rate out" term that is determined solely by the local radial pressure drop Δp_r and the geometry of the slit. Thus we have reduced the number of state variables from four (three velocities and pressure) to only one (i.e., pressure) and the number of independent variables from three to one, the axial distance.

  We note that here again the model equation (6.94c) is identical to that obtained for the catalyst pellet and the heat exchanger fin. However, the final solution, equation (6.96), differs in form from the solutions of Examples 6.7 and 6.8. This is entirely due to a difference in the boundary conditions.

◆ EXAMPLE 6.10 The Streeter–Phelps River Pollution Model: The Oxygen Sag Curve

In this classic 1925 study, probably the first attempt to model the fate of a chemical in the environment, Streeter and Phelps derived an equation that describes the oxygen profile in

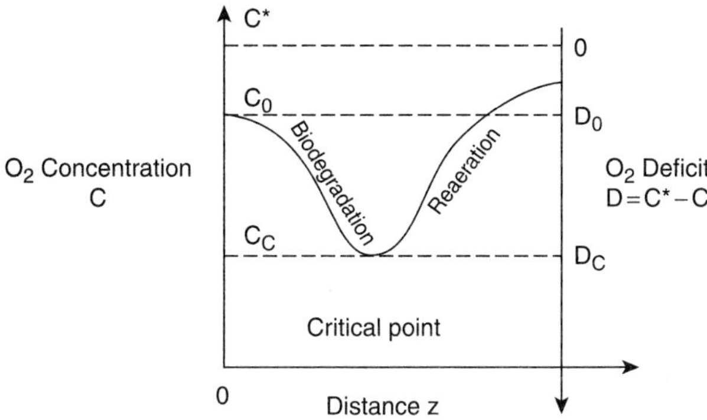

Figure 6.9 Dissolved oxygen profiles in a river with a steady influx of a pollutant: the oxygen sag curve.

a river that receives a steady influx of pollutant at some point upstream. Initially, biodegradation of the pollutant causes a decline in the dissolved oxygen concentration C or, viewed slightly differently, an increase in the oxygen deficit $D = C^* - C$, where C^* is the equilibrium solubility of oxygen in water. As the pollutant concentration L decreases through biodegradation, the decline in oxygen concentration C slows and ultimately passes through a minimum, the so-called critical point, as oxygen supply from the atmosphere replenishes and revives the river. Further "reaeration" restores the oxygen concentration to full saturation levels C^*. The concentration profile that is the result of these processes is termed *the oxygen sag curve*, shown in Fig. 6.9.

In their model, Streeter and Phelps did not consider pollutant adsorption on river sediment, an important removal mechanism. This is taken up in Practice Problem 6.9. Also neglected were the effects of runoff and respiration by algae. Thus, only biodegradation and reaeration rates need to be considered, both of which were assumed to be first order in concentration. The model equations are then as follows.

Oxygen Mass Balance

$$\text{Rate of oxygen in} \qquad - \text{Rate of oxygen out} = 0$$

$$\begin{bmatrix} QC|_z \\ +k_L a(C^* - C)_{avg} A_C \Delta z \end{bmatrix} - \begin{bmatrix} QC|_{z+\Delta z} \\ +k_r L A_C \Delta z \end{bmatrix} = 0 \tag{6.99a}$$

which becomes, after division by $A_c \Delta z$ and letting $\Delta z \to 0$,

$$v\frac{dC}{dz} + k_r L - k_L a(C^* - C) = 0 \tag{6.99b}$$

where $v = Q/A_c$ = average river velocity, k_r = pollutant degradation rate constant, k_L = oxygen mass transfer coefficient (m/s), and a = specific surface area of river (m^2/m^3 river volume).

Alternatively, we can write in terms of the oxygen deficit $D = C^* - C$

$$v\frac{dD}{dz} - k_r L + k_L a D = 0 \tag{6.99c}$$

Since we have two dependent variables D and L, a second equation is required, which is given by a pollutant mass balance:

Pollutant Mass Balance

$$\text{Rate of pollutant in} - \text{Rate of pollutant out} = 0$$

$$[QL|_z] \quad - \quad \begin{bmatrix} QL|_{z+\Delta z} \\ + k_r A_C L \Delta Z \end{bmatrix} \quad = 0 \tag{6.100a}$$

which yields the ODE

$$v\frac{dL}{dz} + k_r L = 0 \tag{6.100b}$$

The latter can be immediately integrated by separation of variables and yields

$$L = L_0 \exp\left(-\frac{k_r}{v}a\right) \tag{6.100c}$$

where L_0 = pollutant concentration at the point of influx. This intermediate result gives the pollutant concentration profile in the river. Substitution of L into equation (6.99b) then leads to an ODE in the oxygen deficit:

$$v\frac{dD}{dz} + k_L a D = k_r L_0 \exp\left(-\frac{k_r}{v}z\right) \tag{6.101}$$

This equation is of the form

$$y' + f(x)y = g(x) \tag{6.102a}$$

which has the solution (see Item 6 Table 2.6)

$$y = \exp\int -f(x)\,dx \left[\int g(x)\exp\int f(x)\,dx\,dx + K\right] \tag{6.102b}$$

Upon evaluation of the integrals we obtain

$$D = \exp\left(-\frac{k_L a}{v}z\right)\left[\frac{k_r L_0}{k_L a - k_r}\exp\left(\frac{k_L a - k_r}{v}\right)z + K\right] \tag{6.103a}$$

Using the boundary condition $D = D_0$ at $z = 0$ to evaluate the integration constant K, we finally find

$$D = \left[D_0 - \frac{k_r L_0}{k_L a - k_r}\right]\exp\left(-\frac{k_L a}{v}z\right) + \frac{k_r L_0}{k_L a - k_r}\exp\left(-\frac{k_r}{v}z\right) \tag{6.103b}$$

This is the equation for the oxygen deficit profile shown in Fig. 6.9.

Equation (6.103b) is often cast into the more convenient form

$$D = \left(D_0 - \frac{L_0}{1-f}\right)\exp\left(-\frac{fk_r}{v}z\right) + \frac{L_0}{1-f}\exp\left(-\frac{k_r}{v}z\right) \tag{6.103c}$$

where $f = k_L a/k_r$ is the so-called self-purification rate (dimensionless).

The critical point or minimum shown in Fig. 6.9 is evaluated by setting $dD/dz = 0$ in equation (6.103c). There results

Critical Distance

$$z_C = \frac{v}{k_r(f-1)} \ln f\left[1-(f-1)\frac{D_0}{L_0}\right] \tag{6.104a}$$

and

Critical Oxygen Deficit

$$D_C = \left(D_0 - \frac{L_0}{f-1}\right)\left\{f\left[1-(f-1)\frac{D_0}{L_0}\right]\right\}^{f/(1-f)} + \frac{L_0}{f-1}\left\{f\left[1-(f-1)\frac{D_0}{L_0}\right]\right\}^{f/(1-f)} \tag{6.104b}$$

These two equations are of importance because they pinpoint the region of the river's greatest vulnerability and give the extent of the oxygen deficit at that location.

COMMENTS

We have here yet another example of a highly complex process being reduced to manageable form by making suitably simplifying assumptions. Let us note what these assumptions are.

- We assumed the river to be uniformly mixed in the vertical direction; that is, there is no variation of oxygen concentration with depth. This assumption has the effect of reducing what would otherwise be a PDE to an ODE format. The assumption holds well for fast, shallow rivers, less so for deep and sluggish waterways.
- Biodegradation of the pollutant was assumed to follow a first-order rate law. Biodegradations are commonly taken to obey Monod kinetics, which at low substrate (pollutant) concentration reduce to first-order kinetics (see Example 5.3). The assumption is therefore not unreasonable.
- Finally, the oxygen transfer from the atmosphere was expressed in terms of the usual combination of a mass transfer coefficient and linear concentration driving force. This, too, is commonly accepted as a reasonable simplification.

◆ EXAMPLE 6.11 Conduction in a Thin Wire Carrying an Electric Current

In this first of two examples involving electrical concepts, we consider the temperature profile that arises when a thin wire is heated by an electrical current. We specify a thin wire, to confine temperature variations to the axial distance $T(x)$. In other words, temperature in the radial direction is assumed to be uniform. We further specify that the wire loses heat to the surroundings at the usual convective rate $q = hA(T - T_a)$,where T_a = ambient temperature.

Before proceeding to set up the model, we briefly address two electrical expressions involved in the problem. The first deals with the heat generated by an electrical current i. We draw for this purpose on the elementary relation which stating that the power generated by an electrical current P (J/s) equals the product of voltage V and current i.

$$P = iV \tag{6.105a}$$

which becomes, after the introduction of Ohm's law, $V = i/R_e$,

$$P = i^2 R_e \tag{6.105b}$$

where R_e is the electrical resistance of the entire wire length. Since we anticipate that an energy balance over the increment Δx will have to be set up to describe the system, we require a second expression relating R_e to length L or distance x. This is given by the fundamental definition of resistance,

$$R_e = \frac{SL}{\pi R^2} \tag{6.105c}$$

where S is the specific resistivity of the material in units of ohm-meters (Ω m).

With the ground thus prepared, we can proceed with the setting up of the energy balance. We write

$$\text{Rate of energy in} \quad - \quad \text{Rate of energy out} \quad = 0$$

$$\left[\begin{matrix} k\pi R^2 \left.\dfrac{dT}{dx}\right|_{x+\Delta x} \\ + \dfrac{Si^2}{\pi R^2}\Delta x \end{matrix} \right] - \left[\begin{matrix} k\pi R^2 \left.\dfrac{dT}{dx}\right|_{x} \\ + h2\pi R\Delta x(T - T_a) \end{matrix} \right] = 0 \tag{6.106a}$$

Note that conduction is in the opposite direction to x, hence takes a positive sign.

Dividing by $\pi R^2 \Delta x$ and letting $\Delta x \to 0$, we obtain

$$k\frac{d^2T}{dx^2} + \frac{Si^2}{(\pi R^2)^2} - \frac{2h}{R}(T - T_a) = 0 \tag{6.106b}$$

or in the equivalent form

$$\frac{d^2\theta}{dx} - \alpha^2\theta = -\beta \tag{6.106c}$$

where $\theta = T - T_a, \alpha^2 = 2h/kR$ and $\beta = Si^2/h(\pi R^2)^2$. This is a second-order ODE with constant coefficients, which can be solved by the D-operator method but will require as well the evaluation of a particular integral because of the nonhomogeneous term $-\beta$. For the homogeneous part of the solution, termed the complementary function θ_C, we obtain

$$(D^2 - \alpha^2)\theta_C = 0 \tag{6.106d}$$

and hence (see Table 2.9)

$$\theta_C = A \sinh \alpha x + B \cosh \alpha x \tag{6.106e}$$

To evaluate the particular integral, we draw on Table 2.10 and obtain

$$\theta_P = \frac{\beta}{\alpha} \tag{6.106f}$$

so that the full solution now becomes

$$\theta = \theta_C + \theta_P = A \sinh \alpha x + B \cosh \alpha x + \frac{\beta}{\alpha} \tag{6.106g}$$

The boundary conditions we need for the evaluation of A and B are somewhat ambiguous. We argue that since the temperature will steadily rise in the x direction, the temperature at the inlet can be assumed to equal that of the surroundings [i.e., $T(0) = T_a$], hence $\theta(0) = 0$. At $x = L$, on the other hand, the argument must be that heat losses are negligible; hence the gradient vanishes at this point, $(d\theta/dx)_{x=L} = 0$. The integration constants then become

From BC 1

$$0 = A \sinh(0) + B \cosh 0 + \frac{\beta}{\alpha} \tag{6.107a}$$

and consequently

$$B = -\frac{\beta}{\alpha} \tag{6.107b}$$

From BC 2

$$\left.\frac{d\theta}{dx}\right|_{x=L} = A \cosh \alpha L + B \sinh \alpha L = 0 \tag{6.107c}$$

and consequently

$$A = \frac{\beta}{\alpha} \tanh \alpha L \tag{6.107d}$$

Hence the full solution is given by the expression

$$\theta = T - T_a = \frac{\beta}{\alpha}(\sinh \alpha x \tanh \alpha L - \cosh \alpha x + 1) \tag{6.107e}$$

This is the temperature distribution in the wire in question.

COMMENTS

- The temperature distribution, while of intrinsic interest, can be cast into a more useful form by using it to calculate the total heat transmitted to the surroundings. This problem is left to the exercises (Practice Problem 6.10). We further note that when the wire temperature rises to substantial levels, say about 150 °C, radiant as well as convective heat losses to the surroundings must be taken into account. The rate of radiant heat transfer has a fourth-power dependence on temperature and is given by

$$q_r = h_r A(T^4 - T_a^4) \tag{6.108}$$

The model equation (6.106b) then becomes

$$k\frac{d^2T}{dx^2} + \frac{Si^2}{(\pi R^2)^2} - \frac{2h_C}{R}(T - T_a) - \frac{2h_r}{R}(T^4 - T_a^4) = 0 \tag{6.109}$$

The expression is now nonlinear because of the radiation term and can no longer be solved by the D-operator method, or indeed by any other analytical procedure. Thus we will be obliged to resort to numerical methods. We note that since the associated boundary conditions are specified at two different locations, special procedures must be implemented for the numerical solution of equation (6.109). These are now available in standard numerical packages, but reaching for the correct one evidently requires an awareness by the user that it is a boundary value problem that is to be solved.

◆ EXAMPLE 6.12 Electrical Potential Due to a Charged Disk

In our second example of an electrical nature, we consider a charged disk, shown in Fig. 6.10, carrying a charge density σ (C/m^2). We set as our task the derivation of an expression for the potential, or voltage at a point P located on the axis of the disk. As a preamble to this task, we define potential, then relate it to the charge q and its associated electrical field **E**. The latter was briefly addressed in Example 1.4, and we recapitulate the results for convenience.

The electrical field **E** is a vector and is defined as the force per unit positive charge q_0 when that charge is placed in an electrical field as a result of another charge q. We have

$$\mathbf{E} = \frac{\mathbf{F}}{q_0} \tag{1.30}$$

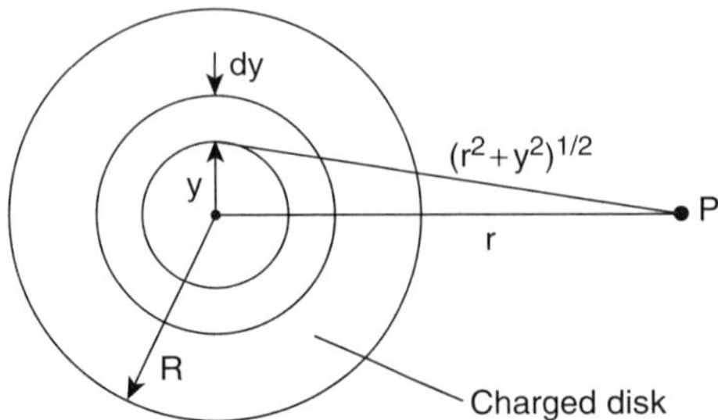

Figure 6.10 Potential due to a charged disk.

If we introduce Coulomb's law, equation (1.9), to express **E** in terms of q, we obtain for the magnitude of the electrical field

$$\mathbf{E} = \frac{1}{4\pi\varepsilon_0}\frac{q}{r^2} \tag{1.31}$$

We next turn to the concept of potential V, which is defined as the *work* required to move the test charge from infinity to some position r. Since work done on a system is taken to be a negative quantity, we obtain

$$W = V_B - V_A = -\int_{r_A}^{r_B} E\,dr \tag{6.110a}$$

$$V_B - V_A = -\int_{r_A}^{r_B} \frac{1}{4\pi\varepsilon_0}\frac{q}{r^2}\,dr = \frac{1}{4\pi\varepsilon_0}\left(\frac{1}{r_B} - \frac{1}{r_A}\right) \tag{6.110b}$$

Letting $r_A \to \infty$ and choosing $V_A = 0$ at this position, we obtain

$$V = \frac{1}{4\pi\varepsilon_0}\frac{q}{r} \tag{6.110c}$$

This is the desired relation between potential and charge, which we need as a starting tool.

We now turn to the principal task of determining the potential at the point P shown in Fig. 6.10. The procedure is to start with the potential due to the differential element of thickness dy and integrate it over the entire disk. The differential charge dq contained in this element is given by

$$dq = \sigma 2\pi y\,dy \tag{6.111}$$

Noting from Fig. 6.10 that the distance from the differential element to the point P is given by $(r^2 + y^2)^{1/2}$, we obtain from equations (6.110c) and (6.111)

$$dV = \frac{1}{4\pi\varepsilon_0}\frac{dq}{r} = \frac{1}{4\pi\varepsilon_0}\frac{\sigma 2\pi y\,dy}{(r^2 + y^2)^{1/2}} \tag{6.112a}$$

$$V = \int dV = \frac{\sigma}{2\varepsilon_0}\int_0^R (r^2 + y^2)^{-1/2} y\,dy \tag{6.112b}$$

Evaluation of the integral is by writing $y\,dy = \frac{1}{2}dy^2$, so that equation (6.112b) becomes

$$V = \frac{\sigma}{2\varepsilon_0}(r^2 + y^2)^{1/2}\Big|_0^R = \frac{\sigma}{2\varepsilon_0}\left[(R^2 + r^2)^{1/2} - r\right] \tag{6.112c}$$

This expression gives us the potential distribution V(r) along the axis of the disk.

COMMENTS

- Before leaving the expression (6.112c), we undertake what we referred to in Chapter 1 as "closure": that is, we analyze the result for potentially interesting features. This includes in particular an examination of limiting cases. Let us do this by considering the two limits $r \to 0$ and $r \to \infty$. For $r \to 0$ we obtain

$$V = \frac{\sigma}{2\varepsilon_0} R \tag{6.113a}$$

which indicates that the potential on the disk itself varies directly with the charge density σ and the disk radius R. This makes eminent physical sense.

To investigate the second case, we expand the square root in equation (6.112c) by means of the binomial theorem and obtain

$$(R^2 + r^2)^{1/2} = r\left(1 + \frac{R^2}{r^2}\right)^{1/2} = r\left(1 + \frac{1}{2}\frac{R^2}{r^2} - \frac{1}{8}\frac{R^4}{r^4} + \cdots\right) \tag{6.113b}$$

Since $R^2/r^2 \ll 1$ as $r \to \infty$, we can truncate the series after the second term and obtain for V

$$V = \frac{\sigma}{2\varepsilon_0}\left(r + \frac{R^2}{2r} - r\right) = \frac{\sigma\pi R^2}{4\pi\varepsilon_0 r} \tag{6.114a}$$

But $\sigma\pi R^2$ is the total charge q on the disk. Hence

$$V_{r\to\infty} = \frac{1}{4\pi\varepsilon_0}\frac{q}{r} \tag{6.114b}$$

which is identical to equation (6.110c) for the potential of a point charge. In other words, at distances sufficiently far from the charged disk, that disk acts like a point charge. This is again in agreement with our expectations on physical grounds.

◆ EXAMPLE 6.13 Production of Silicon Crystals: Getting Lost and Staging a Recovery

An important step in the production of semiconductor devices is the growth of single silicon crystals. In the Czochralski grower (Fig. 6.11), which is the most commonly used device, the charge material is placed in a crucible and melted. A seed crystal is attached to a vertical pull rod, lowered until it touches the melt, and then raised slowly so that crystallization proceeds from the seed crystal. The crystal is rotated as it is pulled, and the crystal diameter is controlled by adjusting pull rate and heat input to the melt.

An important parameter to be established in the preliminary design of these devices is the speed with which the crystal is to be pulled. Evidently, enough time must be allowed for the heat liberated by the solidification process ΔH_s to be conducted away through the solidified crystal. If this is not done, the crystal will melt and break off. The question is then: How slow should the pull rate be to prevent this from happening?

Since the temperature T_s is distributed in the solidified crystal, a first temptation is to make a differential energy balance within the solid. This balance would include energy (enthalpy) entering and leaving the differential element at a velocity v, but it results in too

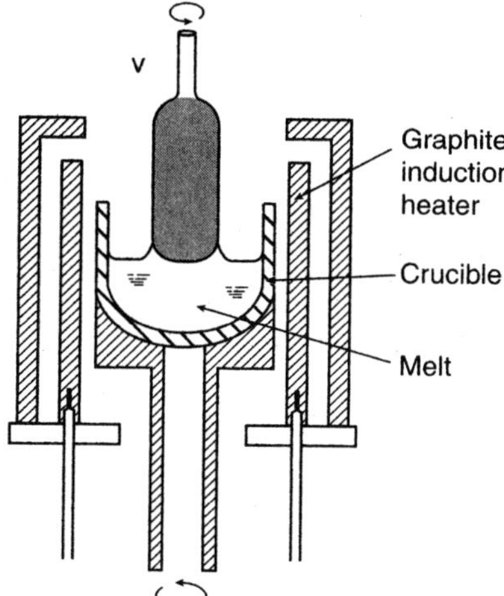

Figure 6.11 Production of silicon crystals in the Czochralski grower.

many unknowns (T_s and v) and does not include the enthalpy of solidification, which is clearly of importance here.

A more promising approach is to make an energy balance at a location where ΔH_s is liberated, that is, around the liquid–solid interface, which we denote by $x = 0$. Setting $x > 0$ in the solid, and $x < 0$ in the liquid, we obtain the energy balance

$$\text{Rate of energy in} \quad - \quad \text{Rate of energy out} = 0$$

$$\left[k_L A \frac{dT_L}{dx} + \rho v A H_L\right] - \left[k_s A \frac{dT_s}{dx} + \rho v A H_s\right] = 0 \tag{6.115a}$$

and consequently

$$k_L \frac{dT_L}{dx} + \rho v \Delta H_s = k_s \frac{dT_s}{dx} \tag{6.115b}$$

where ΔH_s =heat of solidification. This equation in essence states that the heat arriving by conduction from the liquid side and the heat liberated by solidification must equal the heat conducted away from the interface into the solid.

Solving for v yields

$$v = \frac{k_s(dT_s/dx_s) - k_L(dT_L/dx)}{\rho \Delta H_f} \tag{6.115c}$$

from which the velocity of pull can in principle be established.

Here we find ourselves at a seeming impasse, since both derivatives in this expression are unknown. We can, however, resolve the problem by making use of the stipulation, prescribed by industrial practice, that dT_s/dx not exceed 5 K/mm.

This requirement must be met to prevent thermally induced dislocations the silicon crystal structure. Furthermore, for v to be positive, we must have from equation (6.115c)

$$K_s\left(\frac{dT_s}{dx}\right) > k_L\left(\frac{dT_L}{dx}\right) \tag{6.115d}$$

These two conditions enable us to arrive at an appraisal of the velocity v. We can make quantitative estimates by drawing on the following data for silicon:

$$K_s = 31\,\mathrm{J\,s^{-1}m^{-1}K^{-1}}$$

$$K_L = 50\,\mathrm{J\,s^{-1}m^{-1}K^{-1}}$$

$$\Delta H_s = 1.8 \times 10^6\,\mathrm{J/kg}$$

$$\rho = 2300\,\mathrm{kg/m^3}$$

It follows that for v to be positive we must have

$$\frac{dT_L}{dx} < \frac{31}{50}5 = 3.1\,\mathrm{K/mm} \tag{6.115e}$$

Let us use a safe value of 2 K/mm for dT_L/dx. It then follows from equation (6.115c) that

$$v = \frac{(31 \times 5000 - 50 \times 2000)}{2300 \times 1.8 \times 10^{-6}} = 13.3 \times 10^{-6}\,\mathrm{m/s} = 13.3\,\mu\mathrm{m/s} \tag{6.115f}$$

This is the upper velocity limit required to maintain a solid crystal. Commercial silicon crystals are in fact grown at the rate of 14 to 20 μm/s, in agreement with our findings.

COMMENTS

- The reader may feel at this point that after being hopelessly stuck, we were rescued only by invoking an unsuspected and obscure industrial requirement. It is nevertheless a real one, which must be met in the production of high quality crystals. The surprise here is that we were able to accommodate this requirement, as well as the complexities of the system in the single equation (6.115c). The result we obtained is an important one because it allows the manufacturer to operate at maximum production rates without jeopardizing the quality of the crystals.

  The somewhat unstructured and seemingly inspired approach we have taken here may not sit well with all readers. It is, however, a common occurrence in modeling, albeit one that takes some getting used to.

PRACTICE PROBLEMS

6.1 Hypsometric formulas revisited

(a) Derive a hypsometric formula based on the assumption that the adiabatic p-ρ relation given by equation (6.3) applies.

(b) Compare the result with that obtained from equation (6.6) for an altitude of 3000 m, using a value of $\gamma = 1.4$.

ANSWERS (a) $z_2 - z_1 = \frac{P_1\gamma}{\rho_1 g(\gamma - 1)}\left[1 - \left(\frac{P_2}{P_1}\right)^{\frac{\gamma-1}{\gamma}}\right]$

(b) 73 kPa

6.2 **Laminar flow in a slit** Derive the expression for flow rate Q as a function of pressure drop ΔP for laminar flow between two parallel planes held a distance d apart and having a width W.

ANSWER $Q = \frac{1}{12}\frac{\Delta p d^3 W}{\mu \Delta z}$

6.3 **Optimum insulation thickness** As noted in Example 6.4, pipes carrying hot fluids are usually insulated to prevent heat loss to the surroundings and an attendant drop in energy content of the fluid. In the case of steam, this would result in undesirable condensation. Pipes carrying cold fluids, for example, cryogenic liquids such as liquefied natural gas (CH_4) or oxygen, are likewise insulated, in this instance to prevent or minimize evaporation by heat flow from the surroundings. A superficial look at the process would seem to indicate that the thicker the insulation (i.e., the larger the external radius r_0), the less the heat is transmitted through it. A closer inspection of the relevant relations, in this instance equation (6.31), indicates that an increase in the insulation thickness yields diminishing returns and ultimately becomes counterproductive. This is because heat loss is proportional to the external area $2\pi r_0 L$, while resistance to it varies inversely with $\ln(r_0/r_i)$. At low values of r_0 the logarithmic term in (6.31) predominates and heat flow decreases with increasing thickness. With further increase in r_0, the area term $2\pi r_0 L$ becomes more dominant and the heat flow ultimately passes through a minimum. There consequently exists an optimum outer radius, which will minimize heat flow through the insulation. Show that this radius is given by

$$r_{opt} = \frac{k}{h_0}$$

(*Hint*: Neglect internal film resistance and the resistance due to the pipe itself.)

6.4 **A biomedical problem: the Krogh cylinder** Nutrients from flowing blood pass radially into the surrounding tissue, diffusing through it and at the same time undergoing metabolic consumption at a constant rate R (mol/m^3 s). The tissue region is represented by a cylindrical shell, the outer surface of which is impermeable while the inner surface is at a constant concentration. Derive the concentration distribution in the tissue, C(r), (*Hint*: Use the methods developed in Example 6.4.)

ANSWER $C - C_0 = \frac{R}{D}\left(\frac{r^2 - r_i^2}{2} - \frac{r_0^2}{2}\ln\frac{r}{r_i}\right)$

6.5 **Analogy between countercurrent gas absorption and heat exchange** In gas absorption, or gas scrubbing as it is also termed, a gas stream containing a valuable or objectionable substance is contacted with a liquid solvent in which that substance is preferentially absorbed. The operation is frequently carried out in upright cylindrical vessels filled with formed ceramic or plastic particles, called packing, typically having a nominal size of 1 to 3 cm and a high surface area/volume ratio. Solvent is introduced at the top of the vessel and trickles down over the packing, where it comes into intimate contact with the gas. The gas stream to be purified is introduced at the bottom and exits at the top, in countercurrent flow with the liquid stream.

The substance to be recovered is absorbed in the liquid at a rate given in kilograms per second by

$$N = K_Y a(Y - Y^*)V$$

where a is the surface area of the packing per unit volume occupied, K_Y is an overall mass transfer coefficient, Y is the concentration of the substance in the gas phase, and Y^* the concentration, which is hypothetically in equilibrium with the concentration in the liquid phase X. In the present problem we assume the equilibrium relation to be linear, that is,

$$Y^* = mX$$

We came across a similar expression in Example 5.7, dealing with the controlled release of a drug across a two-film resistance, the difference there being that one of the films varied in thickness with time.

In the present case, film thickness, hence $K_Y a$, is assumed to be constant both with time and with distance along the vessel. Gas and liquid flow rates G and S are given in units of kilograms per second–meter squared of cross-sectional area and are taken to be constant because of the low concentration of the substance or solute to be recovered. Concentration units are in kilograms of solute per kilograms of gas or liquid.

(a) Set up the two differential mass balances pertinent to the system.

(b) Without solving the equations, show how the solution for the heat exchanger, equation (6.72), can be transformed into a corresponding expression for countercurrent gas scrubbing. (*Hint*: Compare the two sets of differential balances.)

6.6 Heat effects in a catalyst pellet Derive the model equations whose solution would yield the temperature profile in a spherical catalyst pellet undergoing an exothermic first-order reaction.

6.7 The radial heat exchanger fin Consider a doughnut-shaped fin that extends radially outward from the heat exchanger tube, with inner and outer radii r_i and r_o. Derive the model equation and boundary conditions for this case, but do not attempt a solution.

6.8 The critical point in the oxygen sag curve Derive equations (6.104a) and (6.104b).

6.9 Contamination of a river bed Suppose a contaminant of concentration Y_0 (kg/kg water) enters the river at a point $z = 0$ and we wish to calculate the length of river bed L, which has been contaminated after a time t. This is a fairly complex problem, calling in principle for the use of two PDEs in the two concentrations (water and sediment) and the two independent variables (t and z). It can be reduced to manageable proportions by

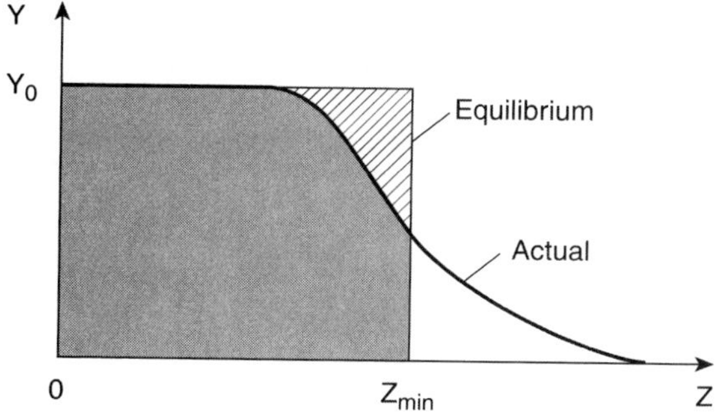

Figure P6.9 River bed contamination.

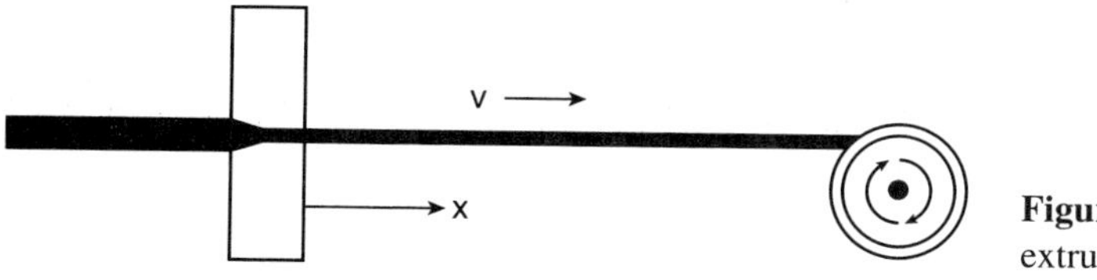

Figure P6.10 Wire extrusion and winding.

assuming the contaminant to be everywhere in equilibrium with the sediment. For trace contaminants, the equilibrium is a linear one, $q = HY$, where q is the sediment phase concentration (kg/kg sediment) and H is the Henry's constant, also termed the partition coefficient. The distance of contamination we obtain in this fashion is a minimum one, z_{min}, corresponding to a rectangular propagation front that results in this case (see Fig. P6.9). In an actual process, the mass transfer resistance between the water and sediment phases slows down the transfer of contaminant and a drawn-out, S-shaped concentration profile results. Nevertheless, z_{min} is a highly useful parameter to have because it informs us of the minimum damage to be expected. Derive an expression for z_{min} as a function of river velocity v, elapsed time t, and pertinent properties of the sediment and water phases. (*Hint*: Make a cumulative contaminant balance.)

6.10 **Heat loss from an electrically heated wire** Derive an expression for the total heat q_{tot} transmitted to the surroundings by an electrically heated wire.

ANSWER $q_{tot} = h2\pi R\frac{\beta}{\alpha}(L - \alpha \tanh \alpha L)$

6.11 **Another wire problem: extrusion and subsequent cooling by the atmosphere** Consider a wire being extruded at a fixed temperature T_0 and velocity v through a die (Fig. P6.10). The wire then passes through air for a long distance before being rolled onto large spools. What distance L will allow sufficient cooling to take place for the wire to be safely wound onto the drums?

Thus the task is to set up and solve a model that will yield the wire temperature as a function of distance. The solution can then be used to establish the distance L required to reduce the temperature to a prescribed safe level.

CHAPTER 7

Some Simple Networks

We use the term "network modeling" to describe actual networks (e.g., electrical circuits), as well as systems that are linear in configuration (i.e., devoid of branches and loops) but lead to sets of simultaneous equations. The networks thus defined, which we consider here, are thermal, chemical, hydraulic, mechanical, and electrical. Some of them occur more than once, and the mathematical tools we use are matrix methods and Laplace transformations as well as standard ODE solution procedures.

Example 7.1 considers a network made up of a stirred tank and a heat exchanger through which the contents of the tank are circulated. The system is analogous to that of the artificial kidney, that is to say the process of dialysis. In Example 7.2 we analyze the behavior of a radioactive decay series using the Laplace transformation. Hydraulic networks are taken up in Example 7.3, and since the systems are taken to be at steady state, the solution method is strictly algebraic. In Example 7.4 we turn to the first of two electrical networks for which we apply the Laplace transformation. Some difficulties arise in the solution of this problem, and the reader is introduced to the notion of "going around the brick wall" rather than confronting the difficulty head on. Example 7.5 considers a mechanical system consisting of vibrating masses, which are again analyzed by means of the Laplace transformation.

Examples 7.6 to 7.9 all use matrix methods as the analytical tool. The first two of these consider simultaneous linear algebraic equations in the context of chemical reactions and plant operations. The former leads to an overdefined system, which is reduced by means of row operations to yield a unique solution. Example 7.7, on the other hand, is neither over- nor underdefined but nevertheless fails to yield a unique solution. The difficulty is overcome by suitably revising the underlying model. Examples 7.8 and 7.9 use matrix multiplication to describe, respectively manufacturing operations and another electrical network.

Finally, Example 7.10 uses yet another electrical network to mimic the transport of carbon dioxide in a leaf. Although an escalation in complexity might have been anticipated, this does not in fact materialize, and the treatment is kept throughout at an elementary and understandable level.

Additional examples dealing with networks and simultaneous equations appear in Chapter 8.

◆ EXAMPLE 7.1 A Thermal Network: External Heating of a Stirred Tank and Analogy to the Artificial Kidney (Dialysis)

It frequently happens in the process industries that the contents of a tank must be either heated or cooled to achieve a certain desirable result. The reasons for doing this are manifold and varied. In the case of a stirred tank reactor, for example, it is common to cool the contents in the case of exothermic (heat-producing) reactions or to provide heat when the reaction is endothermic (heat consuming). Heating may also be used for the simple reason that reactions proceed faster at higher temperature. In Example 3.11, we emphasized that these processes frequently give rise to multiple steady states and pointed out that the analyst must then, by proper manipulation of the heat transfer rate, endeavor to achieve the one steady state that leads to optimum conversion.

Heating may also be used to maintain a substance in its liquid form or to promote the rate of dissolution of a solid. Conversely, when the process involves the crystallization of a substance from its mother liquor, the common practice is to cool the solution to reduce the solubility and induce crystal formation. In yet another application, heating may be used to bring a solution to a boil, while cooling can be applied to bring about the condensation of vapors. These two processes are common adjuncts in the operation of distillation columns.

The heating of the contents of a tank may be brought about in several ways. One can use electrical heaters, which are usually immersed in the tank, or one can choose steam as the heating medium, which is more commonplace. We have sketched three different ways in which this can be done in Fig. 7.1. In case A, steam is passed through an internal coil, which is immersed in the tank. Here, as well as in the other two depicted cases, steam is assumed to condense isothermally at a temperature T_S. In case B, the steam is introduced into an annular space built around the tank. Such jacketed vessels find frequent application. In case C, the entire heat transfer process takes place external to the tank in a specially provided heat exchanger. That exchanger is usually of the shell-and-tube type, which we encountered in Example 6.6. Liquid from the tank is continuously removed from the tank and pumped through the heat exchanger, whereupon the heated liquid is returned to the tank, where we assume that it is instantly and uniformly dispersed throughout the tank contents. This closed-loop operation represents a mini-thermal network, which we wish to address here.

A first examination of this system reveals that we are dealing with two units: one, the tank, is a compartment with assumed uniform contents, while the other is a distributed unit in which the state variable, temperature, varies with distance. A second aspect to the heat exchanger, however, complicates matters. This is because the tubular temperature varies not only with distance, but also with time. To model this situation we would have to resort to a partial differential equation, which would be coupled to the unsteady ODE describing the tank temperature. There is a way of bypassing this difficulty. For this purpose we draw on a concept we established in Example 5.5, that of the quasi-steady state. To apply this concept to the present system, we argue as follows. If the tank contents are reasonably large and the circulation rate moderate, it will take an appreciable length of time to heat the liquid to the desired level, say 60 to 70 °C, starting from room temperature.

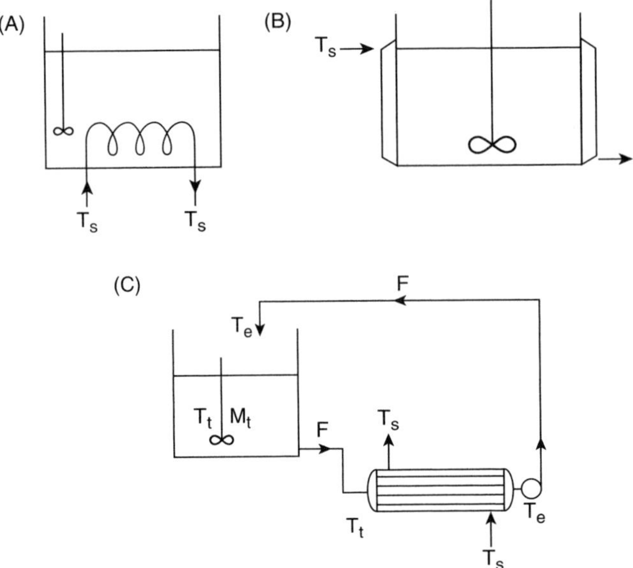

Figure 7.1 Modes of steam heating a tank: (A) coil, (B) jacketed, and (C) external heat exchanger.

It will certainly be of the order of many minutes, not seconds. The residence time of the liquid in the heat exchanger, on the other hand, *will* be of the order of seconds, since the movement of liquids in a typical plant is generally of the order of 0.5 to 1 m/s. It follows that the temperature of the stream entering the exchanger will change only very slowly with time, allowing one to assume that the operation of the heat exchanger will at any instant be at a steady state. The energy balance we have to apply to the heat exchanger is then of the type we established in Example 6.6, with the difference that the shell-side temperature here is constant (isothermally condensing steam). We write for a difference element

Heat Exchanger Energy Balance

$$\begin{array}{ccc} \text{Rate of energy in} - \text{Rate of energy out} = 0 \\ [H|_z + q_{avg}] \quad - \quad [H|_{z+\Delta z}] \quad = 0 \end{array} \tag{7.1}$$

Drawing on the same auxiliary relations used in Example 6.6, equation (7.1) becomes

$$FC_p(T - T_{ref})z + Un\pi d(T_s - T)_{avg}\Delta z - FC_p(T - T_{ref})_{z+\Delta z} = 0 \tag{7.2}$$

where d, n are the diameter and number of tubes. Dividing by Δz and letting $\Delta z \to 0$ then yields the ODE

$$FC_p \frac{dT_t}{dz} - Un\pi d(T_s - T_t) = 0 \tag{7.3a}$$

or equivalently

$$FC_p \frac{d(T_s - T_t)}{dz} - Un\pi d(T_s - T_t) = 0 \tag{7.3b}$$

This expression can be immediately integrated by separation of variables, giving

$$\ln \frac{T_s - T_t}{T_s - T_e} = \frac{UA}{FC_p} \tag{7.4a}$$

where $A = n\pi dL$ = total heat transfer area of the exchanger, and T_e = temperature of the liquid leaving the exchanger. Solving for T_e then yields

$$T_e = T_s - (T_s - T_t)\exp\left(-\frac{UA}{FC_p}\right) \tag{7.4b}$$

This expression is used to derive the desired variation of tank temperature T_t with time. Using an envelope around the entire tank, we write

Tank Energy Balance

$$\begin{array}{ccccc} \text{Rate of energy in} & - & \text{Rate of energy out} & = & \begin{array}{c}\text{Rate of change of}\\ \text{energy contents}\end{array} \\ FC_p(T_e - T_r) & - & FC_p(T_t - T_r) & = & mC_p \dfrac{d(T_t - T_r)}{dt} \end{array} \tag{7.5a}$$

which becomes

$$F(T_e - T_t) = m\frac{dT_t}{dt} \tag{7.5b}$$

Substituting equation (7.4b) into (7.5b) and rearranging leads to

$$F(T_s - T_t)\left[1 - \exp\left(-\frac{UA}{FC_p}\right)\right] = m\frac{dT_t}{dt} \tag{7.6a}$$

or equivalently

$$F(T_s - T_t)\left[1 - \exp\left(-\frac{UA}{FC_p}\right)\right] = -m\frac{d(T_s - T_t)}{dt} \tag{7.6b}$$

This is again easily integrated by separation of variables, yielding

$$\ln\frac{T_s - T_t^0}{T_s - T_t} = \frac{F}{m}\left[1 - \exp\left(-\frac{UA}{FC_p}\right)\right]t \tag{7.7a}$$

Solving for t, we obtain the following equation:

$$t = \frac{m}{F}\left[1 - \exp\left(-\frac{UA}{FC_p}\right)\right]^{-1} \ln\frac{T_s - T_t^0}{T_s - T_t} \tag{7.7b}$$

This expression gives us the time necessary to heat the tank contents from an initial temperature T_t^0 to some final value T_t.

An interesting application of this model in the realm of mass transfer occurs in the operation of so-called hemodialyzers or "artificial kidneys" used to remove accumulated toxic substances from blood. Here the human body takes the place of the tank, with a typical blood volume of 5 L. The blood is continuously circulated through a bundle of hollow fibres that remove the toxins by permeation through the tubular walls into the shell-side saline solution with essentially zero toxin concentration. A typical membrane area is 1 m^2, the permeability K, which takes the place of the heat transfer coefficient, is the order 10^{-4} cm/s, and blood flow rate $Q \cong 5$ cm^3/s. It is left to the ingenuity of the reader to transform equation 7.7b in such a way as to yield the time required to reduce the toxin concentration to, say, 10% of the incoming feed.

COMMENTS

- Use of the quasi-steady-state assumption was the crucial step in arriving at the simple analytical expression (7.7b). Without it, the model would have led to a partial differential equation for the heat exchanger coupled to the ODE resulting from the energy balance around the tank. Although analytical solutions for the system can be derived by the method of Laplace transformation, the task is a fairly onerous one that does not result in sufficient refinement to justify its use.

◆ EXAMPLE 7.2 A Chemical Reaction Network: The Radioactive Decay Series

The decay of radioactive elements is the classic example of a reaction that is accurately described by a first-order rate law. There are two types of radioactive decay: natural and artificial. In both, the disintegration occurs through the emission of α particles, which are helium nuclei, and β particles, which are electrons. In the former, decay occurs spontaneously without outside intervention, while in the case of artificial radioactivity, decay is induced by bombardment with elementary particles, principally neutrons, which are not repelled by the nucleus. There are three naturally occurring decay series, that of uranium-238, and those of the elements actinium and thorium. The uranium decay is composed of 14 steps leading to the stable element lead.

We shall consider a simpler series composed of three steps and endeavor to derive the time dependence of the concentration of each species. This simple network is represented by the following scheme:

$$N_1 \xrightarrow{k_1} N_2 \xrightarrow{k_2} N_3 \xrightarrow{k_3} N_4$$

where the N_i are the number of atoms of element i, and k_i are the associated first-order rate constants. The reaction rates are then represented by

$$\frac{dN_1}{dt} = -k_1 N_1 \tag{7.8a}$$

$$\frac{dN_2}{dt} = -k_2 N_2 + k_1 N_1 \tag{7.8b}$$

$$\frac{dN_3}{dt} = -k_3 N_3 + k_2 N_2 \tag{7.8c}$$

$$\frac{dN_4}{dt} = -k_3 N_3 \tag{7.8d}$$

Equation (7.8a) can be immediately integrated by elementary methods.

The result is substituted in equation (7b) and the process continued down the line. This results in equations (7.8b) to (7.8d) becoming first-order *nonhomogeneous* equations in time. Such equations can in principle be solved by the D-operator method but would in each case require the evaluation of a particular integral, which is somewhat onerous. We draw instead on the alternative method of the Laplace transformation, which is particularly well suited for systems of this type and can be applied in a straightforward manner. Let us see how this works out for the four ODEs in question.

We postulate initial conditions of $N_1(0) = N_1^0$ and $N_2(0) = N_3(0) = N_4(0) = 0$; that is, only the first species is assumed to be present. The procedure we use is to transform all four ODEs, and then apply the same common inversion method, that of the Heaviside expansion, item 8 of Table 2.11. We obtain in the first instance for the first ODE:

$$\frac{dN_1}{dt} = -k_a N_1 \tag{7.9a}$$

Transforming both sides leads to

$$s\overline{N}_1 - N_1^0 = -k_1 \overline{N}_1 \tag{7.9b}$$

and consequently

$$\overline{N}_1 = \frac{N_1^0}{s + k_1} \tag{7.9c}$$

Repeating the procedure for the second ODE gives us

$$s\overline{N}_2 = -k_2 \overline{N}_2 + k_1 \overline{N}_1 \tag{7.10a}$$

and upon introducing of the previous result, equation (7.9b),

$$s\overline{N}_2 = -k_2 \overline{N}_2 + k_1 \frac{N_1^0}{s + k_1} \tag{7.10b}$$

which can be solved for $\overline{N}_2$ to give

$$\overline{N}_2 = \frac{k_1 N_1^0}{(s + k_1)(s + k_2)} \tag{7.10c}$$

Using the same approach for the last two ODEs yields the corresponding expressions for $\overline{N}_3$ and $\overline{N}_4$:

$$\overline{N}_3 = \frac{k_1 k_2 N_1^0}{(s + k_1)(s + k_2)(s + k_3)} \tag{7.11}$$

and

$$\overline{N}_4 = \frac{k_1 k_2 k_3 N_1^0}{s(s + k_1)(s + k_2)(s + k_3)} \tag{7.12}$$

We now apply the Heaviside inversion formula, which is given by

$$L^{-1} \frac{p(s)}{q(s)} = \sum_{i=1}^{4} \frac{(s - a_i)p(a_i)}{q(a_i)} e^{a_i t} \tag{7.13}$$

where L^{-1} is the symbol for the inversion.

The procedure consists of multiplying the right side of equations (7.9) to (7.12) in turn by each factor contained in the denominator and substituting each root of those factors into equation (7.13). This leads to the following result for equation (7.9c):

$$L^{-1} \left\{ \frac{N_1^0}{s + k_1} \right\} = N_1^0 \frac{s + k_1}{s + k_1} e^{-k_1 t} = N_i e^{-k_1 t} \tag{7.14}$$

The same result is also obtained by separation of variables.

For equation (7.10c), the corresponding expression is given by

$$L^{-1} \left\{ \frac{k_1 N_1^0}{(s + k_1)(s + k_2)} \right\} = k_1 N_1^0 \left[\frac{1}{k_2 - k_1} e^{-k_1 t} + \frac{1}{k_1 - k_2} e^{-k_2 t} \right] \tag{7.15a}$$

or

$$\boxed{N_2 = \frac{k_1 N_1^0}{(k_2 - k_1)} \left(e^{-k_1 t} - e^{-k_2 t} \right)} \tag{7.15b}$$

Similarly

$$L^{-1} \left\{ \frac{k_1 k_2 N_1^0}{(s + k_1)(s + k_2)(s + k_3)} \right\} =$$

$$\boxed{N_3 = k_1 k_2 N_1^0 \left[\frac{e^{-k_1 t}}{(k_2 - k_1)(k_3 - k_1)} + \frac{e^{-k_{21} t}}{(k_1 - k_2)(k_3 - k_2)} + \frac{e^{-k_3 t}}{(k_1 - k_3)(k_2 - k_3)} \right]} \tag{7.16}$$

Finally, for the last transform, equation (7.12), we obtain

$$L^{-1}\left\{\frac{k_1k_2k_3N_1^0}{s(s+k_1)(s+k_2)(s+k_3)}\right\}$$

$$= N_4 = k_1k_2k_3N_1^0\left[\frac{e^{0t}}{k_1k_2k_3} + \frac{e^{-k_1t}}{k_1(k_2-k_1)(k_3-k_1)} + \frac{e^{-k_2t}}{k_2(k_1-k_2)(k_3-k_2)} + \frac{e^{-k_3t}}{k_3(k_1-k_3)(k_2-k_3)}\right] \quad (7.17a)$$

This expression can be cast in the alternative form

$$\boxed{\frac{N_4}{N_1^0} = 1 - \frac{k_2k_3e^{-k_1t}}{(k_1-k_2)(k_3-k_1)} - \frac{k_1k_3e^{-k_2t}}{(k_2-k_1)(k_3-k_2)} - \frac{k_1k_2e^{-k_3t}}{(k_2-k_1)(k_3-k_2)}} \quad (7.17b)$$

COMMENTS

- We see here that the Laplace transform yields in fairly straightforward fashion the solution to a set of four linear ODEs. The D-operator method, which can also be applied in principle, was deemed too cumbersome because it involves the repeated evaluation of particular integrals. The Laplace transform, on the other hand, required only a single inversion formula, which could be applied to all four transforms in a routine way to arrive at the final desired result. There also exist matrix methods for the solution of systems of linear ODEs, but these are beyond the scope of the present work. Texts dealing with matrix operations can be found readily.

◆ EXAMPLE 7.3 Hydraulic Networks

Pipe networks, termed hydraulic if they carry water, are a frequent occurrence in plants of all descriptions, in municipal water systems, and on a smaller scale, in localized entities that convey liquids between various sites. These entities can evidently assume a host of different configurations. We have singled out two commonly occurring geometries (Fig. 7.2), those of pipes in series and pipes in parallel. We wish to address the former first, leaving the second one as an exercise (Practice Problem 7.3). We do, however, establish here the relevant equations applicable to both cases.

Two basic relations enter the picture. One is the mass balance or continuity equation, as it is termed in flowing systems. This relation was introduced in Example 4.7. For the present two cases it takes the forms

Pipes in Series

$$Q_1 = Q_2 = Q_3 = \text{constant} \quad (7.18a)$$

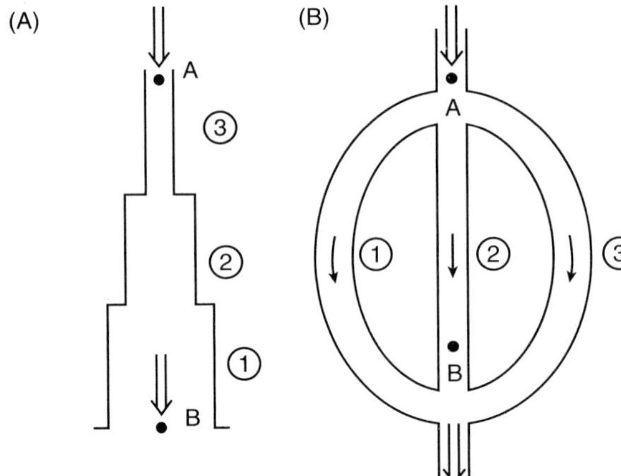

Figure 7.2 Hydraulic networks in series (A) and in parallel (B).

or alternatively, in terms of velocities and diameters,

$$v_1 d_1^2 = v_2 d_2^2 = v_3 d_3^2 = \text{constant} \tag{7.18b}$$

Pipes in Parallel

Here the continuity equation yields the relation

$$Q_1 + Q_2 + Q_3 = Q_{tot} = \text{constant} \tag{7.19a}$$

or alternatively

$$v_1 d_1^2 + v_2 d_2^2 + v_3 d_3^2 = \frac{4}{\pi} Q_{tot} \tag{7.19b}$$

The second relation is a force balance that leads to the Bernoulli equation also given in Example 4.7 [equation (4.88)]. When friction effects arise, a term called the friction head Δh_f is added to that equation, which now becomes

$$\frac{\Delta p}{\rho} + g\Delta z + \Delta h_f = 0 \tag{7.20}$$

where we have omitted kinetic energy changes, which are negligible here.

For the two pipe systems in question, we obtain first

Pipes in Series

$$(\Delta h_f)_{tot} = \Delta h_{f_1} + \Delta h_{f_2} + \Delta h_{f_3} \tag{7.21}$$

For the second case, Δp and Δz must be the same for each pipe in series. We therefore have

Table 7.1 System of Three Pipes in Series

Pipe	L (m)	d (cm)	f
1	100	8	0.0072
2	150	6	0.0065
3	80	4	0.0080

Pipes in Parallel

$$\Delta h_{f_1} = \Delta h_{f_2} = \Delta h_{f_3} \tag{7.22}$$

Let us apply some numerical values to these equations. Suppose the pipe system in series is described by the data given in Table 7.1.

We further assume a change in elevation of 5 m and a total pressure drop $p_A - p_B = 150\,\text{kPa}$. The fluid is water with a density of $\rho = 1000\,\text{kg/m}^3$. The task is to calculate the flow rate that results in the system.

From the equations (7.20) and (7.21) we then obtain

$$\frac{\Delta p}{\rho g} + \Delta z + \frac{4}{2g}\left(f_1 v_1^2 \frac{L_1}{d_1} + f_2 v_2^2 \frac{L_2}{d_2} + f_3 v_3^2 \frac{L_3}{d_3}\right) = 0 \tag{7.22$'$}$$

where Δ denotes the difference between final and initial states, and the expression for the friction head $\Delta h_f = \Delta p_f/\rho$ comes from equation (1.19) of Table 1.3.

We start by expressing the velocities v_2 and v_3 in terms of a single velocity v_1. We do this by using the continuity equation (7.18b) and obtain

$$v_2 = \frac{d_1^2}{d_2^2} v_1 = \frac{8^2}{6^2} v_1 = \frac{16}{9} v_1 \tag{7.23}$$

and

$$v_3 = \frac{d_1^2}{d_3^2} v_1 = \frac{8^2}{4^2} v_1 = 4 v_1$$

Equation (7.23) consequently becomes

$$\frac{\Delta p}{\rho g} + \Delta z + \frac{4 v_1^2}{2g}\left[f_1 \frac{L_1}{d_1} + \left(\frac{16}{9}\right)^2 f_2 \frac{L_2}{d_2} + 4^2 f_3 \frac{L_3}{d_3}\right] = 0 \tag{7.24}$$

We follow this up with the substitution of numerical values and obtain

$$-\frac{150{,}000}{1000 \times 9.81} - 5 + \frac{2 v_1^2}{9.81}\left[0.0072 \frac{100}{0.08} + \left(\frac{16}{9}\right)^2 0.0065 \frac{150}{0.06} + 4^2 0.008 \frac{80}{0.04}\right] = 0 \tag{7.25}$$

This is solved for v_1 to yield $v_1 = 0.565$ m/s, which is used to calculate the desired flow rate Q:

$$Q = \frac{\pi d_1^2}{4} v_1 = \frac{\pi (0.08)^2}{4} 0.565 \times 3600$$

$$\boxed{Q = 10.2 \text{ m}^3/\text{h}} \tag{7.26}$$

A more complex network, calling for the use of numerical methods, is taken up in Chapter 8 (Practice Problem 8.8).

◆ EXAMPLE 7.4 An Electrical Network: Hitting a Brick Wall and Going Around It

In Example 2.15 we introduced the concept of a basic electrical network termed an RLC circuit. The purpose there was to demonstrate the equivalence of such circuits, composed of a resistance R, a coil with inductance L, and a capacitor with capacitance C, to an analogous mechanical system composed of a mass, a spring, and a damper. We return here to the RLC circuit and consider somewhat more complex networks composed of two RLC circuits, joined together as indicated in Fig. 7.3. We want to solve the ODEs that describe these systems, that is, to establish the time variations of the capacitance charge q and the resulting currents in the circuits i. The procedure for setting up the ODEs outlined in Example 2.15, consists, in the first place, of establishing the voltage drop across each of the elements, which are known from elementary electricity theory and were given in Chapter 2 as follows:

$$V_L = L\frac{di}{dt} \tag{2.100a}$$

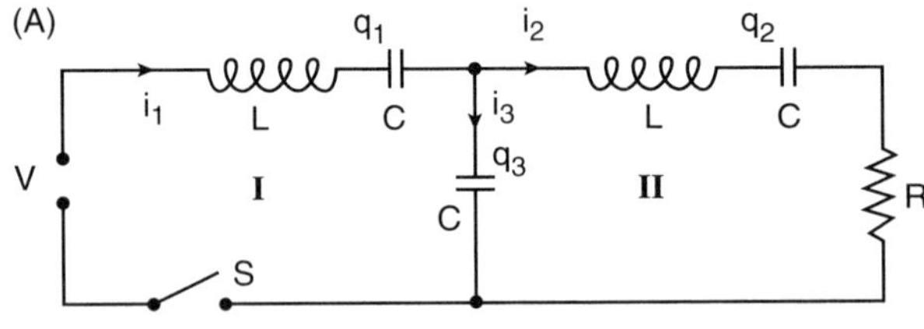

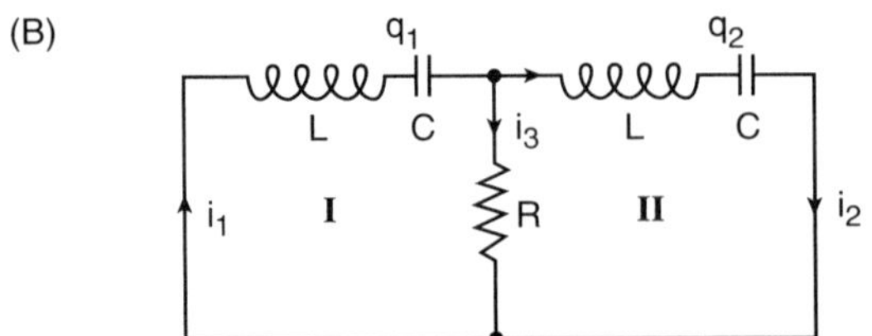

Figure 7.3 Two electrical networks composed of RLC circuits.

$$V_R = iR \tag{2.100b}$$

$$V_C = \frac{q}{C} \tag{2.100c}$$

This is followed by an application of Kirchhoff's second rule [equation (1.8)], which states that the sum of the voltage drops across each of the elements plus any externally applied voltage in the circuit must equal zero. Let us see how this works out for the circuit shown in Fig. 7.3A. We assume the imposed voltage to be of a sinusoidal form $V = V_0 \sin \omega t$. We obtain

For Circuit I

$$L\frac{di_1}{dt} + \frac{1}{C}q_1 + \frac{1}{C}q_3 = V_0 \sin \omega t \tag{7.27a}$$

For Circuit II

$$L\frac{di_2}{dt} + \frac{1}{C}q_2 + Ri_2 + \frac{1}{C}q_3 = 0 \tag{7.27b}$$

Since the current i equals the rate of change of the charge dq/dt, we can write the two equations in the equivalent form

$$Lq_1'' + \frac{1}{C}q_1 + \frac{1}{C}q_3 = V_0 \sin \omega t \tag{7.28a}$$

$$Lq_2'' + \frac{1}{C}q_2 + Rq_2' + \frac{1}{C}q_3 = 0 \tag{7.28b}$$

These are two coupled ODEs in the three dependent variables q_1, q_2, and q_3. A third relation is therefore required, which is provided by Kirchhoff's first rule [equation (1.7)]. That rule in essence states that the algebraic sum of the currents entering and leaving a junction must equal zero. Applying this to Fig. 7.3A we obtain

$$i_1 = i_2 + i_3 \tag{7.29a}$$

or equivalently

$$q_1' = q_2' + q_3' \tag{7.29b}$$

The model is now complete and consists of equations (7.28a), (7.28b), and (7.29b). The ultimate aim is to derive, as stated, the time dependence of the charges q_1, q_2, and q_3, and through them the transients of the currents i_1, i_2, and i_3 by using the relation $i = dq/dt$. Let us see how far we can get by applying the Laplace transformation. Initial conditions are set equal to zero, that is, $q_1(0) = q_2(0) = q_3(0) = q_1'(0) = q_2'(0) = q_3'(0) = 0$.

Using item 5 of Table 2.11 and item 5 of Table 2.12, we then obtain for the transformed equations

$$Ls^2\overline{q_1} + \frac{1}{C}\overline{q_1} + \frac{1}{C}\overline{q_3} = \frac{V_0\omega}{s^2 + \omega^2} \tag{7.30a}$$

$$Ls^2\overline{q_2} + \frac{1}{C}\overline{q_2} + Rs\overline{q_2} - \frac{1}{C}\overline{q_3} = 0 \tag{7.30b}$$

$$\overline{q_1} = \overline{q_2} + \overline{q_3} \tag{7.30c}$$

where $\overline{q}$ denotes the transformed variable.

Eliminating $\overline{q_1}$ and $\overline{q_3}$ in succession yields the following expression in $\overline{q_2}$:

$$\frac{RL}{C}s^3\overline{q_2} + 2Ls^2\overline{q_2} + 2\frac{Ls^2}{C}\overline{q_2} + \frac{1}{C}Rs\overline{q_2} + \frac{2}{C}\overline{q_2} + \frac{1}{C^2}\overline{q_2} = \frac{V_0\omega}{s^2+\omega^2} \tag{7.31a}$$

or solving for $\overline{q_2}$,

$$\overline{q_2} = \frac{V_0\omega}{(s^2+\omega^2)(RLs^3/C + 2Ls^2 + 2Ls^2/C + Rs/C + 2/C + 1/C^2)} \tag{7.31b}$$

We see immediately that to invert this expression by the Heaviside expansion (item 8 of Table 2.11), we must find the roots of the cubic polynomial in the denominator of equation (7.31b), which can only be done numerically. We are consequently unable to obtain by Laplace transformation explicit expressions for the variables $q_1(t)$, $q_2(t)$, and $q_3(t)$. The proceedings have thus seemingly come to a halt: we have hit a brick wall.

All is not lost, however. In the Preface and in Chapter 1 we warned readers that they must often content themselves with limiting solutions, or with upper and lower bounds to the solution being sought. Such a limiting solution arises in the present case if we let $t \to \infty$. This yields the ultimate steady-state solution for the system, which is often of greater importance and significance to the analyst than the transients that precede it. It tells us what the ultimate long-term behavior of the system will be.

The steady-state solutions can be obtained in a deceptively simple fashion without having to invert the expression (7.31b). We merely return to the original ODEs (7.28a) and (7.28b) and set the time derivatives q'' and q' equal to zero. This is the condition that holds at the steady state. We obtain

$$(q_2)_{ss} = -(q_3)_{ss} \tag{7.32a}$$

and

$$(q_1)_{ss} - (q_2)_{ss} = CV_0 \sin \omega t \tag{7.32b}$$

Introducing the Kirchhoff relation (7.29b) into equation (7.32b) then gives us

$$\boxed{(q_3')_{ss} = (i_3)_{ss} = V_0C\omega \cos \omega t} \tag{7.33a}$$

and from it

$$\boxed{(i_2)_{ss} = -V_0C\omega \cos \omega t} \tag{7.33b}$$

and

$$\boxed{(i_1)_{ss} = V_0C(\sin \omega t - \omega \cos \omega t)} \tag{7.33c}$$

These are the steady-state currents that result in the three branches of the system shown in Fig. 7.3A.

COMMENTS

- We have here an example of how a seeming impasse (the brick wall) can be overcome by looking for limiting solutions (going around the wall). It also teaches us the importance of contenting ourselves with partial answers. As this example has shown, these are often not trivial and can in fact be more revealing than a full solution to the problem.

◆ EXAMPLE 7.5 A Mechanical Network: Resonance of Two Vibrating Masses

Figure 7.4 shows a system of two masses attached to two springs and acted on by a periodic force. In Example 2.19 we introduced the concept of resonance, which arises when the natural frequency of a system equals that of an imposed periodic force. It was seen there that the natural frequency was obtained from the roots of the transform of the vibrating system. We now wish to examine the conditions under which resonance will arise in a system of two vibrating masses. That is, under what conditions will the natural frequency of the system equal ν: Specifically, we consider a system with the masses $m_1 = 2$ kg and $m_2 = 1$ kg and the spring constants $k_1 = 4$ kg/m and $k_2 = 2$ kg/m. We proceed to write the pertinent force balances, which are as follows:

$$m_1 \frac{d^2X_1}{dt^2} + k_1X_1 - k_2(X_2 - X_1) = 0 \tag{7.34a}$$

$$m_2 \frac{d^2X_2}{dt^2} + k_2(X_2 - X_1) = F_0 \cos 2\pi\nu t \tag{7.34b}$$

We assume zero initial displacements and velocities and obtain by Laplace transformation

$$m_1s^2\overline{X_1} + k_1\overline{X_1} - k_2(\overline{X_2} - \overline{X_1}) = 0 \tag{7.35a}$$

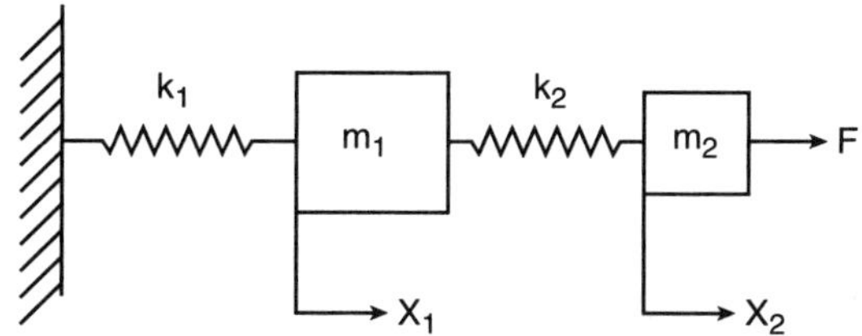

Figure 7.4 A system of two vibrating masses.

$$m_2 s^2 \overline{X_2} + k_2(\overline{X_1} - \overline{X_1}) = \frac{F_0}{s^2 + (2\pi v)^2} \tag{7.35b}$$

Solving equation (7.35a) for $\overline{X_2}$ yields

$$\overline{X_2} = \left(\frac{k_1}{k_2} + 1 + \frac{m_1 s^2}{k_2^2}\right)\overline{X_1} \tag{7.36}$$

which upon substitution into equation (7.35b) leads to the result

$$\overline{X_1} = -\frac{2\pi\nu F_0}{[(s^2 + (2\pi\nu)^2]\left[m_1 m_2 s^4/k_2 + (k_1/k_2 + m_1 + m_2)s^2 + k_1\right]} \tag{7.37}$$

To obtain the roots of the quadratic in the denominator of equation (7.37), we write

$$\frac{m_1 m_2}{k_2}y^2 + \left(\frac{k_1}{k_2} + m_2 + m_1\right)y + k_1 = 0 \tag{7.38a}$$

or numerically

$$\frac{2 \times 1}{2}y^2 - \left(\frac{4}{2} + 1 + 2\right)y + 4 = 0 \tag{7.38b}$$

where $y = s^2$.

This equation has the roots

$$y_1 = -4 \qquad \text{and} \qquad y_2 = -1 \tag{7.39}$$

so that equation (7.37) now becomes

$$\overline{X_1} = \frac{2\pi\nu F_0}{[s^2 + (2\pi\nu)^2](s^2 + 4)(s^2 + 1)} \tag{7.40}$$

As we saw in Example 2.19, for resonance to occur, the roots of the quadratic factors in the denominator must be identical. In other words, we must have

$$(2\pi\nu)^2 = 4 \tag{7.41a}$$

and

$$(2\pi\nu)^2 = 1 \tag{7.41b}$$

so that

$$\nu_1 = \frac{1}{\pi} \tag{7.41c}$$

$$\nu_2 = \frac{1}{2\pi} \tag{7.41d}$$

Thus if the forcing function $F = F_0 \cos 2\pi\nu t$ has either one of these two frequencies, the system will be in resonance. In practical terms this means that the masses will vibrate with ever increasing amplitudes, ultimately resulting in rupture.

COMMENTS

- The reader may find it rather cavalier that we did not bother to invert the expression (7.40) but relied instead on the results of Example 2.19 to prove our point. This approach is termed Laplace domain analysis, in contrast to time domain analysis, which requires inversion of the transform. Laplace domain analysis is often resorted to when qualitative insight rather than quantitative results are being sought and is a time-saving device of considerable power as well as elegance. Here we are able to obtain both insights and results. There may still be some uneasiness as to the validity of this approach, since the equation (7.40) contains an additional factor in the denominator compared to equation (2.155) of Example (2.19). To show that this does not alter the basic result, we argue as follows.

  In Example 2.19, the factor $1/(s^2 + \omega_0^2)^2$ gave rise to the inverse factor $t\cos\omega_0 t$. The effect of extending this expression into the form $1/[(s^2+1)(s^2+\omega_0^2)^2]$ can be deduced by invoking the convolution integral, item 7 of Table 2.11. The inversion of the extended expression is then given by

$$L^{-1}\frac{1}{(s^2+1)(s^2+\omega_0^2)^2} = \int_0^t \sin(t-\tau)t\cos\omega_0 t\,dt \tag{7.42}$$

  Simple integration by parts can be used to show that the factor $t\cos\omega_0 t$, signaling an increase in amplitude with time, not only remains in being but is joined by additional time-dependent factors. We have

$$\int_0^t \sin(t-\tau)t\cos\omega_0 t\,dt = t\cos\omega_0 t\int \sin(t-\tau)\,dt\,\Big|_0^t - \int_0^t \sin(t-\tau)d(t\cos\omega_0 t) \tag{7.43}$$

  from which it is seen that at least one time-dependent amplitude will arise in this system.

  It is natural for the reader to balk at these seemingly inspired shortcuts. They are, however, part and parcel of successful modeling and require skill and vision. Given patience and sustained practice, these skills, too, can ultimately be acquired.

◆ EXAMPLE 7.6 Application of Matrix Methods to Stoichiometric Calculations

The following problem arises frequently in chemical process calculation: Given a number n of chemical compounds, how many independent reactions can be formulated that will satisfy the usual requirements of an atom balance?

To lead into this topic and the attendant matrix operations, consider the following set of algebraic equations:

$$\begin{aligned} 2x_1 + 4x_2 - 2x_3 + 2x_4 &= 0 \\ x_1 + \tfrac{3}{2}x_2 &= 0 \\ 3x_1 + 5x_2 - x_3 + x_4 &= 0 \end{aligned} \tag{7.44}$$

The coefficients in this set can be viewed as the number of a particular atomic species in a compound, and the unknowns x_i as equaling the number of moles of a particular compound participating in a reaction. The three equations together can then be taken to represent atomic species balances (e.g., a carbon balance, an oxygen balance, and a hydrogen balance).

To solve this set by Gaussian elimination, described in Chapter 2, we construct the coefficient array as follows:

$$\begin{matrix} x_1 & x_2 & x_3 & x_4 & \text{RHS} \end{matrix}$$

$$\begin{pmatrix} 2 & 4 & -2 & 2 & 0 \\ 1 & \frac{3}{2} & 0 & 0 & 0 \\ 3 & 5 & -1 & 1 & 0 \end{pmatrix} \tag{A}$$

In a chemical context, this array would represent the atomic species matrix, denoting, for example, that compound x_1 contains two carbon atoms, one oxygen atom, and three hydrogen atoms.

We then proceed to reduce the array by row operations to a normalized upper triangular form, that is, to an array that has coefficients of 1 on the diagonal and 0 entries below it. This is done as follows.

1. Divide each entry of the first row by 2. We obtain

$$\begin{pmatrix} 1 & 2 & -1 & 1 & 0 \end{pmatrix} \tag{B}$$

2. Add -1 times this row to the second row and -3 times this row to the third row, which gives the result

$$\begin{pmatrix} 1 & 2 & -1 & 1 & 0 \\ 0 & -\frac{1}{2} & 1 & -1 & 0 \\ 0 & -1 & 2 & -2 & 0 \end{pmatrix} \tag{C}$$

3. Divide each entry of the second row by $-\frac{1}{2}$ to obtain:

$$\begin{pmatrix} 0 & 1 & -2 & 2 & 0 \end{pmatrix} \tag{D}$$

4. Add -2 times this row to the first row and 1 times this row to the third row. This yields

$$\begin{pmatrix} 1 & 0 & 3 & -3 & 0 \\ 0 & 1 & -2 & 2 & 0 \\ 0 & 0 & 0 & 0 & 0 \end{pmatrix} \tag{E}$$

where we have accomplished the desired reduction to a normalized upper triangular form. Since the third row is identically zero, the highest order nonvanishing determinant of this array will have order 2, and it follows immediately that only two of the three equations of our original set (7.44) are independent: that is, the set has rank 2. We use the coefficients of the reduced array to give these equations as follows:

$$\begin{aligned} x_1 &= -3x_3 + 3x_4 \\ x_2 &= 2x_3 - 2x_4 \end{aligned} \tag{7.45}$$

If we now add the two trivial equations

$$\begin{aligned} x_3 &= 1x_3 \\ x_4 &= 1x_4 \end{aligned} \tag{7.46}$$

we can write the solution $\mathbf{x}$ to our set in the vector form

$$\mathbf{x} = \begin{pmatrix} x_1 \\ x_2 \\ x_3 \\ x_4 \end{pmatrix} = \begin{pmatrix} -3 \\ 2 \\ 1 \\ 0 \end{pmatrix} x_3 + \begin{pmatrix} 3 \\ -2 \\ 0 \\ 1 \end{pmatrix} x_4 \tag{7.47}$$

This is precisely the solution we were looking for, since in the context of chemical reaction stoichiometry, the solutions x_i would represent the number of moles or the stoichiometric coefficients needed to write the reaction equations.

Let us summarize the procedure that led to this result, using the context of a system of reactions.

- Write the atom matrix array corresponding to the array (A).
- Reduce by means of elementary row operations to the usual form consisting of an identity matrix I, a block of nonreduced columns C, and one or more rows of zeros:

$$\begin{pmatrix} \mathrm{I} & \mathrm{C} \\ 0 & 0 \end{pmatrix} \tag{F}$$

  This corresponds to the array (E), with the last column omitted.
- Form an array composed of $-\mathrm{C}$ and an identity matrix of size equal to the number of columns of C. This leads to

$$\begin{pmatrix} -\mathrm{C} \\ \mathrm{I} \end{pmatrix} \tag{G}$$

  or in terms of the result we obtained in equation (7.47),

$$\begin{pmatrix} -3 & 3 \\ 2 & -2 \\ 1 & 0 \\ 0 & 1 \end{pmatrix} \tag{H}$$

  This is the array that would yield the stoichiometric coefficients of the reaction equations, which will here be two in number. Let us illustrate the procedure with an actual example.

Suppose we wish to determine the number of independent chemical reactions involving five species, shown in the following atom matrix:

$$\begin{array}{c} \\ H \\ C \\ O \end{array} \begin{array}{c} \begin{array}{ccccc} H_2 & CH_4 & CO & CH_3OH & H_2O \end{array} \\ \begin{pmatrix} 2 & 4 & 0 & 4 & 2 \\ 0 & 1 & 1 & 1 & 0 \\ 0 & 0 & 1 & 1 & 1 \end{pmatrix} \end{array} \tag{A'}$$

We proceed to reduce it to the form of array (F) by using the following operations.

1. Multiply row 1 by $\frac{1}{2}$ and subtract from the result twice row 2. This yields

$$\begin{pmatrix} 1 & 0 & -2 & 0 & 1 \\ 0 & 1 & 1 & 1 & 0 \\ 0 & 0 & 1 & 1 & 1 \end{pmatrix} \tag{B'}$$

2. Add twice row 3 to row 1 and subtract row 3 from row 2, which gives the desired result of the form of array (F):

$$\begin{pmatrix} 1 & 0 & 0 & 2 & 3 \\ 0 & 1 & 0 & 0 & -1 \\ 0 & 0 & 1 & 1 & 1 \end{pmatrix} \tag{C'}$$

Cast this result in configuration (G), which yields the array

$$\begin{pmatrix} -2 & -3 \\ 0 & 1 \\ -1 & -1 \\ 1 & 0 \\ 0 & 1 \end{pmatrix} \begin{array}{c} H_2 \\ CH_4 \\ CO \\ CH_3OH \\ H_2O \end{array} \tag{D'}$$

$$R \times n\,1 \quad R \times n\,2$$

We note that this results in two independent reactions, with the species coefficients given by the numbers in the array. These reactions are identified as

$$2H_2 + CO = CH_3OH \tag{E'}$$

$$3H_2 + CO = CH_4 + H_2O \tag{F'}$$

COMMENTS

- The procedure we have outlined essentially consists of solving a set of equations whose coefficients are obtained from atom balances, while the unknowns consist of the stoichiometric coefficients being sought. The set has more unknowns than equations, which is called overdetermined. Such sets can still yield unique solutions as we indicated in Chapter 2. These solutions are the independent reaction equations just derived. This is yet another example of the power of matrix methods to handle and analyze arrays of linear algebraic equations, or networks, as we refer to them here.

◆ EXAMPLE 7.7 Diagnosis of a Plant Flow Sheet

Plant operations can often be represented schematically in terms of so-called flow sheets, an example of which appears in Fig. 7.5. Such diagrams are composed of a network of units dedicated to certain operations, such as the mechanical or chemical transformation of materials. The assembly line of an automobile plant can be thought of as one such network. Another example is provided by chemical plants in which raw materials are fed to a reactor and the products are subsequently separated and purified. One feature that frequently makes an appearance in such diagrams is that of recycle streams or stream loops. In other words, the operation is not a linear, straight-through one but either feeds part of a product forward to a later unit or returns it to a unit located upstream. Suppose, for example, that bolts of two sizes are produced in a particular unit. The two sets could then be sent to two different locations downstream, where they are needed for two different operations, resulting in a looped pathway. Streams 2 and 3 in Fig. 7.5 typify such an arrangement. In chemical plants, part of the raw materials fed to a reactor remain unreacted. They can be separated out in a subsequent unit and recycled back into the feed to the reactor. Streams 6 and 7 show this type of arrangement.

Networks of the type just described are customarily analyzed by means of material inventories or mass balances. To do this, some knowledge is required of the percentage of material that is fed forward from a particular unit or recycled back to an earlier stage. Suppose we stipulate for the example shown in Fig. 7.5 that 30% of stream 1 is to go to stream 2 and the remaining 70% to stream 3, while 50% of stream 5 enters stream 6. Let us analyze the resulting mass balances, where x_i denotes the mass flow rates in kilograms per second:

Stage A: $-x_1 + 0x_2 + 0x_3 + 0x_4 + 0x_5 + x_6 + 0x_7 = -100$

Stage B: $x_1 - x_2 - x_3 + 0x_4 + 0x_5 + 0x_6 + 0x_7 = 0$

Stage C: $0x_1 + x_2 + 0x_3 - x_4 + 0x_5 + 0x_6 + 0x_7 = 0$

Stage D: $0x_1 + 0x_2 + x_3 + x_4 - x_5 + 0x_6 + 0x_7 = 0$

Stage E: $0x_1 + 0x_2 + 0x_3 + 0x_4 + x_5 - x_6 - x_7 = 100$

Stream 1 split: $0.3x_1 - x_2 + 0x_3 + 0x_4 + 0x_5 + 0x_6 + 0x_7 = 0$

Stream 5 split: $0x_1 + 0x_2 + 0x_3 + 0x_4 + 0.5x_5 - x_6 + 0x_7 = 0$

This set of equations can be written in matrix form as follows

$$\mathbf{Ax} = \mathbf{B} \tag{7.48}$$

where **A** is the coefficient matrix and **x** and **B** are the row and column vectors representing the unknowns x_i and the nonhomogeneous terms, respectively.

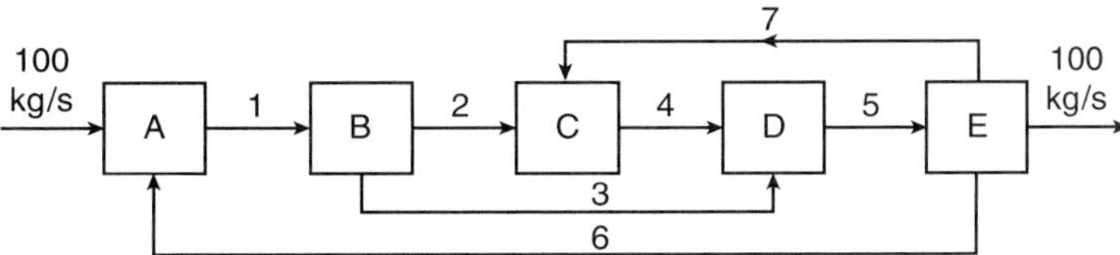

Figure 7.5 A plant flow sheet.

The augmented coefficient matrix **A : B**, which we require for the Gaussian elimination we intend to use, is then given by the array

$$\begin{bmatrix} -1 & 0 & 0 & 0 & 0 & 1 & 0 & \vdots & -100 \\ 1 & -1 & -1 & 0 & 0 & 0 & 0 & \vdots & 0 \\ 0 & 1 & 0 & -1 & 0 & 0 & 1 & \vdots & 0 \\ 0 & 0 & 1 & 1 & -1 & 0 & 0 & \vdots & 0 \\ 0 & 0 & 0 & 0 & 1 & -1 & -1 & \vdots & 100 \\ 0.3 & -1 & 0 & 0 & 0 & 0 & 0 & \vdots & 0 \\ 0 & 0 & 0 & 0 & 0.5 & -1 & 0 & \vdots & 0 \end{bmatrix} \tag{A}$$

We start with a superficial inspection of this array and note that a complete cancellation of all nonzero terms occurs when we add the first four rows of the assembly. That is, we obtain for the sum $R_1 + R_2 + R_3 + R_4$ the zero-row vector

$$0 \quad 0 \quad 0 \quad 0 \quad 0 \quad 0 \quad 0 \quad \vdots \quad 0 \tag{B}$$

Whatever the outcome of further steps in the Gaussian elimination, this first result indicates that the rank of the coefficient and augmented matrices will be *less* than the number of unknowns, here 7. It will consequently fail to meet the criterion for a unique solution, which as stated in Table 2.5 requires that this rank, or equivalently the highest order of the determinant of the coefficient matrix be *equal* to the number of unknowns. We conclude that the array, as it stands, contains at least one inconsistent equation, namely, the mass balance for the stage E of Fig. 7.5. One way to rectify this situation is to drop the offending equation and replace it by a simpler and viable alternative. Our proposal is to substitute the condition that 40% of stream 5 is to enter stream 7. The revised array then becomes

$$\begin{bmatrix} -1 & 0 & 0 & 0 & 0 & 1 & 0 & \vdots & -100 \\ 1 & -1 & -1 & 0 & 0 & 0 & 0 & \vdots & 0 \\ 0 & 1 & 0 & -1 & 0 & 0 & 1 & \vdots & 0 \\ 0 & 0 & 1 & 1 & -1 & 0 & 0 & \vdots & 0 \\ 0 & 0 & 0 & 0 & 0.4 & 0 & -1 & \vdots & 0 \\ 0.3 & -1 & 0 & 0 & 0 & 0 & 0 & \vdots & 0 \\ 0 & 0 & 0 & 0 & 0.5 & -1 & 0 & \vdots & 0 \end{bmatrix} \tag{C}$$

A cursory inspection of array (C) shows that simple additive or multiplicative row operations do not lead to matrix singularity. We can therefore with some confidence proceed to the solution of this set by Gaussian elimination. This will be left to the exercises (Practice Problem 7.6).

◆ EXAMPLE 7.8 Manufacturing Costs: Use of Matrix–Vector Products

The material costs involved in manufacturing products composed of subsidiary units of a given price can be tersely and conveniently expressed in the form of a matrix–vector product. Consider as an example the following case.

An electronics manufacturer produces two types of circuit boards, whose major components are diodes, transistors, and resistors. Let y_1, y_2, y_3 denote the cost of diodes, transistors, and resistors, respectively. Then the materials cost of the circuit boards, excluding the cost of assembling the units, is given by

$$\begin{aligned} x_1 &= b_{11}y_1 + b_{12}y_2 + b_{13}y_3 \\ x_2 &= b_{21}y_1 + b_{22}y_2 + b_{23}y_3 \end{aligned} \tag{7.49}$$

where b_{ij} gives the number of components j used to produce the circuit board i. We can associate with this system a component cost vector **y**, a board cost vector **x**, and the quantity matrix **B**, so that

$$\mathbf{y} = \begin{bmatrix} y_1 \\ y_2 \\ y_3 \end{bmatrix} \quad \mathbf{x} = \begin{bmatrix} x_1 \\ x_2 \end{bmatrix} \quad \mathbf{B} = \begin{bmatrix} b_{11} & b_{12} & b_{13} \\ b_{21} & b_{22} & b_{23} \end{bmatrix} \tag{7.50}$$

We now recall the definition of matrix multiplication given in Chapter 2, of which the multiplication of a matrix by a column vector was a special case. That definition stated that the product **AB** of two matrices is a new matrix **C** whose components C_{ij} are composed of the sum of the products of the ith row of matrix **A** with the jth column of the matrix **B**. For the case given here, the matrix **A** is a column vector composed of only one column, while the matrix **B** is composed of two rows. It follows that the entries for the product **By** will be composed of a *single* entry composed of the sum of the products $b_{ij}y_j$ and that there will be two such sums, resulting in a new column vector. But a closer inspection of equations (7.49) reveals that the sums given on the right of those expressions are precisely the entries that result from the multiplication of **B** by **y**. We can therefore write

$$\mathbf{x} = \mathbf{B}\mathbf{y} \tag{7.50$'$}$$

Let us demonstrate the use of that expression with a numerical example. Suppose we are given

$$\mathbf{B} = \begin{bmatrix} 1 & 1 & 2 \\ 3 & 0 & 1 \end{bmatrix} \qquad \mathbf{y} = \begin{bmatrix} 1 \\ 2 \\ 3 \end{bmatrix}$$

Then

$$\mathbf{x} = \begin{bmatrix} 1 \times 1 + 1 \times 2 + 2 \times 3 \\ 3 \times 1 + 0 \times 2 + 1 \times 3 \end{bmatrix} = \begin{bmatrix} 9 \\ 6 \end{bmatrix}$$

That is, board 1 costs 9 monetary units and board 2 costs 6 such units. Note again that the assembly cost is not included. An extension of this problem, the consideration of the cost of making a product with these circuit boards, is taken up in Practice Problem 7.7.

◆ EXAMPLE 7.9 More About Electrical Circuits: The Electrical Ladder Networks

Electrical ladder networks consist of circuits in which resistors are wired in series or as shunts, as shown in Fig. 7.6A, where i_1 and V_1 are the input current and voltage, and i_5 and V_5 are the corresponding outputs. The task we set ourselves is to relate the output current and voltage to the input values, taking into account the arrangement of resistors. This can be conveniently done by a repeated application of matrix products. To show this, we first consider the current–voltage relations that apply to individual resistors wired either in series or as shunts. These are depicted in Fig. 7.6B and 7.6C. Consider first the single resistor wired as a shunt (Fig. 7.6C). Two basic relations of circuit theory apply here. From Ohm's law we obtain

$$i = \frac{V_1}{R} = \frac{V_2}{R} \tag{7.51a}$$

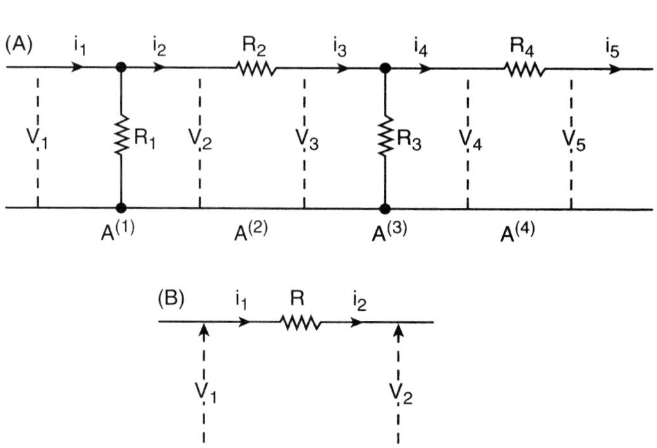

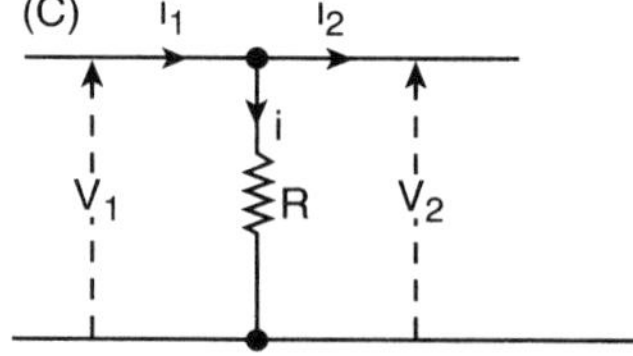

Figure 7.6 Electrical ladder networks: (A) a sample network, (B) resistor in series, and (C) shunt.

and hence

$$V_1 = V_2 \tag{7.51b}$$

The second relation comes from Kirchhoff's junction law, which here reads

$$i_1 - i - i_2 = 0 \tag{7.52a}$$

and can be combined with Ohm's law to yield

$$i_2 = -i + i_1 = -\frac{V}{R} + i_1 \tag{7.52b}$$

If we now define an input and output vector as follows

$$\mathbf{x}^{(1)} = \begin{bmatrix} V_1 \\ i_1 \end{bmatrix} \quad \text{and} \quad \mathbf{x}^{(2)} = \begin{bmatrix} V_2 \\ i_2 \end{bmatrix} \tag{7.53}$$

Then by the relations (7.51b) and (7.52b) we obtain

$$\mathbf{x}^{(2)} = \mathbf{A}\mathbf{x}^{(1)} \tag{7.54}$$

where **A** is referred to as the transfer matrix given by

$$\mathbf{A} = \begin{bmatrix} 1 & 0 \\ -1/R & 1 \end{bmatrix} \tag{7.55}$$

Equivalence of the matrix relation (7.54) and the scalar equations (7.51b) and (7.52b) is easily shown by using the usual rules of matrix–vector multiplication to decompose the matrix (7.54).

Consider next the series resistor shown in Fig. 7.6B. Here we have the relations

$$i_1 = i_2 \tag{7.56a}$$

and by Ohm's law

$$V_2 = V_1 - Ri_1 \tag{7.56b}$$

We can Use the same input/output vectors as before to recast these two relations in matrix–vector form to yield

$$\mathbf{x}^{(2)} = \mathbf{A}\mathbf{x}^{(1)} \tag{7.57}$$

where $\mathbf{A} = \begin{bmatrix} 1 & -R \\ 0 & 1 \end{bmatrix}$.

It is left to the reader to show the equivalence of (7.57) and the equations (7.56).

Let us now examine how the two matrix–vector relations can be used to establish the relation between input and output currents and voltages for the network shown in Fig. 7.6A. Starting on the extreme right we have

$$\begin{bmatrix} V_5 \\ i_5 \end{bmatrix} = \begin{bmatrix} V_4 \\ i_4 \end{bmatrix} \tag{7.58}$$

Applying this relation repeatedly and in succession to each element of the ladder leads to the sequence

$$\begin{bmatrix} V_5 \\ i_5 \end{bmatrix} = \mathbf{A}^{(4)} \begin{bmatrix} V_4 \\ i_4 \end{bmatrix} = \mathbf{A}^{(4)}\mathbf{A}^{(3)} \begin{bmatrix} V_3 \\ i_3 \end{bmatrix} = \cdots = \mathbf{A}^{(4)}\mathbf{A}^{(3)}\mathbf{A}^{(2)}\mathbf{A}^{(1)} \begin{bmatrix} V_1 \\ i_1 \end{bmatrix} \tag{7.59}$$

Therefore, the transfer matrix for any ladder network may be obtained by multiplying the transfer matrices of the basic shunt and series resistors.

Application of equation (7.59) to a numerical example is left to the exercises (Practice Problem 7.8).

◆ EXAMPLE 7.10 Photosynthesis and Respiration of a Plant: An Electrical Analogue for the CO_2 Pathway

Carbon dioxide plays a central role in the life and growth of plants. It is, on the one hand, taken up from the surrounding air and converted to organic compounds and oxygen, which is released to the atmosphere. This process is known as photosynthesis and requires light for its sustenance. Plants also *produce* carbon dioxide, much as humans do, by metabolizing or "burning" nutrients. In this process, known as respiration, oxygen from the air is a reactant, while carbon dioxide is a product that is released to the atmosphere. Respiration is thus the reverse of photosynthesis but plays a more subdued role, leaving the plant as a net consumer of carbon dioxide.

The transport of carbon dioxide and oxygen occurs principally within the foliage of a plant, and it is here also that the process of photosynthesis takes place. To gain an understanding of this important event, we examine briefly the internal structure and cell organization of the foliage.

Figure 7.7A shows the cell structure of a typical leaf. Transport of gases takes place mainly through openings termed stomatal pores (stoma: Greek for mouth), which are located on the underside of the leaf. Stomata have a typical length of 10 to 20 μm and an average radius of 5 to 10 μm; they cover a fractional area ranging from 0.002 to 0.02 (0.2–2%). The remainder of the leaf surface is covered by a layer called cuticle, which is essentially impermeable to gases (Fig. 7.7A).

The stomatal pores are flanked on either side by guard cells, which control the size of the opening by expanding or contracting in response to external stimuli. In the dark, the guard cells are triggered to expand, thus partially closing the stomatal pores. Very little photosynthesis takes place under these conditions, and the principal remaining process is that of respiration, which releases CO_2 to the atmosphere.

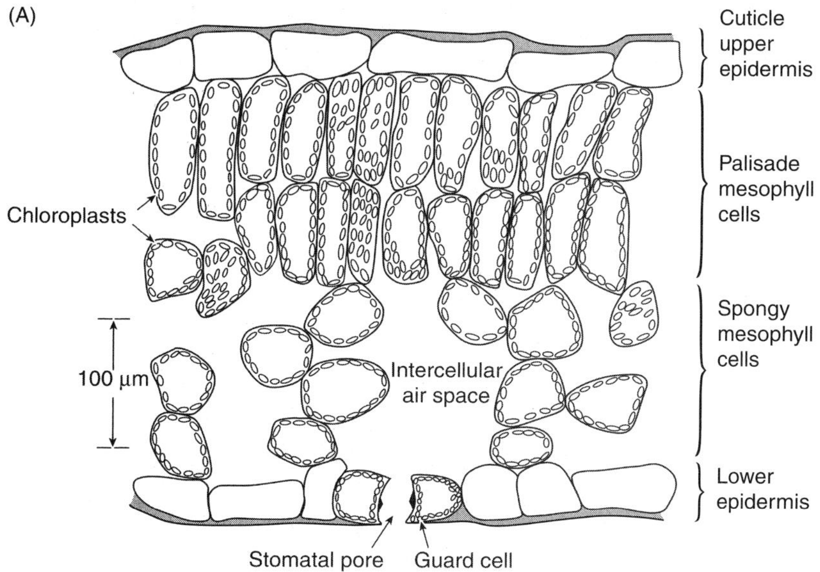

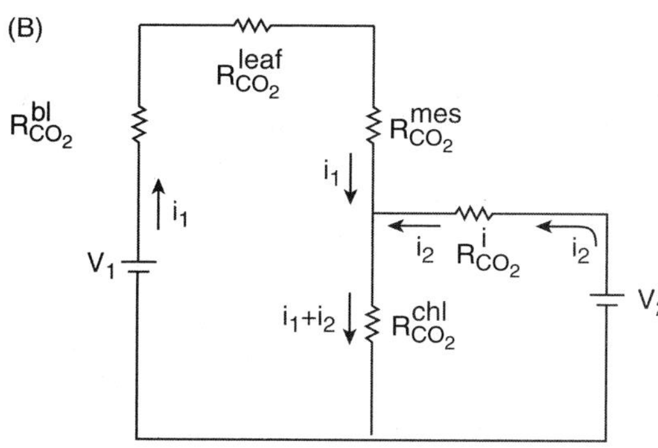

Figure 7.7 Plant physiology and photosynthesis: (A) structure of a leaf and (B) electrical network mimicking photosynthesis and respiration (bl, boundary layer; leaf, leaf surface; mes, mesophyllic cells; chl, chloroplasts).

The interior of the leaves contains the so-called mesomorphic cells, which are arranged either in a loosely packed, "spongy" configuration or in the denser "palisade" form. Intercellular air fills the intervening spaces. These mesomorphic cells contain smaller cellular entities termed chloroplasts, which carry chlorophyll, the principal substance responsible for photosynthesis.

The task we set ourselves is to calculate the net carbon dioxide uptake by a tree of modest size, bearing 10^4 leaves, each with an average area of 25×10^{-4} m^2. The course of CO_2 transport and reaction is evidently a highly intricate one. It involves diffusion through a stagnant boundary layer and the stomatal openings, penetration of the intercellular air space, passage into the mesophyllic and chloroplast cells, and, finally, reaction by photosynthesis. Because of the complexity of these processes, it has become customary to express the overall uptake in terms of a CO_2 concentration driving force ΔC and a series of resistances, $\sum R$, each one of which describes a particular step in the overall process.

Table 7.2 Representative Values of Resistances for CO_2 Diffusion into Leaves

Component	Resistance ($s\,m^{-1}$)
Leaves, open stomata	R^{leaf} 500–2500
Mesophyll cells	R^{mes} 200–800
Chloroplasts	R^{chl} < 400

We have

$$\frac{N}{A} = \frac{\Delta C}{\sum R} \tag{7.60a}$$

where N denotes moles of CO_2 taken up per second and A is the area of the underside of a leaf.

The reader will note the similarity to the conduction of heat we considered in Example 6.4, and electrical engineers will of course recognize its equivalence to Ohm's law. In fact, we shall shortly make use of an electrical analogue to describe the dual effect of photosynthesis and respiration.

Early work dealing with CO_2 uptake had shown that photosynthesis is fast in comparison to the rate of transport to and within the leaf. That process consequently becomes the rate-determining step and comprises resistances due to the boundary layer R^{bl}, resistance at the stomatal leaf surface R^{leaf}, resistance of the mesophyllic cells R^{mes}, and resistance due to the chloroplasts R^{chl}. These resistances have been measured experimentally and are summarized in Table 7.2. They are the principal resistances to carbon dioxide uptake, those due to the boundary layer and the intercellular air space being negligible in comparison.

We now turn to the task of representing the dual processes of photosynthesis and respiration by means of an electrical analog (Fig. 7.7B), where two interconnected loops are composed of resistances in series and each is fed by a battery with voltages V_1 and V_2, respectively. The left loop represents photosynthesis and the right loop the respiration process, with the current i_1 mimicking CO_2 intake N/A and i_2 the amount of carbon dioxide produced by respiration, N^r/A. We now apply Kirchhoff's second rule to the left loop, which implies that the overall changes in potential in going completely around that loop must be zero (see Table 1.2, item 7). We then use Ohm's law as an adjunct and, neglecting boundary layer resistance, we obtain

$$V_i - i_1(R^{leaf} + R^{mes} + R^{chl}) - (i_1 + i_2)R^{chl} = 0 \tag{7.60b}$$

This becomes, in terms of CO_2 transport

$$\Delta C - \frac{N}{A(R^{leaf} + R^{mes})} - \left(\frac{N}{A} + \frac{N^r}{A}\right) R^{chl} = 0 \tag{7.60c}$$

where ΔC has replaced the electrical driving force V_1.

We can recast this expression into the form given by equation (7.56) and obtain

$$\frac{N}{A} = \frac{\Delta C}{\left(R^{leaf} + R^{mes} + R^{chl} + \frac{N^r/A}{N/A} R^{chl}\right)} \tag{7.60d}$$

Let us use this relation to calculate some values for N/A for the modest-sized tree we had chosen before, with a total leaf area of 25 m^2. The CO_2 content of air is currently at an approximate and average level of 370 parts per million (ppm), that is, 370×10^{-6} mol CO_2 per mole of air and is growing at the rate of 2 ppm/year. An equivalent way of expressing this is to say that the carbon dioxide exerts a pressure equal to 370×10^{-6} times the total atmospheric pressure of 10^5 Pa. We can then write the ideal gas law [equation (4.45)] in the form

$$C = \frac{n}{V} = \frac{p}{RT} \tag{7.61a}$$

$$(C_{CO_2})_{air} = \frac{370 \times 10^{-6} \times 10^5}{8.314 \times 298} = 0.015\ \text{mol/m}^3 \tag{7.61b}$$

This gives us the molar concentration of carbon dioxide in the air for use in equation (7.58). If we further assume that the CO_2 level in chloroplasts is near zero because of the high rate of conversion by photosynthesis, we obtain

$$\Delta C \cong 0.015\ \text{mol/m}^3 \tag{7.61c}$$

Finally, we use the experimental finding that during daylight, here taken to be an average of 10 effective hours, the ratio $(N^r/A)/(N/A)$ is approximately 0.3; that is, the amount of carbon dioxide produced by respiration amounts to roughly 30% of the total uptake from the atmosphere.

With these numbers in place, and using the lower values of the resistances cited in Table 7.2, we obtain for the daily production of CO_2

$$N = A\frac{\Delta C}{\sum R} 3600 \times 10 \tag{7.62a}$$

$$= 25\frac{0.015}{500 + 200 + 200 + (0.3 \times 200)} 3600 \times 10 \tag{7.62b}$$

$$\boxed{N = 14.1\ \text{mol/day}}$$

COMMENTS

- We have here a fruitful application of simple modeling and experimentation to a problem of considerable complexity. Although Fick's law could in principle be applied to each segment of the CO_2 pathway and the resulting theoretical resistances summed, the complex geometry of the system, consisting as it does of irregular arrays of cells, would quickly defeat this approach. One is therefore forced to combine modeling with experiment, a not unusual occurrence in the simulation of complex processes.

  It is of some interest to examine how well the earth's vegetation copes with the CO_2 emissions of its surroundings. We refer the reader to Practice Problem 7.9, which asks for an estimate of the daily emissions of a typical passenger car. The startling result emerges that some 27 trees of the modest size proposed here are needed to compensate for the carbon dioxide production of a typical passenger car. This shows the limited capacity of our vegetation to act as a carbon sink and underlines the necessity to reduce emissions to sustainable levels.

PRACTICE PROBLEMS

7.1 The hairpin heat exchanger Derive the model equations and the boundary conditions which describe the performance of a hairpin heat exchanger, shown in Fig. P7.1. Do not solve.

7.2 Determining the age of the Shroud of Turin: radiocarbon dating The earth's atmosphere is constantly being bombarded by cosmic rays that originate in outer space and consist of electrons, neutrons, and atomic nuclei. One of the important reactions between the atmosphere and cosmic rays is the capture of neutrons by atmospheric nitrogen to produce the radioactive carbon-14 isotope and hydrogen. The unstable carbon atoms eventually form $^{14}CO_2$, which mixes with the ordinary carbon dioxide in the air, and enters the biosphere when CO_2 is taken up in plant photosynthesis. When an individual plant or animal dies, the carbon-14 isotope on it is no longer replenished, so the ratio of ^{14}C to ^{12}C diminishes as ^{14}C decays. The same change occurs when carbon atoms are trapped in coal and petroleum, or in wood preserved underground.

In 1955 Willard F. Libby suggested that this fact could be used to estimate the length of time the carbon-14 isotope of a particular specimen has been decaying without replenishment. The technique was subsequently used to determine the age of a host of ancient artifacts, including the Shroud of Turin, which had been purported to be the burial cloth of Jesus. Since in fresh samples the ratio $^{14}C/^{12}C$ is about $1/10^{12}$, and in older samples even less, the equipment used to measure decay must be extremely sensitive.

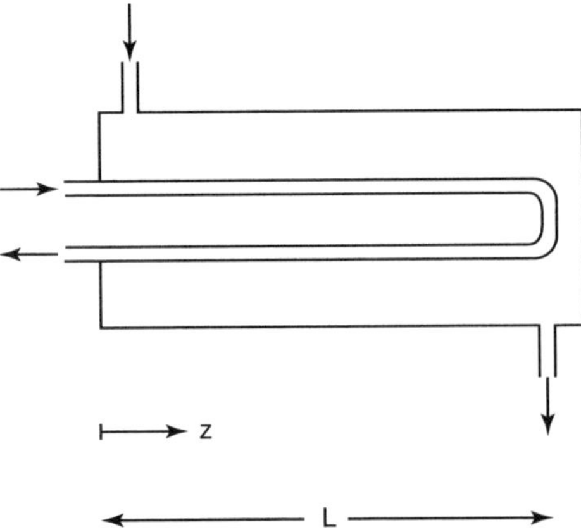

Figure P7.1 The hairpin heat exchanger.

In 1988, three laboratories in Europe and the United States, working on less than 50 mg samples, independently determined the age of the Shroud of Turin. In one of the measurements, the value of $^{14}C/^{12}C$ was found to be $0.99912/10^{12}$. Using the fact that the half-life of ^{14}C is 5.73×10^3 years, estimate the year the shroud was made.

7.3 **The parallel pipe network** Use the pipe lengths and diameters given in Table 7.1 to calculate the flow rates Q_1, Q_2, Q_3 in each branch of the pipe network shown in Fig. 7.2B. Use the following friction factor values arrived at after two iterations, using the values of Fig. 7.2B as starting values:

$$f_1 = 0.0067 \qquad f_2 = 0.0062 \qquad f_3 = 0.0076$$

ANSWER

$$Q_1 = 62.5\,m^3/h$$
$$Q_2 = 25.9\,m^3/h$$
$$Q_3 = 11.4\,m^3/h$$

7.4 **Electrical network without an imposed voltage** Derive the time dependence of the currents i_1 and i_2 in the circuit shown in Fig. 7.3B, assuming $i_1(0) = i_2(0) = 0$ and $q_1(0) = q_2(0) = q_0$.

7.5 **More about stoichiometry** In a particular industrial chemical process, acetone $[(CH_3)_2CO]$ is thermally decomposed to produce ketene (CH_2CO), which is the starting material for acetic anhydride, an important chemical. Besides ketene, the thermal process itself generates methane (CH_4), ethylene (C_2H_4), hydrogen (H_2), and carbon monoxide (CO), as by-products. Thus this initial process involves the six species CO, $(CH_3)CO$, CH_2CO, CH_4, C_2H_4, and H_2. Establish the type and number of independent reactions that can take place in the reactor.

7.6 **More Gaussian elimination** Solve the equations represented by array (C) of Example 7.7.

ANSWER

$$x_1 = 600$$
$$x_2 = 0$$
$$x_3 = 600$$
$$x_4 = 400$$
$$x_5 = 1000$$
$$x_6 = 500$$
$$x_7 = 400$$

7.7 **More about manufacturing costs**

We pursue Example 7.8 by considering that the circuit boards are used further to manufacture stereo sets, of which there are three models: Standard, Superior, and Deluxe. Let z_1, z_2, and z_3 denote the costs of producing these three models. Then

$$\begin{aligned} z_1 &= a_{11}x_1 + a_{12}x_2 \\ z_2 &= a_{21}x_1 + a_{22}x_2 \\ z_3 &= a_{31}x_1 + a_{32}x_2 \end{aligned} \qquad \text{(A)}$$

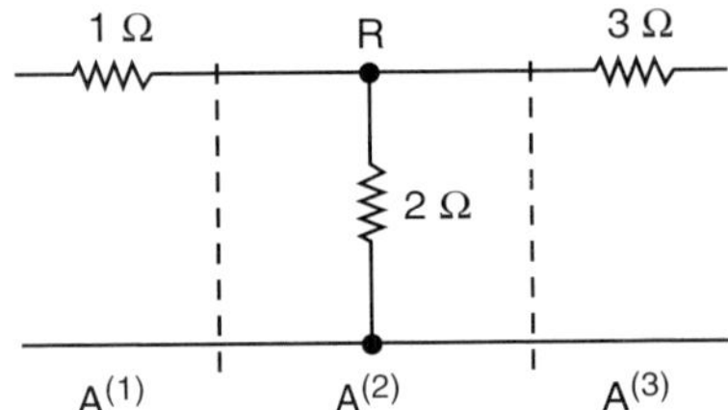

Figure P7.8

where a_{ij} gives the number of circuit boards of type j used to produce the stereo model i. For example the Deluxe model (model 3) requires a_{32} circuit boards of type 2. It then follows from the rules of matrix–vector multiplication discussed in Example 7.8 that set (A) can be written in matrix–vector notation as follows

$$\mathbf{z} = \mathbf{A}\mathbf{x} \tag{B}$$

where

$$\mathbf{z} = \begin{bmatrix} z_1 \\ z_2 \\ z_3 \end{bmatrix} \qquad \mathbf{A} = \begin{bmatrix} a_{11} & a_{12} \\ a_{21} & a_{22} \\ a_{31} & a_{32} \end{bmatrix} \tag{C}$$

We wish to establish the cost **z** in terms of the cost of diodes, transistors, and resistors given by the vector **y**. Show that this is given by the relation $\mathbf{z} = \mathbf{A}(\mathbf{B}\mathbf{y}) = \mathbf{C}\mathbf{y}$.

7.8 **A simple ladder network** Consider the simple ladder network shown in Fig. P7.8. Show that for an input

$$\begin{bmatrix} V_1 \\ i_1 \end{bmatrix} = \begin{bmatrix} 40 \\ 10 \end{bmatrix}$$

the output is given by

$$\begin{bmatrix} V_4 \\ i_4 \end{bmatrix} = \frac{1}{2}\begin{bmatrix} 90 \\ -10 \end{bmatrix}$$

7.9 **How many trees to a car?** Consider a car that is driven an average of 100 km a day and consumes a liter of gasoline for every 15 km. Calculate the daily production of CO_2, assuming the fuel to consist entirely of octane (C_8H_{18}), with a density of 800 g/L. From this, and the result of Example 7.10, estimate the number of small trees that will compensate for the CO_2 emission of a single car.

ANSWER 27

CHAPTER 8

More Mathematical Tools: Dimensional Analysis and Numerical Methods

8.1 Dimensional Analysis

8.1.1 Introduction

The section of Chapter 1 entitled 'When Not to Model' drew attention to the difficulties inherent in modeling highly complex systems and the need in these cases for at least some experimentation to derive the desired information. Heat and mass transfer coefficients in turbulent flow were two cases in which our advice was to set aside modeling, at least in part, in favor of experimentation. It was also pointed out that this task can be considerably eased by combining the pertinent physical parameters of the system into dimensionless groups, thus reducing the number of variables to be dealt with.

Suppose, for example, that six physical parameters affect the behavior of a system. In the case of heat transfer in turbulent flow in a pipe, these might be flow velocity v, pipe diameter d, and the physical properties of the fluid (density, heat capacity, viscosity, and thermal conductivity). Suppose further that we wish to run tests at 10 values of each parameter. The total number of experiments required would then amount to 10^6. If, on the other hand, we were able to combine the parameters into only three dimensionless groups, the number of tests would drop to 10^3. An experienced graduate student, running one test a day, would thus require only 3 years to complete his Ph.D. Without the use of dimensionless groups, this number would balloon to 3000 years. This process of expressing the behavior of a physical system in terms of dimensionless groups or parameters is known as dimensional analysis. It is a powerful tool for arriving at a first qualitative description of complex systems and eases enormously the experimental work required to quantify that relationship.

The process of dimensional analysis can be carried out in three deceptively simple steps:

1. List all the variables that affect the system behavior.
2. Write down the dimensions of these variables.
3. Combine the variables into a functional relationship involving dimensionless groups or some other dimensionally consistent form.

Some comments are necessary to ensure the proper application of this scheme.

Step 1. This is by far the most important and difficult step and confronts the user with the task of deciding which independent variables to include in the analysis. There is no easy recipe for carrying out this step, but the following suggestions may be found useful:

(a) Apply physical reasoning to identify the pertinent variables.

(b) Use experimental observations to amplify this list and to verify the final result.

(c) If no functional relationship of dimensional consistency results, reexamine the situation by adding or omitting parameters.

(d) Make sure that only truly independent variables are included. Thus if mass and acceleration are chosen as parameters, force cannot be added to the list because it is related to the former through Newton's law.

We shall have occasion to practice this important step in the illustrations that follow and the reader will find that, as often as not, some inspired guesswork will be involved.

Step 2. The basic dimensions commonly chosen in dimensional analysis are mass (M), length (L), time (θ), and temperature (T). All other quantities are expressed in terms of these fundamental dimensions. Thus force has the units of $ML\theta^{-2}$ by virtue of Newton's law. The joule (J) is not a fundamental dimension but is instead expressed as $ML^2\theta^{-2}$. To aid the reader in implementing the analysis, a short list of the dimensions of important variables is presented in Table 8.1.

Step 3. The ultimate aim of this step, and of the analysis as a whole, is to express the system behavior in terms of the functional relationship

$$F(\pi_1, \pi_2, \ldots, \pi_p) = 0 \tag{8.1}$$

where π is the symbol commonly used for a dimensionless group. If $x_2 \cdots x_p$ are independent variables, and x_1 the dependent variable, the dimensionless groups can be represented in the form

$$\pi_1 = \frac{x_1}{x_2^{a_1} x_3^{1_2} \cdots} \qquad \pi_2 = \frac{x_i}{x_2^{\beta_1} x_3^{\beta_2} \cdots}, \cdots \tag{8.2}$$

Often it is convenient to solve this relationship for the dependent variable. We then obtain

$$x_1 = x_2^{a_1} x_3^{a_2} \cdots F(\pi_2, \pi_3, \ldots, \pi_n) \tag{8.3}$$

The use of these relations will become clearer in the examples that follow.

We start our examples with an analysis of the simple pendulum; only one dimensionless group results, but we nevertheless manage to illuminate several important aspects of dimensional analysis. We follow this up with a second simple example, that of the vibration of a

Table 8.1 Variables and Their Dimensions

Variable	Symbol	Dimension
Fundamental		
Mass	m	M
Length	l, d	L
Time	t	θ
Temperature	T	T
Mechanical		
Velocity	v	$L\theta^{-1}$
Acceleration	a	$L\theta^{-2}$
Volumetric flow rate	Q	$L^{-3}\theta^{-1}$
Density	ρ	ML^{-3}
Viscosity	μ	$ML^{-3}\theta^{-1}$
Force	F	$ML^{-1}\theta^{-2}$
Pressure	p	$ML^{-3}\theta^{-2}$
Power	P	$ML^{2}\theta^{-3}$
Torque	T	$ML^{2}\theta^{-2}$
Rotational speed	N	θ^{-1}
Frequency	ν	θ^{-1}
Thermal		
Rate of heat flow	q	$ML^{2}\theta^{-3}$
Thermal conductivity	k	$ML\theta^{-3}T^{-1}$
Specific heat	C_p	$L^{2}\theta^{-2}T^{-1}$
Heat transfer coefficient	h	$M\theta^{-3}T^{-1}$
Diffusional		
Concentration	C	ML^{-3}
Rate of mass flow	N	$M\theta^{-1}$
Diffusivity	D	$L^{2}\theta^{-1}$
Mass transfer coefficient	k_C	$L\theta^{-1}$

one-dimensional structure, which likewise results in a single dimensionless group. Both these examples lead to a formal generalization of dimensional analysis known as Buckingham's π theorem. This theorem is taken up in Section 8.1.2.

◆ EXAMPLE 8.1 Time of Swing of a Simple Pendulum

The simple pendulum is defined as one in which frictional effects are neglected and the angle of deflection is small (see Fig. 8.1). The reasons for these assumptions will be made clear in our subsequent comments.

Suppose we wish to find a relation between the time of swing, here taken to be the dependent variable, and the physical parameters of the system. We start by establishing these parameters, which must be independent (step 1), and argue as follows.

Since gravity is the driving force, we take both mass m and the gravitational acceleration g to be relevant independent variables. We also know, either through physical reasoning or by experimental observation, that the time of swing depends on the length of the pendulum. In the device known as a metronome, for example, a rhythmic beat is sounded out with each swing of a pendulum, which is provided with a sliding mass. The frequency of the beat is adjusted by moving the mass up or down, thereby shortening or lengthening the pendulum. Short lengths result in a fast beat, greater lengths in a slow beat.

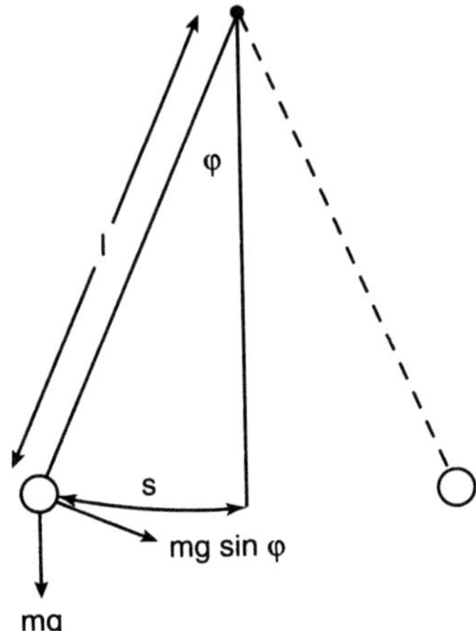

Figure 8.1 The simple pendulum.

Table 8.2 Dimensions for Example 8.1

Variable	Symbol	Dimension
Time	t	θ
Length	l	L
Mass	m	M
Gravitational Acceleration	g	$L\theta^{-2}$

We also know from physical observation that the time of swing does not depend on material properties. The period is the same, irrespective of whether the mass is suspended by a nylon thread, a cotton string, or a piano wire.

With this information in hand, we proceed to step 2 and write out the dimensions of the parameters in accordance with the conventions of Table 8.2.

We next compose a functional relation in the form of a *product of powers* of the variables. This procedure, known as the Rayleigh method, can be used when the number of independent variables equals the number of unknowns. When this is no longer the case, the method is replaced by the Buckingham's π theorem, to be taken up later. For the present, we write

$$t = \text{const} \times l^{\alpha} m^{\beta} g^{\gamma} \tag{8.4a}$$

or in terms of dimensions

$$M^0 L^0 \theta = L^{\alpha} M^{\beta} (L\theta^{-2})^{\gamma} \tag{8.4b}$$

A solution for the unknowns α, β, γ is found by equating the exponents. Thus

$$\alpha + \gamma = 0 \quad \text{Condition on L} \tag{8.4c}$$

$$\beta = 0 \quad \text{Condition on M} \tag{8.4d}$$

$$-2\gamma = 1 \quad \text{Condition on } \theta \tag{8.4e}$$

Solving the equations, we obtain

$$\alpha = \tfrac{1}{2} \tag{8.4f}$$

$$\gamma = -\tfrac{1}{2} \tag{8.4g}$$

and hence by substitution into equation (8.4a)

$$t = \text{const}\left(\frac{l}{g}\right)^{1/2} \tag{8.4h}$$

This expression can also be written in an equivalent dimensionless form:

$$t\left(\frac{g}{l}\right)^{1/2} = \text{const} \tag{8.4i}$$

or

$$f(\pi) = 0 \tag{8.4j}$$

where π is the symbol for the dimensionless group $t(g/l)^{1/2}$, known as the time constant of the pendulum. The relation does not allow us to calculate t, but it does tell us that if we increase the length of the pendulum, the time of swing will rise proportionately. This is useful and important information.

COMMENTS

- The first feature of note in this result is the disappearance of mass as a variable. We observed a similar phenomenon in modeling the trajectory of a projectile (Example 4.5) and noted the power of modeling to reveal the unexpected. The same result emerges if we write out a quantitative model for the system. We have from Newton's law

$$F = ma \tag{8.5a}$$

or equivalently (see Fig. 8.1)

$$-mg\sin\varphi = m\frac{d^2s}{dt^2} = ml\frac{d^2\varphi}{dt^2} \tag{8.5b}$$

where we have replaced the arc length s by the equivalent product φl. Note that the mass cancels out, in agreement with the result of dimensional analysis.

For small angles of deflection we can further write $\sin\varphi \approx \varphi$ and consequently obtain

$$\frac{d^2\varphi}{dt^2} + \frac{g}{l}\varphi = 0 \tag{8.5c}$$

This is the equation that describes the motion of a simple pendulum. Its solution is left to the exercises (Practice Problem 8.1), but we note that one can already see here lurking the same ratio g/l that made its appearance in equation (8.4h).

A second feature worth noting is the square-root dependence of time of swing on length. Doubling the length will quadruple t. This was not expected on physical grounds but is nicely revealed by the simple analysis given here. Experimentation or an analytical solution is still necessary to quantify the relation (8.4i), but the important qualitative features of the system are firmly in place.

We finally note that both qualitative and quantitative results can, in this instance, be derived from the model equation (8.5c). Such model equations are not always available, as we shall show in subsequent applications. It is here that dimensional analysis comes into its own and proves itself as not only a powerful tool but as the *only* tool for the description of complex systems.

◆ EXAMPLE 8.2 Vibration of a One-Dimensional Structure

In this example we wish to derive a qualitative expression for the frequency of vibration ν that results when a momentary force is applied to a one-dimensional structure. Examples of this process include the plucking of a guitar string and the striking of a note on a piano or a xylophone. To formulate step 1 of our procedure, we must again draw to some extent on experimental evidence. We know from our experience that pitch, and therefore frequency, depends on the length l of the "structure," which in the case of a guitar string is varied by applying finger pressure at various positions. Piano keys are attached to wires, and xylophones have metal slabs that vary in length. Density ρ of the material is expected to play an additional role, since heavy materials produce a deeper sound (lower frequency).

We expect frequency to be also affected by the elasticity of the material. Our experience here is that a string that resists displacement will vibrate at a higher frequency on being momentarily displaced. To describe this phenomenon, we draw on the elasticity modulus, or Young's modulus E, first seen in Chapter 4 (Example 4.1). We know that E, which has the units of pressure (i.e., $ML^{-1}\theta^{-2}$), varies directly with the resistance to displacement. Thus, materials with high values of E will produce a high pitch or frequency. We conclude that the system will be described by the four variables ν, ρ, l, and E.

Let us now proceed to steps 2 and 3 of the analysis. The dimensions of the variables are given in the Table 8.3. Assuming that the relation is of the form

$$\nu = \text{const} \times l^{\alpha} \times E^{\beta}\rho^{\gamma} \tag{8.6a}$$

and upon replacing the variables with their corresponding dimensions, one obtains

$$\theta^{-1} = L^{\alpha}(ML^{-1}\theta^{-2})^{\beta}(ML^{-3})^{\gamma} \tag{8.6b}$$

Table 8.3 Dimensions for Example 8.2

Variable	Symbol	Dimensions
Frequency	ν	θ^{-1}
Length	l	L
Density	ρ	ML^{-3}
Elasticity	E	$ML^{-1}\theta^{-2}$

or equivalently

$$M^0L^0\theta^{-1} = L^{\alpha-\beta-3\gamma}M^{\beta+\gamma}\theta^{-2\beta} \tag{8.6c}$$

Equating the indices gives

$$\alpha - \beta - 3\gamma = 0 \qquad \text{Condition on L} \tag{8.6d}$$

$$\beta + \gamma = 0 \qquad \text{Condition on M} \tag{8.6e}$$

$$-2\beta = -1 \qquad \text{Condition on } \theta \tag{8.6f}$$

Solving the equations then gives

$$\alpha = -1 \qquad \beta = \tfrac{1}{2} \qquad \gamma = -\tfrac{1}{2} \tag{8.6g}$$

With these values, equation (8.6a) becomes

$$\nu = \text{const} \times \frac{(E/\rho)^{1.2}}{l} \tag{8.6h}$$

with alternative formulations given by

$$\frac{\nu l}{(E/\rho)^{1/2}} = \text{const} \tag{8.6i}$$

or

$$F(\pi) = F\left[\frac{\nu l}{(E/\rho)^{1/2}}\right] = 0 \tag{8.6j}$$

These are the relations that express, albeit only in semiquantitative fashion, the dependence of vibrational frequency ν on the system parameters.

COMMENTS

- We start by noting that the qualitative dependence of frequency on the system parameters observed experimentally is confirmed by the expression (8.6h). Thus frequency varies inversely with length: to double it, we must reduce the structure to half its length. We term this process "scaling" and note that it constitutes an important end use of the results obtained by dimensional analysis. It allows us to translate the observations made on a small model to predict the performance of a much larger unit, termed a prototype.

  Frequency also varies as expected directly with E or "stiffness" and inversely with ρ or "inertia." The dependence, however, is relatively weak, as nicely revealed by the square-root relation in equation (8.6h).

  We emphasize that the square root of elasticity to density equals the velocity of sound c in the material, that is,

$$c = \left(\frac{E}{\rho}\right)^{1/2} \tag{8.7}$$

Thus strings made of aluminum or steel, which produce approximately the same velocity of sound, will somewhat surprisingly have the same vibrational frequency.

This second simple example is again amenable, like the first one, to a quantitative analytical treatment. The governing expression here is the so-called wave equation, which is a partial differential equation. Its solution is more challenging than what we have seen so far and requires considerable mathematical background. Dimensional analysis avoids this difficulty and manages at the same time to reveal the important qualitative features of the system.

8.1.2 Systems with More Variables than Dimensions: The Buckingham π Theorem

The preceding example problems were solved with relative ease because the number of independent variables fortuitously equaled the number of fundamental dimensions. The resulting algebraic equations [(8.4c–e) and (8.6d–f)] could consequently be solved in straightforward fashion and the result expressed in terms of a *single* dimensionless group. When the number of variables exceeds that of the dimensions, more than one dimensionless group will necessarily be involved. Then we must determine how many such groups are to be brought in and how they are to be composed. These problems were first addressed by Buckingham and resolved in his famous π theorem. Simply stated, that theorem reads as follows.

Given p variables $x_1, x_2, \ldots, x_p$ related to a physical phenomenon that can be expressed in terms of r fundamental dimensions: these variables, which include the dependent variable, can be gathered into p − r dimensionless groups π and cast in the functional form

$$F(\pi_1, \pi_2, \ldots, \pi_{p-r}) = 0 \tag{8.8a}$$

This relationship now takes the place of the Rayleigh method we had used earlier, which resulted in the form

$$x_1 = \text{const}\; x_2^{\alpha} x_3^{\beta} \cdots x_p^{\pi} \tag{8.8b}$$

In other words, we have managed to replace a functional relation that involves p variables by one that involves only p − r variables. This constitutes a considerable simplification.

To implement this extended version of dimensional analysis, we retain the three steps we had formulated earlier in the introduction but amplify step 3 in the following fashion.

Step 3a: Select a set of variables that equals the number of dimensions and does not include the dependent variable. Raise each variable to some unknown power α, β, ... and form a product of these variables. This product is placed in the *denominator* of each dimensionless group π.

Step 3b: Place the remaining p − r variables, which now include the dependent variable, in the *numerators* of the p − r dimensionless groups. This results in the set of π's shown in equation (8.2).

Step 3c: Equate the dimensions in the numerator and denominator; evaluate the unknown powers of the denominator by requiring π to be dimensionless.

We demonstrate the use of this procedure in Example 8.3.

◆ EXAMPLE 8.3 Heat Transfer to a Fluid in Turbulent Flow

In Example 6.6, we modeled a heat exchanger in which heat was released by condensing steam and passed to a fluid in turbulent tubular flow. This process involved the use of a heat transfer coefficient, which is a complex function of the system variables. We had indicated in Chapter 1 that the evaluation of this function was beyond the reach of strictly analytical models and that it would be necessary to resort to a combination of dimensional analysis and experimentation. In what follows we first turn to the task of applying dimensional analysis to establish the functional form (8.8a) and then report the results of experiments that have led to a numerical quantification of this relation.

To implement step 1 of our procedure, we start by noting that both fluid mechanical and thermal properties are expected to affect the heat transfer coefficients. The former determine the degree of turbulence or ability to form eddies, hence the rate at which the hot fluid near the tubular wall is transported into the colder core of the flowing fluid. The thermal properties, on the other hand, dictate the rate at which heat is conducted and absorbed in the fluid being heated. We propose to use fluid velocity v, density ρ, and viscosity μ as the parameters that affect the degree of turbulence in the flow, and heat capacity C_p and thermal conductivity k as a measure of its capacity to store and transmit heat. We expect the pipe diameter to play a role as well since it determines the distance over which heat has to be transmitted. We can now proceed to step 2 and list in Table 8.4 the dimensions of these variables, some of which also can be found in Table 8.1.

We note that there are four fundamental units (i.e., M, L, θ, T), and we therefore require, for step 3a, four independent variables to be placed in the denominator of the dimensionless group. We choose v, d, ρ, and k, and proceed to form the first π by placing the dependent variable h in the numerator (step 3b). Hence

$$\pi_1 = \frac{h}{v^{\alpha_1} d^{\alpha_2} \rho^{\alpha_3} k^{\alpha_4}} \Rightarrow \frac{M\theta^{-3}T^{-1}}{(L\theta^{-1})^{\alpha_1}(L)^{\alpha_2}(ML^{-3})^{\alpha_3}(ML\theta^{-3}T^{-1})^{\alpha_4}} \tag{8.9a}$$

Equating the dimensions in the numerator and denominator as required by step 3, we obtain

$$1 = \alpha_3 + \alpha_4 \quad \text{Condition on M} \tag{8.9b}$$

$$0 = \alpha_1 + \alpha_2 - 3\alpha_3 + \alpha_4 \quad \text{Condition on L} \tag{8.9c}$$

Table 8.4 Dimensions for Example 8.3

Variable	Symbol	Dimension
Velocity	v	$L\theta^{-1}$
Density	ρ	ML^{-3}
Viscosity	μ	$ML^{-3}\theta^{-1}$
Thermal conductivity	k	$ML\theta^{-3}T$
Heat capacity	C_p	$L^2\theta^{-2}T^{-1}$
Heat transfer coefficient	h	$M\theta^{-3}T^{-1}$
Diameter	d	L

$$-3 = -\alpha_1 - 3\alpha_4 \qquad \text{Condition on } \theta \tag{8.9d}$$

$$-1 = -\alpha_4 \qquad \text{Condition on T} \tag{8.9e}$$

Solving these four equations simultaneously yields $\alpha_1 = \alpha_3 = 0, \alpha_2 = -1$ and $\alpha_4 = 1$, and consequently from equation (8.9a)

$$\pi_1 = \frac{hd}{k} = Nu \tag{8.9f}$$

This group is termed the Nusselt number and is given the symbol Nu.

Placing next the viscosity μ with dimensions $ML^{-1}\theta^{-1}$ in the numerator, altering the left side of conditions (8.9b–e) accordingly, replacing the exponent α with β, we obtain

$$1 = \beta_3 + \beta_4 \qquad \text{Condition on M} \tag{8.10a}$$

$$-1 = \beta_1 + \beta_2 - 3\beta_3 + \beta_4 \qquad \text{Condition on L} \tag{8.10b}$$

$$-1 = -\beta_1 - 3\beta_4 \qquad \text{Condition on } \theta \tag{8.10c}$$

$$0 = -\beta_4 \qquad \text{Condition on T} \tag{8.10d}$$

and consequently $\beta_1 = \beta_2 = \beta_3 = 1, \beta_4 = 0$. Thus the second dimensionless group becomes

$$\pi_2 = \frac{\mu}{vd\rho} \tag{8.10e}$$

The inverse of this group is the celebrated Reynolds number, which is given the symbol Re.

The final placement is that of the heat capacity C_p in the numerator of the third dimensionless group π_3. We obtain the equations

$$0 = \gamma_3 + \gamma_4 \qquad \text{Condition on M} \tag{8.11a}$$

$$2 = \gamma_1 + \gamma_2 - 3\gamma_3 + \gamma_4 \qquad \text{Condition on L} \tag{8.11b}$$

$$-2 = -\gamma_1 - 3\gamma_4 \qquad \text{Condition on } \theta \tag{8.11c}$$

$$-1 = -\gamma_4 \qquad \text{Condition on T} \tag{8.11d}$$

with the result $\gamma_1 = \gamma_2 = \gamma_3 = -11, \gamma_4 = 1$, so that

$$\pi_3 = \frac{C_p vd\rho}{k} \tag{8.11e}$$

Substitution of equation (8.10e) into (8.11e) yields the alternative formulation

$$\pi_3 = \frac{C_p\mu}{k} Re \tag{8.11f}$$

where the group $C_p\mu/k$ is termed the Prandtl number and given the symbol Pr.

Following equation (8.8a), the sum total of these results can be expressed in the form

$$f(\mathrm{Nu}, \mathrm{Pr}, \mathrm{Re}) = 0 \quad (8.12a)$$

or alternatively as

$$\mathrm{Nu} = F(\mathrm{Re}, \mathrm{Pr}) \quad (8.12b)$$

This is the expression that is used to correlate experimental measurements of heat transfer coefficients in cylindrical tubes as well as in a host of other geometries. Note that we have managed to reduce the number of variables from seven to three.

COMMENTS

- Evidently some guesswork was involved in the choice of independent parameters, but the variables ultimately chosen make good physical sense. Thus the formation of eddies, which affects proportionately the magnitude of the heat transfer coefficient, will rise with increasing velocity but will be resisted by highly viscous fluids. On the other hand, both the capacity to transmit heat and the rate at which heat is transmitted will be affected by heat capacity, thermal conductivity, and density of the fluid. This reassures us that we have made a reasonable choice of variables. There is, however, still an element of inspired guesswork involved.

  The second point of note is the designation of the dimensionless groups after prominent contributors to the field. Table 8.5 gives a short list of such dimensionless numbers drawn from the fields of fluid mechanics and heat and mass transfer. The Mach number, which is the velocity of the fluid divided by the velocity of sound, is associated with supersonic gas flow, while the Froude number arises in the formation of waves and is used in ship design. In these expressions, l is the principal linear dimension.

Table 8.5 Dimensionless Numbers

Name	Definition
Fluid flow	
Reynolds number	$\mathrm{Re} = \frac{lv\rho}{\mu}$
Mach number	$\mathrm{Ma} = \frac{v}{C}$
Froude number	$\mathrm{Fr} = \frac{v^2}{lg}$
Heat transfer	
Nusselt number	$\mathrm{Nu} = \frac{hl}{k}$
Prandtl number	$\mathrm{Pr} = \frac{C_p\mu}{k}$
Stanton number	$\mathrm{St} = \frac{h}{v\rho C_p} = \frac{\mathrm{Nu}}{\mathrm{RePr}}$
Mass transfer	
Sherwood number	$\frac{k_C l}{D}$
Schmidt number	$\frac{\mu}{\rho D}$

Table 8.6 Correlations for Drag and Heat Transfer Coefficients

Drag Coefficients		Heat Transfer Coefficients	
Flat plate[a]		Flate plate	
$Re < 5 \times 10^5$	$C_D = \frac{1.33}{Re^{1/2}}$	$Re < 10^5$	$St = 0.66(Re)^{-1/2}(Pr)^{-2/3}$
$Re > 10^7$	$C_D = \frac{0.072}{Re^{1/5}}$	$Re > 10^5$	$St = 0.036(Re)^{-0.2}(Pr)^{-2/3}$
Sphere		Sphere	
$Re < 0.1$	$C_D = \frac{24}{Re}$		$Nu = 2.0 + 0.6\,(Re)^{1/2}(Pr)^{1/3}$
$Re > 10^5$	$C_D = 0.22$		
Cylinder[b]		Cylinder	
$Re < 10$	$C_D = \frac{0.33}{Re}$	$1 < Re < 4000$	$Nu = 0.43 + 0.53\,Re^{0.5}\,Pr^{0.31}$
$Re > 10^5$	$C_D = 0.36$		
		Inside tubes	
		$Re > 2100$	$Nu = 0.026(Re)^{0.89}(Pr)^{1/3}$

[a]Body parallel to flow.
[b]Axis normal to flow.

Finally we remind the reader that dimensional analysis merely reduces the number of variables that need to be examined for a particular process. It does not yield a quantitative relation between the variables. That relation must be established experimentally, but the effort is now much less than it would have been in the absence of dimensional analysis. Some correlations derived from these experiments listed in Table 8.6, include drag coefficients, which are taken up in the Example 8.4, as well as heat transfer coefficients.

◆ EXAMPLE 8.4 Drag on Submerged Bodies, Horsepower of a Car

Any relative motion between a fluid and a body fully or partially submerged in the fluid causes a drag force on the body. We have already alluded to this force as a factor in determining the trajectory of a projectile (recall Example 4.5 and Practice Problem 4.8). There is a vast array of similar situations in which drag arises and plays a crucial role in determining the system behavior. A body falling through air or a diver rising through water experiences a drag force that, along with the gravity and buoyancy forces, determine the velocity of the entity. Aircraft in flight must overcome the drag force due to their own motion as well as that of the surrounding air. Stationary bodies such as buildings, or moored marine structures such as piers or offshore oil rigs, are similarly exposed to drag forces caused by wind and waves.

When the velocities involved are low—that is, when flow is in the so-called laminar regime—and the body shape is simple (sphere, flat plate), a fully analytical solution is possible, although only with considerable difficulty. More complex shapes in laminar flow can nowadays be handled by numerical methods collectively known as computational fluid dynamics (CFD). In most other cases, however, particularly those that give rise to turbulent conditions, one still must resort to a combination of experimentation and dimensional analysis. We address this latter aspect here.

As on earlier occasions we start our deliberations by determining the variables involved in the process. Drag force F_D is the chosen dependent variable, while relative velocity v, viscosity μ, density ρ, and some linear dimension of the body, l, are the principal independent variables. The role of the first three parameters in determining the intensity of turbulence was discussed in the preceding example. The choice of l arises from the intuitive notion that the size of the body will have something to do with the magnitude of the drag force. Whether it enters the final functional relation as a volume, an area, or merely a linear dimension is ultimately determined by the dimensional analysis. Note, however, that we have specified only *one* such dimension; that is, we have limited ourselves to simple forms such as a flat plate, a sphere, or a cylinder. Shape and orientation of the body in the flow field will evidently also play a role, but we sidestep this issue for the time being by confining our analysis to a *particular* form and orientation, (e.g., a flat plate parallel to the direction of flow). We shall have more to say about these factors in the Comments section.

Let us choose the three variables ρ, l, and v for the denominator, as well as drag force F_D and viscosity μ for the numerator, giving us the following dimensionless groups:

$$\pi_1 = \frac{F_D}{\rho^{\alpha_1} l^{\alpha_2} v^{\alpha_3}} \quad \text{and} \quad \pi_2 = \frac{\mu}{\rho^{\beta_1} l^{\beta_2} v^{\beta_3}} \tag{8.13a}$$

Dimensionally, these parameters become

$$[\pi_1] = \frac{ML\theta^{-2}}{(ML^{-3})^{\alpha_1}(L)^{\alpha_2}(L\theta^{-1})^{\alpha_3}} \quad \text{and} \quad [\pi_2] = \frac{ML^{-1}\theta^{-1}}{(ML^{-3})^{\beta_1}(L)^{\beta_2}(L\theta^{-1})^{\beta_3}} \tag{8.13b}$$

Equating the corresponding exponents in numerator and denominator, and solving the resulting systems of equations, we obtain

$$\begin{aligned} \alpha_1 &= 1 & \beta_1 &= 1 \\ \alpha_1 &= 2 & \beta_2 &= 1 \\ \alpha_3 &= 2 & \beta_3 &= 1 \end{aligned} \tag{8.13c}$$

These values are substituted into equation (8.13a) and the two groups gathered together into the functional relationship (8.8a):

$$f\left(\frac{F_D}{\rho l^2 v^2}, \frac{\mu}{\rho l v}\right) = 0 \tag{8.13d}$$

where we recognize the second group as the inverse of the Reynolds number Re we encountered in Example 8.3. We can therefore write equation (8.13d) in the alternative form

$$F_D = \rho l^2 v^2 F(\text{Re}) \tag{8.13e}$$

We note that the drag force varies with the square of the linear dimension l. The resulting area A_C is taken to be the projected cross-sectional area of the body normal to the flow, except in the case of the flat plate, where it equals the lateral area parallel to the direction of flow. The product ρv^2 is seen to be proportional to the kinetic energy of the fluid. We emphasize this by dividing the right side of equation (8.13e) by 2 and denote the residual functional form as the "drag coefficient" C_D. There results

$$F_D = C_D(Re)\tfrac{1}{2}\rho v^2 A_C \tag{8.13f}$$

Let us use this expression to calculate the drag force required to move objects through air at a speed of 100 km/h. Air has a normal density ρ of 1.18 kg/m^3 and a viscosity of 1.84×10^{-5} Pa s. Choosing first a sphere of diameter 1 m, we obtain a Reynolds number of

$$Re = \frac{\rho l v}{\mu} = \frac{1.18 \times 1 \times 10^5/3600}{1.84 \times 10^{-3}} = 1.8 \times 10^6 \tag{8.14a}$$

so that (see Table 8.6)

$$C_D = 0.22 \tag{8.14b}$$

Consequently the drag force is given by

$$F_D = 0.22 \times \frac{1}{2} \times 1.18 \left(\frac{10^5}{3600}\right)^2 \frac{\pi \times 1^2}{4} = 80\,\text{N} \tag{8.14c}$$

For motor vehicles, the drag coefficient is somewhat higher: $C_D \approx 0.3$ for passenger cars and $C_D \approx 1$ for tractor-trailer trucks. Let us assume a projected area of 3 m^2 for an average modern passenger car and calculate both drag force and the power required to overcome the air resistance. We have

$$F_D = 0.3 \times \frac{1}{2} \times 1.18 \left(\frac{10^5}{3600}\right)^2 3 = 410\,\text{N} \tag{8.15a}$$

Since 1 kJ/s = 1 kW = 0.7846 hp (horsepower), we have

$$P = \frac{11.3}{0.746} = 15.2\ \text{hp} \tag{8.15b}$$

This is the power required to overcome air resistance when driving at 100 km/h. Now the resistance due to the rolling friction varies linearly with velocity and equals that caused by air drag at about 100 km/h (~60 mph). Thus to drive at this speed, approximately 30 hp is required to overcome air resistance and rolling friction. At 150 km/h this rises to $15 \times 1.5 + 15 \times 1.5^2 = 58$ hp. Even higher speed and rapid acceleration add to these requirements. Consequently passenger car engines are usually designed to generate about 100 to 150 hp.

COMMENTS

- The reader will have noted from Table 8.6 and our earlier comments that both shape and orientation play a role in determining the drag force. Thus equation (8.13f) should actually read

$$F_D = C_D(Re, \text{shape, orientation})\tfrac{1}{2}\rho v^2 A \tag{8.16}$$

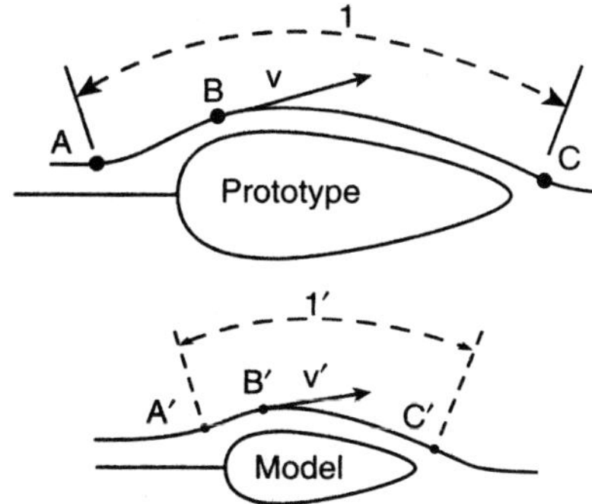

Figure 8.2 Similarity of geometry and orientation.

This complication can be avoided, and the same equation (8.13f) applied to different bodies, provided they are "similar" both in geometry and in orientation. Figure 8.2 shows two such similar bodies, a small-sized model, and its larger counterpart, the prototype. The two are seen to be perfect replicas, and thus both satisfy geometric similarity. This implies, for example, that if the cross-sectional areas of the bodies at some point one-quarter of the way along the axis are in the ratio 1:10, then the areas at the midpoint or three-quarters of the way to the rear tip must equally be in the ratio 1:10. In the same way, similarity in orientation implies that if the velocities or forces at corresponding points B and B′ are in a particular ratio, then the same ratio applies at all other corresponding points, for example, C and C′. This correspondence is referred to as kinematic and dynamic similarity.

One additional factor that affects the drag force is surface finish. We know intuitive by that a rough surface will give rise to higher drag forces than a smooth one. Geometric similarity can be claimed only if the ratios of roughness (i.e., the height of the protuberances), in the model and in the prototype are identical to the ratio of other dimensions, say the diameter. This similarity is of considerable importance in model studies. Thus while one may be satisfied with a rough casting of a marine propeller or a turbine, any experiment on a scaled-down model must be carried out with a correspondingly smooth surface finish. This frequently entails painstaking work.

We leave the topic of fluid mechanics and turn our attention to an example involving principles of elasticity. We dealt with these principles earlier in the context of solid mechanics (see Examples 4.1 and 4.2) and will now extend them to fluid media. In particular, we shall address the phenomena associated with underwater explosions. The process here is again a highly complex one that does not easily yield to a fully analytical treatment. Dimensional analysis provides us with an avenue by which to arrive at some first qualitative results, which can nevertheless be exploited to give some highly useful answers.

◆ EXAMPLE 8.5 Design of a Depth Charge

When a high explosive is detonated under water, it is converted almost instantaneously into a gas. The initial pressure generated, p_0, depends on the nature of the explosive.

For example, for TNT p_0 is 10^5 atm (10 GPa). Mass does not influence this value but does affect the size of the initial pressure cavity.

The explosion creates a spherical shock wave that is transmitted in the form of a pressure front of magnitude p, where p depends on both the radial distance r and the time t. The underlying model is consequently a partial differential equation (the same wave equation we saw Example 8.2), the solution of which requires considerable experience and effort. We wish to avoid this by making use of dimensional analysis to arrive at a partial quantification of the process.

The problem to be addressed entails a newly developed explosive being considered for use in antisubmarine warfare. In a model test with 0.5 kg of the explosive, a shock pressure of 1000 atm was generated at a distance of 10 m. It is estimated that this is the minimum pressure required to damage or destroy a submarine.

Suppose that the accuracy of delivery places the explosion within 100 m of a hostile vessel. What is the mass of explosive needed for the device to be effective (i.e., for the pressure to be 1000 atm at a distance of 100 m from the explosion)?

It is not a priori known that dimensional analysis will provide that answer. We expect, on physical grounds, that the shock pressure will vary directly with the size of the charge and inversely with distance r, raised to some unknown power. There is some hope of gathering these variables into a single dimensionless ratio, in which case the answer could be found by simple scaling.

The choice of additional variables must start with the realization that pressure is propagated by collisions between neighboring particles. In liquids and solids, that collision, or transition of pressure, is brought about by vibrations of the constituent molecules. The larger the mass of these molecules, the greater their inertia and consequently the smaller the amplitude of the induced vibrations. We tentatively conclude from this that pressure will vary inversely with the density of the material: that heavy materials will dampen the pressure and act as shock absorbers. We also sense that the more closely packed the molecules, the faster the transmission of the pressure pulse. Put another way, materials that are difficult to compress, or are "hard," enhance the speed at which pressure propagates.

This compressibility is usually expressed through the so-called bulk modulus K, which is the proportionality constant relating applied pressure p and relative volume change $\Delta V/V$. Thus

$$p = K\frac{\Delta V}{V} \tag{8.17}$$

where K has the dimension of pressure.

The reader will note the similarity of this expression to equation (4.4), which has the Young modulus E as proportionality constant and relates applied stress to linear elongation. Equation (8.17) can thus be viewed as a three-dimensional analogue to equation (4.4), albeit an inverted one, since it considers contraction rather than expansion.

We find from these deliberations that six parameters are involved in the process: the dependent variable p and the independent variables p_0, K, mass of the explosive m, density of the medium ρ, and radial distance r.

Since these can be expressed in terms of the three basic dimensions M, L, θ, we have, according to Buckingham's theorem, three dimensionless groups, π_1, π_2, π_3.

Although the standard procedure outlined earlier for determining these groups can be applied here as well, we prefer in this instance to use some convenient shortcuts that establish the result in rapid fashion. We argue as follows.

A first dimensionless group can be immediately written down by nondimensionalizing pressure with respect to the detonation pressure p_0. Thus

$$\pi_1 = \frac{p}{p_0} \tag{8.18a}$$

and consequently

$$\frac{p}{p_0} = f(\pi_2, \pi_3) \tag{8.18b}$$

We then place p_0 and m in the numerator (of π_2 and π_3) and the remaining variables in the denominators. Hence

$$\pi_2 = \frac{p_0}{\rho^{\alpha_1} r^{\alpha_2} K^{\alpha_3}} = \frac{ML^{-1}\theta^{-2}}{(ML^{-3})^{\alpha_1}(L)^{\alpha_2}(ML^{-1}\theta^{-2})^{\alpha_3}} \tag{8.18c}$$

and

$$\pi_3 = \frac{m}{\rho^{\beta_1} r^{\beta_2} K^{\beta_3}} = \frac{M}{(ML^{-3})^{\beta_1}(L)^{\beta_2}(ML^{-1}\theta^{-2})^{\beta_3}} \tag{8.18d}$$

A quick inspection of these groups and their dimensions shows that they can be nondimensionalized by setting $\alpha_1 = \alpha_2 = \beta_3 = 0$, $\alpha_3 = \beta_1 = 1$, and $\beta_2 = 3$. We consequently have

$$\frac{p}{p_0} = f\left(\frac{p_0}{K}, \frac{m}{\rho r^3}\right) \tag{8.18e}$$

This result, although arrived at in a somewhat intuitive fashion, nevertheless satisfies all the requirements of the π theorem, namely, the ratios are dimensionless and three in number.

We now turn to a numerical evaluation of the answer being sought. Since the pressure ratio p/p_0 is the same for both model and prototype explosions, the groups on the right must likewise be the same for the two detonations. We have

$$\left(\frac{m}{\rho r^3}\right)_{\text{model}} = \left(\frac{m}{\rho r^3}\right)_{\text{prototype}} \tag{8.18f}$$

or

$$\left(\frac{0.5}{\rho 10^3}\right)_{\text{model}} = \left(\frac{m}{\rho 100^3}\right)_{\text{prototype}} \tag{8.18g}$$

From this it follows that

$$\boxed{(m)_{\text{prototype}} = 500\ \text{kg}} \tag{8.18h}$$

That is, to satisfy the stated conditions, the prototype charge must be one thousand times larger than that used in the model.

COMMENTS

- Here is yet another example of the power of dimensional analysis to resolve complex problems. In this instance a single test was sufficient to arrive at the desired answer, since p/p_0 was kept constant and the result could be obtained by simple scaling of the dimensionless group $m/\rho r^3$. If it had been desired to calculate the pressure generated by a given mass m at a particular location r, we would have had to quantify the expression (8.18e), either by analytical methods or through extensive testing.

8.2 Numerical Methods

8.2.1 Introduction

Thus far we have restricted ourselves to analytical treatments, only occasionally selecting numerical methods (see Examples 3.5 and 3.11) and not elaborating on the procedures involved. We left numerical methods for the last chapter, not out of any wish to minimize their importance but because of the belief that they typically should be introduced only after some experience had been gained in setting up models and solving them by simple analytical methods.

The use of numerical methods is nowadays all pervasive in the scientific and engineering world. Entire disciplines, such as computational fluid dynamics, are devoted to their application and to the development of appropriate software. Much if not most of this work deals with the solution of partial differential equations.

In our treatment we shall limit ourselves to an exposition of the principal numerical tools needed in the solution of the following systems of equations:

Simultaneous linear algebraic equations (SLAE)

Single nonlinear equations (NLE)

Simultaneous nonlinear equations (SNLE)

Ordinary differential equations (ODE)

The methodology for solving these systems is well established and has been incorporated in software packages for scientific and engineering use. We address these briefly before embarking on the actual numerical solution of equations.

8.2.2 Numerical Software Packages

The development of software packages that comprise solution techniques for a range of problems dates back to the late 1960s and early 1970s. An early and popular version was the IMSL package (International Mathematical and Statistical Library), with capabilities for solving algebraic and ordinary differential equations of the initial value type, as well as statistical problems. The relentless drive for greater sophistication and flexibility led to the development, in the early 1980s, of the currently popular packages MATHEMATICA and MAPLE, joined at a later date by the equally popular MATLAB package. Those software packages are being continually refined and expanded, and we give a brief synopsis of their properties and current capabilities.

MATHEMATICA and MAPLE are comparable in performance and capabilities, which include graphics and sound. They provide solutions of algebraic equations and ODEs, and PDEs of both the IVP and BVP types. In addition they are capable of handling AE/ODE/PDE combinations, which often arise in modeling, and they can be linked to other programs written in FORTRAN and C. These standard features can be enhanced by add-on packages such as finite element modeling (MAPLE) and time series analysis (MATHEMATICA).

The outstanding feature of both these packages, and one that distinguishes them from MATLAB, is their ability to handle symbolic operations. This enables them to provide the user with exact analytical solutions, where they exist, of algebraic and differential equations. MATHEMATICA, which was the first on the scene with this capability, still excels in the field of symbolic manipulation, which includes the analytical evaluation of integrals. This dual capability of handling numerical and analytical solutions makes MATHEMATICA the software of choice for many users. Example 8.9 provides an example of this aspect.

The strength of the MATLAB package, a relative newcomer to the scene, lies in matrix-related operations. It includes tools for data acquisition and analysis (e.g., regression and curve fitting) and has capabilities as well for visualization and image processing. Since most solution methods for simultaneous equations involve vector–matrix operations, MATLAB gives good results in handling sets of algebraic and ordinary differential equations. As well, its capabilities can be greatly enhanced by add-on packages, particularly in the fields of control and of finance.

In the sections that follow, we shall describe and apply several numerical solution methods (Gaussian elimination, the Newton and Newton–Raphson algorithms, and the Euler and Runge–Kutta routines). All of these can be implemented by any one of the three software packages we have mentioned. Our own preference is for MATHEMATICA, and it is this package we shall apply in our examples.

8.2.3 Numerical Solution of Simultaneous Linear Algebraic Equations: Gaussian Elimination

In Section 2.2 dealing with matrix operations, we acquainted the reader with the Gaussian elimination method for solving sets of linear algebraic equations. These sets can be represented by the matrix–vector equation

$$\mathbf{AX} = \mathbf{B} \tag{2.53}$$

in which $\mathbf{A}$ denotes the coefficient matrix of the vector $\mathbf{X}$ of the unknowns and $\mathbf{B}$ is the column vector comprising the nonhomogeneous terms. Gaussian elimination was seen as consisting of the systematic reduction, by a series of elementary row operations of the augmented matrix $\mathbf{A:B}$ to echelon form, from which the solutions could be easily extracted by back-substitution. Example 8.6 provides a typical computer output obtained in these operations.

◆ EXAMPLE 8.6 The Global Positioning System Revisited: Using the MATHEMATICA Package for Gaussian Elimination

In Example 3.5 we gave the results of Gaussian elimination in the solution of three equations that arose in connection with the use of the global positioning system (GPS). The computer output obtained in connection with the problem looks like this:

```
GPS
The 3-dimensional coordinates of each satellite are the
    following:
```

x_1 := 2088202.299

y_1 := -22757191.370

z_1 := 25391471.881

x_2 := 11092568.240

y_2 := -14198201.090

z_2 := 21471165.950

x_3 := 35606984.591

y_3 := -4447027.239

z_3 := 9101378.572

x_4 := 3966929.048

y_4 := 7362851.831

z_4 := 26388447.172

The distances registered in the GPS device are:

r_1 := 23204698.51

r_2 := 21585835.37

r_3 := 31364260.01

r_4 := 24966798.73

eqn1 := $2\ (x_2 - x_1)\ x + 2\ (y_2 - y_1)\ y + 2\ (z_2 - z_1)\ z = r_1^2 - r_2^2 - x_1^2 + x_2^2$

eqn2 := $2\ (x_3 - x_1)\ x + 2\ (y_3 - y_1)\ y + 2\ (z_3 - z_1)\ z = r_1^2 - r_3^2 - x_1^2 + x_3^2$

eqn3 := $2\ (x_4 - x_1)\ x + 2\ (y_4 - y_1)\ y + 2\ (z_4 - z_1)\ z = r_1^2 - r_4^2 - x_1^2 + x_4^2$

The equations with numerical coefficients are:

$1.80087 \times 10^7\ x - 4.88202 \times 10^6\ y - 7.84061 \times 10^6\ z$ == 1.91194×10^{14}

$6.70376 \times 10^7\ x + 1.46203 \times 10^7\ y - 3.25802 \times 10^7\ z$ == 8.18238×10^{14}

$3.75745 \times 10^6\ x + 3.82401 \times 10^7\ y + 1.99395 \times 10^6\ z$ == -7.35071×10^{13}

In[721] := Solve[{eqn1, eqn2, eqn3}, {x, y, z}]

Out[721] = {{x → -3.0168×10^6, y → 7231.03, z → -3.13188×10^7}}

COMMENTS

- The striking feature in this output is the terseness of the statements implementing the solution. The bulk of the output is taken up with the description of the system to be solved and a listing of the numerical coefficients of the equations involved. Implementation takes up only two lines and is obtained with the command In[721], where Solve is the code for the Gaussian elimination routine and Out[721] is the result of its application. The entire procedure is thus hidden in a "black box," which can be cracked open to provide more solution details, albeit at the cost of some additional effort. We do this shortly, in Example 8.7.

Black boxes containing standard AE or ODE solution methods often run to several hundred pages of instructions or more, which the typical user never sees. We have tried to alleviate the discomfort this creates by providing a fairly detailed synopsis of the solution methods involved (see also Chapter 2). This makes for greater confidence in applying these methods and enables us as well both to interpret the reasons for termination that occasionally appear in the output and to take appropriate countermeasures. An example of this will be seen in Example 8.8.

8.2.4 Numerical Solution of Single Nonlinear Equations: Newton's Method

The solution of single nonlinear equations is a frequent requirement in the sciences and engineering, although it is somewhat overshadowed by the even greater need to solve *sets* of such equations. The methodology for solving simultaneous equations is often derived from solution methods for single equations, and it is therefore appropriate to start our deliberations with this simpler but nevertheless important case.

The most frequently encountered form of nonlinear equation is the polynomial equation, followed by trigonometric and other transcendental expressions. The roots may be real, complex, or multiple, or infinite in number. Some of these more complex cases are addressed in our comments at the end of the section. Our principal concern will be to establish the methodology for locating *single real* roots.

Among the earliest attempts to solve nonlinear equations numerically was the procedure known as Newton's method. It is based on a simple and elegant application of calculus and has survived the passage of time remarkably well. Contemporary refinements often use the method as a starting point, and the extension to simultaneous equations known as the Newton–Raphson method is based on the concept of Newton's original procedure.

The derivation of Newton's method is best described graphically, as shown in Fig. 8.3A.

We start by casting the equation in the form

$$f(x) = 0 \tag{8.19a}$$

and choose a point x_0 on the abscissa so that in general

$$f(x_0) \neq 0 \tag{8.19b}$$

We next draw a tangent to the curve f(x) at the point x_0, and find its intersection x_1 with the abscissa. The procedure is repeated, using x_1 and subsequent values $x_2, x_3, \ldots,$ as the new starting points until the root x_r is reached. This completes the solution.

To cast the procedure in mathematical form, we use the fact that the slope of the tangent at $x = x_0$ equals the derivative $f'(x_0)$, which in turn is equal to the ratio of the ordinate $f(x_0)$ to

(A)

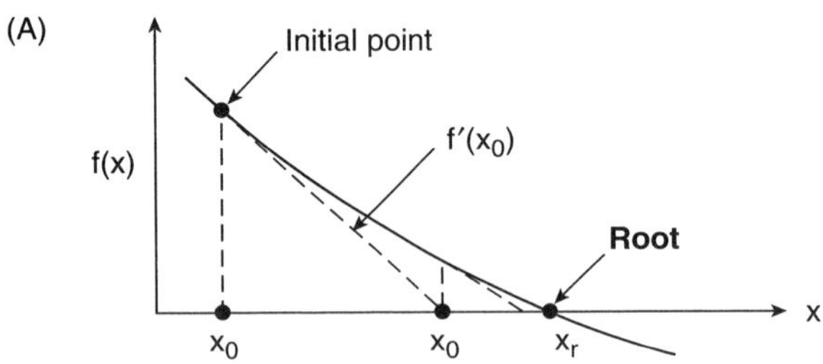

(B)

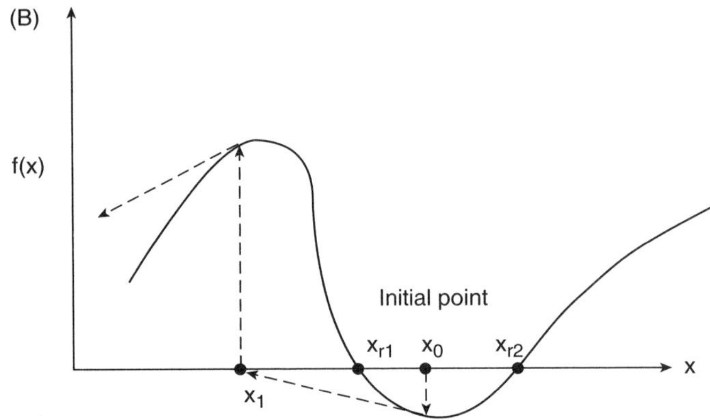

Figure 8.3 Newton's method: (A) convergence to root and (B) divergence.

the abscissa difference $x_1 - x_0$. We have

$$\frac{f(x_0)}{x_1 - x_0} = -\left(\frac{df}{dx}\right)_{x=x_0} = -f'(x_0) \tag{8.19c}$$

Solving for $x_1 - x_0$ and generalizing the procedure, we arrive at the relation

x_{n+1}	$-$	x_n	$=$	$-\dfrac{f(x_n)}{f'(x_n)}$
New value of x closer to the root x_r		Old value of x		Function over derivative of the function at x_n

(8.19d)

This is the algorithm that implements the solution of nonlinear equations by Newton's method. We shall see in the next section that it can be further generalized into a vector–matrix form, which is used as an algorithm in the solution of simultaneous nonlinear equations.

Comments We start by noting that convergence to the root being sought is not automatic. It happens on occasion that the initial value chosen for x_0 results in an overshooting of the root, as shown in Fig. 8.3B, or in an immediate divergence away from the root. This type of instability in the solution of nonlinear equations was among the first sources of chaotic behavior discovered in early studies of the topic. To overcome the difficulty, one uses the simple device of adjusting the initial guess, guided wherever possible by physical reasoning, until convergence is obtained.

To solve for complex roots, a variety of methods are available. A preferred procedure is to extend Newton's method by substitution of the complex variable $z = x + iy$ into the equation to be solved, $f(x) = 0$. We obtain

$$f(z) = u(x, y) + iv(x, y) = 0 \tag{8.20a}$$

and consequently

$$u(x, y) = 0 \tag{8.20b}$$

$$v(x, y) = 0 \tag{8.20c}$$

The problem is thus reduced to the solution of two simultaneous nonlinear equations that can be implemented by the Newton–Raphson method, described in the next section.

Multiple roots arise when $f(x) = 0$ is tangent to the abscissa at x_r. Here Newton's method works in principle but shows very slow convergence, since $f(x)/f'(x) = 0/0$ at the root. Special extensions of Newton's method involving the use of second derivatives have been successful in overcoming this difficulty.

We note that most physical problems involve single real roots, and the complications due to complex or multiple roots do not arise. Furthermore, one can usually find a solution upon setting upper and lower physical bounds that are in fairly close proximity of each other. An initial guess can therefore be made that is within reasonable reach of the solution being sought, resulting in rapid convergence. There are exceptions, however: the polynomials that result from D-operator equations of order greater than 2 or from the Laplace transformation of simultaneous ODEs. Both real and complex roots are of interest here and must therefore be sought out. Problems of this type arise in solid mechanics and in process control, but we do not address them here.

We now turn to an illustration of the application of Newton's method. Our example is based on the concept of chemical equilibrium, which is a rich source of nonlinear algebraic, or polynomial equations. With this example, we introduce the notion of a chemical equilibrium constant that describes these systems and then use it to address a problem of industrial importance.

◆ EXAMPLE 8.7 Chemical Equilibrium: The Synthesis of Ammonia by the Haber Process

In many chemical reactions, the starting materials, or reactants, are not quantitatively converted into product. They proceed instead to an equilibrium position in which the product pressures or concentrations stand in a constant ratio to the pressures or concentration of unreacted starting materials. That ratio is known as the equilibrium constant, which has a fixed numerical value for all concentrations and pressures and varies only with temperature.

Consider the reaction describing the synthesis of ammonia (NH_3), an important industrial chemical, from its constituent elements nitrogen and hydrogen. The reaction equation is given by

$$N_2 + 3H_2 \rightleftarrows 2NH_3 \tag{8.21}$$

where we have used two opposing arrows to indicate that ammonia is produced as well as decomposed, leading to an equilibrium among the pressures of the participating species. That equilibrium is expressed through an equilibrium constant K_p, which has the following form:

$$\frac{p_{NH_3}^2}{p_{N_2} p_{H_2}^3} = K_p(T) \tag{8.22}$$

Here the symbol p is used to denote the so-called partial pressure of a particular species; p can be viewed as the fraction of the total pressure P_T due to the presence of that species. Note that each partial pressure is raised to a power equal to the corresponding coefficient in the reaction equation (8.21). This is a general feature of all equilibrium constants.

Partial pressure and total pressure are related through the expression

$$p_i = \frac{n_i}{n_T} P_T \tag{8.23}$$

where n_i is the number of moles of species I and n_T denotes the total number of moles of all participating gaseous molecules. This relation makes physical sense because it states that the pressure contributed by each species, p_i, varies directly, as it must, with the number of molecules of that species. Introducing the relation into equation (8.22) leads to the following alternative formulation of the equilibrium constant:

$$\frac{n_{NH_3}^2}{n_{N_2} \times n_{H_2}^3} \left(\frac{n_T}{p_T}\right)^2 = K_p(T) \tag{8.24}$$

This is a more convenient expression for practical calculations because it contains the *moles* of each species, which are the quantities of direct interest. Note that n_{N_2} and n_{H_2} are the amounts that prevail at equilibrium and are not to be confused with the initial amounts introduced.

Let us see what use we can make of this expression. Suppose the reaction is to be carried out at a particular total pressure P_T and temperature T and that we wish to calculate the amount of ammonia formed, n_{NH_3}. This is a seemingly impossible task, since the expression (8.24) contains the four unknowns n_{NH_3}, n_{H_2}, n_{N_2} and n_T. These unknowns, however, are not independent variables but are related to one another through the coefficients seen in the chemical reaction equation (8.21). Let us start with one mole of nitrogen and three moles of hydrogen, and let us also denote the amount of ammonia formed by 2x. We can then compose the following "balance sheet" or material balance:

Reaction:	N_2	+	$3H_2$	$\rightleftarrows$	$2NH_3$	
Moles at start:	1		3		0	
Moles at equilibrium:	$1 - x$		$3 - 3x$		$2x$	(8.25)
Total moles at equilibrium:	$n_T = 1 - x + 3 - 3x + 2x = 4 - 2x$					

As a consequence of these calculations, we can express equation (8.24) in terms of the single variable x, as follows:

$$\frac{2x^2}{(1-x)(3-3x)^3}\left(\frac{4-2x}{P_T}\right)^2 = K_p \tag{8.26a}$$

This is the equation that now is to be solved for x, to obtain the number of moles of ammonia formed. Note that this amount represents the maximum that can be produced under the specified conditions of temperature and total pressure. In the industrial production of ammonia, it is impractical to drive the reaction to complete equilibrium, and thus yields lower than the maximum must be accepted.

The reaction rate for ammonia synthesis is extremely low at normal conditions and must be augmented through the use of catalysts and high temperatures to reach useful production levels. High pressures also favor ammonia formation, as can be deduced from equation (8.24). Current practice calls for the use of iron catalysts in combination with certain metal oxides, temperatures of several hundred degrees, and pressures ranging up to 1000 atm. Let us choose a temperature of 500 °C and a pressure of 500 atm and use equation (8.26a) to calculate the amount of ammonia produced under these conditions. The equilibrium constant for the reaction at 500 °C has a value of $K_p = 1.45 \times 10^{-5}$ atm^{-2}. We obtain

$$\frac{8x^2(2-x)^2}{27(1-x)^4} = 1.45 \times 10^{-5} \times 500^2 \tag{8.26b}$$

or alternatively

$$f(x) = \frac{x^2(2-x)^2}{(1-x)^4} - 12.23 = 0 \tag{8.26c}$$

with the range for x given by $1 > x > 0$. We proceed to use the software package MATHEMATICA to solve this equation by Newton's method.

```
        "Ammonia synthesis"

        Ammonia synthesis
In[566] := FindRoot[x^2(2 - x)^2/(1 - x)^4 - 12.23 = 0,
    {x, {0, .999}}]
Out[566]= {x → 0.528446}

        Here are the steps:
        Step 1:
In[567] := f[x_] := x^2(2 - x)^2/(1 - x)^4 - 12.23
In[568] := x0 = 0.5
           X1 = x0 - f[x0]/f'[x0]
Out[569] := 0.533646
        Step 2 :
In[570] := x0 = 0.533646
           x1 = x0 - f[x0]/f'[x0]
```

```
Out[571]= 0.528603
        Step 3 :
In[572]:= x0 = 0.528603
             x1 = x0 - f[x0]/f'[x0]
Out[573] = 0.528446
        Step 4 :
In[574] := x0 = 0.528446
             x1 = x0 - f[x0]/f'[x0]
Out[575] = 0.528446
No change in the value in 6 significant digits.
```

COMMENTS

- It will be noted that the actual solution produced by the package consists of the two terse lines In[566] and Out[566]. The code FindRoot automatically calls for the use of Newton's method and specifies a solution range of (0, 0.999), leaving it to the software to choose an initial trial value within that range. The initial value can also be specified by the user instead. This is the route chosen in the next example, in which we study the production of silicon films by chemical vapor deposition.

  For convenience, we have extracted the steps that lead to the end result. The input of the first step specifies the initial guess, which is typically chosen by the package to be the midpoint of the range stipulated (i.e., $x_0 = 0.5$). This value is also, rather fortuitously, close to the root, so that only a few steps are needed to reach the final destination. Also specified in the input to the first step is the Newton algorithm, equation (8.19d). These inputs, and the corresponding outputs, are repeated a total of four times to obtain the final result of $x = 0.528446$. Since a value of $x = 1$ represents total conversion, the fractional yield in ammonia under the stated conditions is 0.528446, or approximately 53%.

  The synthesis of ammonia was first achieved by F. Haber in 1905 after years of experimenting with literally hundreds of catalysts. It led to the development of synthetic fertilizers, which are today an important component product of the chemical industry.

8.2.5 Numerical Simulation of Simultaneous Nonlinear Equations: The Newton–Raphson Method

We have seen that the Newton method results in an algorithm involving both the functional form of the equation and its derivative as well as an increment in the variable. We reproduce it in terse form:

$$x_{n+1} \quad - \quad x_n \quad = \quad -\frac{f(x_n)}{f'(x_n)} \tag{8.27a}$$

New value   Old value   Function over derivative

For simultaneous equations, it has been shown that the same scheme applies but in vector–matrix form. We have

$$\mathbf{x_{n+1}} \quad - \quad \mathbf{x_n} \quad = \quad -\frac{\mathbf{f(x_n)}}{\mathbf{f'(x_n)}} \tag{8.27b}$$

New vector	Old vector	Vector over matrix

Here both the variables $\mathbf{x}$ and the equation function $\mathbf{f}$ are column vectors, while $\mathbf{f'}$ is a matrix composed of the various partial derivatives of $\mathbf{f}$. Suppose, for example, that we have two simultaneous equations

$$f_1(x_1, x_2) = 0 \tag{8.28a}$$

and

$$f_2(x_1, x_2) = 0 \tag{8.28b}$$

for which we are seeking the roots. Then $\mathbf{f'}$ will be composed of the array

$$\mathbf{f'} = \begin{matrix} \dfrac{\partial f_1}{\partial x_1} & \dfrac{\partial f_1}{\partial x_2} \\ \dfrac{\partial f_2}{\partial x_1} & \dfrac{\partial f_2}{\partial x_2} \end{matrix} \tag{8.28c}$$

This matrix of partial derivatives is called the Jacobian of the function f.

In examining equation (8.27b), we find ourselves at a seeming impasse: $\mathbf{f/f'}$ appears to be an unknown, making it impossible to solve for the principal unknown of the equation, $\mathbf{x_{n+1}}$. We resolve this dilemma by first writing

$$\frac{\mathbf{f(x_n)}}{\mathbf{f'(x_n)}} = \mathbf{f(x_n)} \tag{8.29a}$$

and then transposing to obtain

$$\mathbf{f'}(x_n) \quad \cdot \quad \mathbf{F(x_n)} \quad = \quad \mathbf{f(x_n)} \tag{8.29b}$$

Known matrix of partial derivative	Unknown vector–matrix quotient	Known function vector

Here $\mathbf{F(x_n)}$ is a column vector of unknowns, $\mathbf{f(x_n)}$ is a column vector of nonhomogeneous terms, and $\mathbf{f'(x_n)}$ acts as a coefficient matrix. Both $\mathbf{f(x_n)}$ and $\mathbf{f'(x_n)}$ are known, since they are evaluated at the "old" value x_n. Thus the problem of solving for the quotient $\mathbf{F(x_n)}$ is reduced to *solving a set of linear algebraic equations* in the unknowns $F_1(x_n), F_2(x_n), \ldots, F_m(x_n)$. This can be done by the Gaussian elimination discussed in Chapter 2.

We note that the set of equations must be solved *every time we evaluate new values* $\mathbf{x}_{n+1}$ on our way to finding the roots. Furthermore, if no convergence is obtained, all initial guesses $\mathbf{x}_0$ must be adjusted in principle. Application of the Newton–Raphson method is therefore a considerable undertaking. We demonstrate its use with Example 8.8.

◆ EXAMPLE 8.8 More Chemical Equilibria: Producing Silicon Films by Chemical Vapor Deposition (CVD)

In Example 6.13 we considered the fabrication of silicon chips by drawing crystals from a silicon melt. Another step in the manufacture of semiconductor devices involves the deposition of thin films of silicon onto wafers. This is done by passing the gaseous silicon compound silane (SiH_4) over the wafer at elevated temperatures, causing it to decompose into solid silicon, which is deposited, and hydrogen gas. Thus

$$SiH_4(g) \rightleftarrows Si\,(s) + 2H_2(g) \tag{8.30a}$$

It turns out that the process is accompanied by two side reactions, which take place simultaneously and act to reduce the yield of silicon. They are

$$SiH_4 \rightleftarrows SiH_2 + H_2 \tag{8.30b}$$

and

$$SiH_4 + SiH_2 \rightleftarrows Si_2H_6 \tag{8.30c}$$

The chemical equilibrium constants for these processes at 1000 °C are given by the following relations:

$$\frac{n_{H_2}^2}{n_{SiH_4}}\frac{P_T}{n_T} = K_{p1} = 1.29 \times 10^6 \text{ atm} \tag{8.31a}$$

$$\frac{n_{SiH_2} \times n_{H_2}}{n_{SiH_4}}\frac{P_T}{n_T} = K_{p2} = 8.8 \times 10^{-4} \text{ atm} \tag{8.31b}$$

$$\frac{n_{SiH_6}}{n_{SiH_4} \times n_{SiH_2}}\frac{P_T}{n_T} = K_{p3} = 1.01 \times 10^3 \text{ atm}^{-1} \tag{8.31c}$$

We now wish to establish how much of the SiH_4 is converted to silicon, which is the desired product, and to what extent the side reactions diminish the conversion. This requires the full solution of the equation set (8.31a) to (8.31c).

We note that here again, as in the Example 8.7, we have more variables (four) than equations (three). We must therefore interrelate these variables, as we did before, by means of a "balance sheet," which is in essence a material balance. We write, for each reaction in turn, the initial moles and the moles left over at equilibrium. This yields the following series:

Reaction :	SiH_4	$\rightleftarrows$	Si	+	$2H_2$
Moles at start:	a		0		0
Moles at equilibrium :	$a-x$		x		$2x$
Reaction :	SiH_4	$\rightleftarrows$	SiH_2	+	H_2
Moles at start:	b		0		0
Moles at equilibrium :	$b-y$		y		y
Reaction :	SiH_4	+	SiH_2	$\rightleftarrows$	Si_2H_6
Moles at start:	c		y		0
Moles at equilibrium :	$c-z$		z		z

Consequently we have for the moles of each species at equilibrium

$$n_{SiH_4} = a + b + c - x - y - z \tag{8.32a}$$

$$n_{SiH_2} = y - z \tag{8.32b}$$

$$n_{Si_2H_6} = z \tag{8.32c}$$

$$n_{H_2} = 2x + y \tag{8.32d}$$

and for the total number of moles

$$n_T = a + b + c + x + y - z \tag{8.32e}$$

Let us assume that we start the reaction with one mole of SiH_4 (i.e., $a+b+c = 1$) and set the total pressure at 1 atm. We then obtain by substitution into the equilibrium constants (8.31a) to (8.31c):

$$\frac{(2x+y)^2}{(1-x-y-z)(1+x+y-z)} = 1.29 \times 10^6 \text{ atm} \tag{8.33a}$$

$$\frac{(y-z)(2x+y)}{(1-x-y-z)(1+x+y-z)} = 8.80 \times 10^{-4} \text{ atm} \tag{8.33b}$$

$$\frac{z(1+x+y-z)}{(1-x-y-z)(y-z)} = 1.01 \times 10^3 \text{ atm}^{-1} \tag{8.33c}$$

We now proceed to solve this system by the Newton–Raphson method, using the MATHEMATICA package. To aid in the task of setting initial values, we note that the variables range from zero to one, that is,

$$1 > x > 0 \qquad 1 > y > 0 \qquad 1 > z > 0 \tag{8.34}$$

Input and output then appear as follows.

Chemical Vapor Deposition problem

$$f[x\_, y\_, z\_] := \frac{(2x + y)^2}{(1 - x - y - z)(1 + x + y - z)} - 1290000$$

$$g[x\_, y\_, z\_] := \frac{(y - z)(2x + y)}{(1 - x - y - z)(1 + x + y - z)} - 0.000880$$

$$h[x\_, y\_, z\_] := \frac{z(1 + x + y - z)}{(1 - x - y - z)(y - z)} - 1010$$

```
FindRoot [{f[x, y, z] == 0, g[x, y, z] == 0,
    h[x, y, z] == 0}, {x, 0.4}, {y, 0.3}, {z, 0.2],
    WorkingPrecision → 16, MaxIterations → 25]
FindRoot::jsing : Encountered a singular Jacobian at the point
    {x, y, z} = {6.77512 x 10^11, -3.62964 x 10^11, -3.18306 x
    10^11}. Try perturbing the initial point(s).
FindRoot[{f[x, y, z] == 0, g[x, y, z] == 0, h[x, y, z] == 0},
    {x, 0.4}, {y, 0.3}, {z, 0.2}, WorkingPrecision → 16,
    MaxIterations → 25]
Rearrange the equations to avoid division by zero
f1[x_, y_, z_] := (2x + y)^2- 1290000 (1 - x - y - z)
    (1 + x + y - z)
g1[x_, y_, z_] := (y - z) (2x + y) - 0.000880 (1 - x - y - z)
    (1 + x + y - z)
h1[x_, y_, z_] := z (1 + x + y - z) - 1010
    (1 - x - y - z) (y - z)
FindRoot[{f1[x, y, z] == 0, g1[x, y, z] == 0, h1[x, y, z] == 0},
    {x, 1}, {y, 0}, {z, 0}, WorkingPrecision → 16,
    MaxIterations → 25]
    {x → 0.999998, y → 1.36541 x 10^-9, z → 1.0682 x 10^-12}
```

COMMENTS

- We start by noting that for the purpose of implementing the solution routine, the equations are required to be of the form

$$F(x, y, z) = 0 \tag{8.35}$$

This is reflected in the first three lines. This is followed by a line with the code FindRoot, which calls for the automatic implementation of the Newton–Raphson routine. Failure to converge results in a switch to alternative routines, of which the user is apprised. The solution is then initiated by entering the initial givens (0.4, 0.3, 0.2), the working precision (16), and the maximum number of iterations (25).

The user is next informed that the equations in the form given result in a singular Jacobian; that is, $\mathbf{f}'$ in equation (8.28c) cannot be evaluated. The proceedings have consequently come to a halt, and the system must be examined to determine the culprit. Attention quickly focuses on the denominators of the equations, all of which contain *differences*. This form has been repeatedly brought to the reader's attention as one that harbors the source of potential singularities. The computer output indicates that such singularities had in fact arisen. Some thought will bring us to the conclusion that this particular difficulty can be easily overcome by multiplying each equation by the denominators appearing in them. With this new formulation in place, convergence is rapid, leading to the results given in the last line.

The reader will note from these results that both SiH_4 and Si_2H_6 ($y - z = 1.36541 \times 10^{-9}$, $z = 1.0682 \times 10^{-12}$) are produced in minuscule amounts and that the conversion of SiH_4 to the desired silicon is for all practical purposes complete. Such conclusions might have been inferred from the small values of K_{p2} and K_{p3} compared with K_{p1}, but confirmation by actual calculation was essential. The vapor deposition of silicon can consequently be carried out at the specified temperature without serious interference from the side reactions (8.30b) and (8.30c).

8.2.6 Numerical Solution of Ordinary Differential Equations: The Euler and Runge–Kutta Methods

In dealing with the numerical solution of ODEs, the order of the equations and the type of boundary conditions must be distinguished at the outset. Most numerical packages deal with sets of first-order equations in which all the necessary boundary conditions are specified at the same point in time or space: initial value problems in the nomenclature of Section 2.3.3. Higher order equations can be decomposed into an equivalent set of first-order ODEs. For example, the second-order equation that describes diffusion and reaction in a spherical catalyst pellet is given by

$$\frac{d^2C}{dr^2} + \frac{2}{r}\frac{dC}{dr} + \frac{k_r}{D}C = 0 \tag{8.36a}$$

and can be decomposed into the equivalent first-order set

$$p = \frac{dC}{dr} \tag{8.36b}$$

and

$$\frac{dp}{dr} + \frac{p}{r} + \frac{k}{D}C = 0 \tag{8.36c}$$

However, this model has the additional complication of having its boundary conditions specified at two different locations, the center and the surface of the pellet. Such systems were referred to as boundary value problems in Section 2.3.3. They can in principle be handled by guessing the missing initial conditions and the concentration at the center, then integrating by repeated adjustment of its value until the surface boundary condition is matched.

In what follows, we limit ourselves to first-order ODEs of the initial value type. Let us demonstrate the numerical solution of these equations by considering the following single first-order ODE:

$$\frac{dy}{dx} = f(x, y) \tag{8.37}$$

All numerical integration methods start by putting the equation in the following difference form:

$$\int_y dy = y_{j+1} - y_j = \int_{x_j}^{x_{j+1}} f(x, y)\, dy \cong f(x, y)_{avg} \Delta x \tag{8.38}$$

where the integral in question has been approximated by the expression $f(x, y)_{avg} \Delta x$ and Δx is often referred to as the step size.

It is the choice and form of $f(x, y)_{avg}$ that distinguishes the various numerical integration methods. In the so-called single-step procedure, which includes the Euler and Runge–Kutta methods, both y and f(x, y) are evaluated at the last value of x, x_j. In the so-called open or explicit multistep methods, one reaches further back and evaluates these variables at values of x_j, x_{j-1}, and so on, to get a more accurate description of $f(x, y)_{avg}$. In yet another class of procedures, the so-called closed or implicit multistep method, a forward projection is made by including values of y and f(x, y) evaluated at x_{j+1}. Since this results in the unknown y_{j+1} appearing on both sides of equation (8.38), an iterative solution procedure must be applied. Finally, we note that open and closed methods are usually combined into the so-called predictor–corrector method. The latter is used in the well-known Gear package popular in the 1970s and 1980s, which also has provisions for the automatic adjustment of step size.

Let us start by considering the simplest of these procedures, the Euler method. Here $f(x, y)_{avg}$ is simply evaluated at the last value of the independent variable, x_j, and we have

$$y_{j+1} - y_j = [f(x_j, y_j)](x_{j+1} - x_j) \tag{8.39}$$

Let us see how this relation is applied in practice. Suppose we wish to solve by Euler's method the system

$$\frac{dy}{dx} = f(x, y) = -y^2 \qquad y(0) = 1 \tag{8.40a}$$

which has the analytical solution

$$y = (x + 1)^{-1} \tag{8.40b}$$

Applying the algorithm (8.38) with a step size $\Delta x = 0.1$, we obtain the steps listed in Table 8.7. One notes that there is an increasing divergence from the analytical solution with an increase in the number of steps. This can be partly remedied by choosing a smaller step size, but applying a more sophisticated procedure is often more fruitful.

Table 8.7

	y Values	
	Numerical	Analytical
For the 1st step		
$y_1 - 1 = -(1)^2 \times 0.1 = -0.1$	0.9	0.909
For the 2nd step		
$y_2 - 0.9 = -(0.9)^2 \times 0.1 = -0.081$	0.819	0.833
For the 3rd step		
$y_3 - 0.819 = -(0.819)^2 \times 0.1 = -0.067$	0.752	0.769

One such procedure is the Runge–Kutta or RK method, which seeks to improve the accuracy by *weighting* $f(x, y)_{avg}$ in the following fashion

$$y_{j+1} - y_j = f(x, y)_{avg}\Delta x = \tfrac{1}{6}[K_1 + 2K_2 + 2K_3 + K_4]\Delta x \tag{8.41a}$$

where the K_i are the so-called Runge–Kutta constants, given by

$$K_1 = f(x_j, y_j) \tag{8.41b}$$

$$K_2 = f\left[x_j + \frac{\Delta x}{2}, y_j + \left(\frac{\Delta x}{2}\right)K_1\right] \tag{8.41c}$$

$$K_3 = f\left[x_j + \frac{\Delta x}{2}, y_j + \left(\frac{\Delta x}{2}\right)K_2\right] \tag{8.41d}$$

$$K_4 = f[x_j + \Delta x, y_j + (\Delta x)K_3] \tag{8.41e}$$

This particular version of the RK method is referred to as a fourth-order one because of the number of constants involved. For a first-order RK routine, $K_2 = K_3 = K_4 = 0$ and the coefficient $\frac{1}{6}$ in equation (8.41a) is replaced by 1. The routine then reduces to Euler's method. The procedure in both cases consists of evaluating the constants at the position x_j in accordance with equations (8.41b) to (8.41e). The constants are then substituted into equation (8.41a) to obtain the new value y_{j+1}, which is used in implementing the next step.

When a set of *simultaneous* first-order ODEs is to be solved, there is an *array* of functions f, dependent variables y, and RK constants K_i (but the independent variable x is not present). It then becomes convenient to represent these parameters, as well as the ODEs and their initial conditions, as *vectors*, in place of the single values we had dealt with before. We write for the ODEs and their initial condition

$$\frac{d\mathbf{y}}{dx} = \mathbf{f}(x, \mathbf{y}) \tag{8.42a}$$

and

$$\mathbf{y}(a) = \mathbf{A}$$

where

$$\mathbf{y} = \begin{pmatrix} y_1 \\ y_2 \\ \vdots \\ y_n \end{pmatrix} \tag{8.42b}$$

$$\mathbf{f}(x, \mathbf{y}) = \begin{pmatrix} f_1(x, y_1, y_2, \ldots, y_n) \\ f_2(x, y_1, y_2, \ldots, y_n) \\ \vdots \\ f_n(x, y_1, y_2, \ldots, y_n) \end{pmatrix} \tag{8.42c}$$

$$\mathbf{A} = \begin{pmatrix} A_1 \\ A_2 \\ \vdots \\ A_n \end{pmatrix} \tag{8.42d}$$

In similar fashion one forms vectors of the RK constants, which now become

$$\mathbf{K_{1j}} = f(x_j, \mathbf{y_j}) \tag{8.42e}$$

$$\mathbf{K_{2j}} = \mathbf{f}\left[x_j + \frac{\Delta x}{2}, \mathbf{y_j} + \left(\frac{\Delta x}{2}\right)\mathbf{K_{1j}}\right] \tag{8.42f}$$

$$\mathbf{K_{3j}} = \mathbf{f}\left[x_j + \frac{\Delta x}{2}, \mathbf{y_j} + \left(\frac{\Delta x}{2}\right)\mathbf{K_{2j}}\right] \tag{8.42g}$$

$$\mathbf{K_{4j}} = \mathbf{f}[x_j + \Delta x, \mathbf{y_j} + (\Delta x)\mathbf{K_{3j}}] \tag{8.42h}$$

These constants are then substituted into the vector equivalent of equation (8.41a), namely,

$$\mathbf{y_{j+1}} - \mathbf{y_j} = \tfrac{1}{6}[\mathbf{K_{1j}} + 2\mathbf{K_{2j}} + 2\mathbf{K_{3j}} + \mathbf{K_{4j}}]\Delta x \tag{8.42i}$$

This expression yields the updated array $\mathbf{y_{j+1}}$ of dependent variables, which is used in the next step of the integration procedure.

We note that none of these manipulations involve new types of computation. The basic RK procedure used for single ODEs is retained, with the difference that all variables, with the exception of x, are stored and updated as vector arrays.

It is important to keep in mind that all dependent variables are computed and incremented *simultaneously*. If the model contains algebraic equations that are linked to ODEs, they, too, must be solved and updated simultaneously.

◆ EXAMPLE 8.9 The Effect of Drag on the Trajectory of an Artillery Piece

In Example 4.5 we examined the trajectory of a projectile fired with a muzzle velocity v_0 and an angle of elevation α. By neglecting air resistance, we were able to integrate the two ODEs, which resulted from a simple application of Newton's law and yielded the horizontal and vertical distances x and y as a function of time. From this information we could derive the range of the projectile, the maximum height attained, and the optimum angle of elevation.

We now turn to an examination of the effect of drag. The pertinent equations were given in Practice Problem 4.8 and are repeated here

$$\frac{d^2x}{dt^2} = -k\frac{dx}{dt}\left[\left(\frac{dx}{dt}\right)^2 + \left(\frac{dy}{dt}\right)^2\right]^{1/2} \tag{8.43a}$$

$$\frac{d^2y}{dt^2} = -g - k\frac{dy}{dt}\left[\left(\frac{dy}{dx}\right)^2 + \left(\frac{dy}{dt}\right)^2\right]^{1/2} \tag{8.43b}$$

where $k = \frac{1}{2}(\rho/m)A_C$, ρ = air density, and m and A_C are, respectively, the mass and the maximum cross-sectional area of the projectile. Both equations are nonlinear and must be integrated numerically.

We first note that these expressions are second-order ODEs that must be converted to an equivalent set of first-order equations to enable us to apply the Euler or RK integration procedure. This is easily accomplished by setting

$$u = \frac{dx}{dt} \tag{8.44a}$$

$$v = \frac{dy}{dt} \tag{8.44b}$$

so that equations (8.43a) and (8.43b) become

$$\frac{du}{dt} = -ku[u^2 + v^2]^{1/2} \tag{8.44c}$$

$$\frac{dv}{dt} = -g - kv[u^2 + v^2]^{1/2} \tag{8.44d}$$

It is this set of four simultaneous ODEs that now must be solved numerically. The pertinent initial conditions are given by

$$x(0) = 0 \tag{8.45a}$$

$$y(0) = 0 \tag{8.45b}$$

$$u(0) = v_0 \cos\theta \tag{8.45c}$$

$$v(0) = v_0 \sin\theta \tag{8.45d}$$

where v_0 = muzzle velocity and θ is the angle of elevation. We set v_0 = 400 m/s and k = 0.0001, varying θ over the range of 30° to 60°. The value of k corresponds to a projectile of 10 kg and 10 cm maximum diameter, air density of 1.18 kg/m^3, and C_D = 0.22, a typical value for artillery pieces. The computer input and output using the MATHEMATICA package is then as follows.

```
Projectile paths for varying angles
    (30 to 60 degrees, step 5 degrees) keeping
    the initial velocity constant at 400m/s

F[u_,v_] := -k√(u²+v²) u

G[u_,v_] := -g-k√(u²+v²) v
k := 0.0001
g := 9.81
θ1 := 30
θ2 := 35
θ3 := 40
θ4 := 45
θ5 := 50
θ6 := 55
θ7 := 60

S1=NDSolve[{u'[t]==F[u[t], v[t]], v'[t]==G[u[t], v[t]],
    x'[t]==u[t], Y'[t]==v[t],u[0]==400Cos[Piθ1/180],
    mv[0]==400Sin[Piθ1/180], X[0]==0, y[0]==0},
    {x[t], y[t], u[t], v[t]}, {t, 60}]

S2=NDSolve[{u'[t]==F[u[t], v[t]], v'[t] ==G[u[t], v[t]],
    x'[t]==u[t], Y'[t]==v[t], u[0]==400Cos[Piθ2/180],
    mv[0]== 400Sin[Piθ2/180], X[0]==0, y[0]==0}, {x[t], y[t],
    u[t], v[t]}, {t, 60}]

S3=NDSolve[{u'[t]== F[u[t], v[t]], v'[t]== G[u[t], v[t]],
    x'[t]== u[t], } Y'[t]== v[t], u[0]== 400Cos[Piθ3/180],
    mv[0]== 400Sin[Piθ3/180], X[0]== 0, y[0]== 0}, {x[t], y[t],
    u[t], v[t]}, {t, 60}]

S4=NDSolve[{u'[t]== F[u[t], v[t]], v'[t]== G[u[t], v[t]],
    x'[t]== u[t], Y'[t]== v[t], u[0]== 400Cos[Piθ4/180],
    mv[0]== 400Sin[Piθ4/180], X[0]== 0, y[0]== 0}, {x[t], y[t],
    u[t], v[t]}, {t, 60}]

S5=NDSolve[{u'[t]== F[u[t], v[t]], v'[t]== G[u[t], v[t]],
    x'[t]== u[t], Y'[t]== v[t], u[0]== 400Cos[Piθ5/180],
    mv[0]== 400Sin[Piθ5/180],X[0]== 0, y[0]== 0}, {x[t], y[t],
    u[t], v[t]}, {t, 60}]

S6=NDSolve[{u'[t]== F[u[t], v[t]], v'[t]== G[u[t], v[t]],
    x'[t]== u[t], Y'[t]== v[t], u[0]== 400Cos[Piθ6/180],
```

```
    mv[0]==400Sin[Piθ6/180], X[0]==0, y[0]==0}, {x[t], y[t],
    u[t], v[t]}, {t, 60}]

S7=NDSolve[{u'[t]==F[u[t], v[t]], v'[t]==G[u[t], v[t]],
    x'[t]==u[t], Y'[t]==v[t], u[0]==400Cos[Piθ7/180],
    mv[0]==400Sin[Piθ7/180], X[0]==0, y[0]==0}, {x[t], y[t],
    u[t], v[t]}, {t, 60}]

ParametricPlot[{Evaluate[{x[t],y[t]}/.S1],
    Evaluate[{x[t],y[t]}/.S2],
    Evaluate[{x[t],y[t]}/.S3], Evaluate[{x[t],y[t]}/.S4],
    Evaluate[{x[t],y[t]}/.S5], Evaluate[{x[t],y[t]}/.S6],
    Evaluate[{x[t],y[t]}/.S7], (t,0,60,PlotRange → All]

The closed form solution for the projectile equation in the
    absence of air resistance : H is the height and L is the
    total distance traveled

U := VCos[θ]
V := VSin[θ]--gt
H[T_] := ∫_0^T vdt

H
```

$$-\frac{gT^2}{2}+TVSin[\theta]$$

$$\text{Solve }[-\frac{gT^2}{2}+\text{ T V Sin}[\theta]==0,\{T\}]$$

$$\{\{T \to 0\},\ \{T \to \frac{2VSin[\theta]}{g}\}\}$$

$$t_0=\frac{2V\,Sin[\theta]}{g}$$

$$\frac{2VSin[\theta]}{g}$$

```
L := ut0

L
```

$$\frac{2V^2\,Cos[\theta]\,Sin[\theta]}{g}$$

$$\text{Simplify }[\frac{2V^2Cos[\theta]\,Sin[\theta]}{g}]$$

$$\frac{V^2Sin[\theta]}{g}$$

COMMENTS

- The first 11 lines list the input to the package and include the seven different angles of elevation θ to be tested, as well as the coefficient k set at 0.0001 and the gravitational acceleration. The four equations themselves are solved by the routine NDSolve,

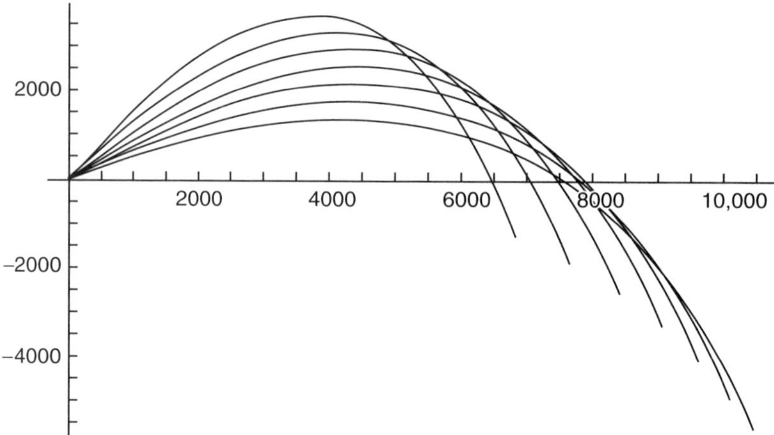

Figure 8.4 Trajectory of an artillery piece with air resistance: effect of elevation angle.

which employs the fourth-order Runge–Kutta method. Here, as elsewhere, the package has provisions to switch automatically to more elaborate routines, such as predictor–corrector methods, if difficulties should arise in the solution of the ODEs. When this happens, the user is apprised.

The NDSolve statements, one for each of the angles specified, list the function to be used in lines 1 and 2, as well as the four required initial conditions [equations (8.45a–d)]. The interval of integration is from 0 to 60 s.

The results are presented in graphical form, generated by the routine Parametric Plot and shown in Fig. 8.4. We note that the maximum range obtained without air resistance, $r_{max} = v_0^2/g = 400^2/9.81 = 16{,}310$ m [equation (4.73)] is more than halved when drag is included. Furthermore, the angle of elevation needed to yield the maximum range is no longer 45° but has dropped to approximately 35°. Thus the results undergo a considerable change when drag is included. The case of no drag nevertheless provides useful upper limits to the height and range attainable.

We have also used this example to provide a glimpse of the analytical capabilities of the MATHEMATICA package. These are shown in the section following the Parametric Plot Statement and include derivation of the height H at any time and the range L that holds for H = 0. The expressions obtained are identical to those derived analytically in Example 4.5. That is, we have

$$y\,(= H) = \frac{gt^2}{2} + (v_0 \sin\alpha)t \tag{4.66}$$

and

$$r\,(= L) = \frac{v_0^2 \sin 2\theta}{g} \tag{4.72}$$

This is a nice example of symbolic manipulation by the MATHEMATICA package.

PRACTICE PROBLEMS

8.1 **Time of swing of a simple pendulum: the analytical solution** Quantify the result (8.4i) by solving equation (8.5c).

8.2 **Doping of a silicon chip** To control the conductivity of a semiconductor device, it is necessary to introduce very small amounts of specific atoms termed *dopants* (e.g., boron), into the base material of the semiconductor. This process is often carried out in two stages. In the first step, the dopant is deposited from the vapor phase onto the semiconductor as a thin surface layer. In the second step, also carried out at high temperatures, the deposit is allowed to diffuse into the matrix interior. This process is termed drive-in diffusion.

For the device to function properly, it is necessary to control both the level and the spatial distribution of the dopant concentration within the semiconductor. In other words, we wish to predict how long the drive-in process should proceed to ensure that, a particular concentration level will be attained at some prescribed location. This generally requires the solution of a PDE called Fick's equation, which is given by

$$D\frac{\partial^2 C}{\partial x^2} = \frac{\partial C}{\partial t}$$

and yields full dopant concentration distribution as a function of time. To avoid this step, we can use test data and dimensional analysis to predict position or time of a particular concentration level.

Suppose that a particular concentration is found to have penetrated to a depth of 15 μm after 10 h of drive-in diffusion. How long would it take for the same concentration to penetrate to a depth of 30 μm?

ANSWER 40 h

8.3 **Mass transfer to a fluid in turbulent flow** In Example 1.7 we dealt with a description of the release of a substance from a tubular wall into a fluid in turbulent flow. The model derived there contained a mass transfer coefficient k_C, which is the counterpart to the heat transfer coefficient we examined in this chapter (see example 8.3). With Example 8.3 and Table 8.5 as guides, use dimensional analysis to derive a dimensionless functional relationship for the mass transfer coefficient.

ANSWER $Sh = f(Re, Sc)$

8.4 **Power requirements of an electric fan** Suppose that in the design of an electric fan, the power required to drive a model fan of a given diameter and at a given number of revolutions per second has been measured. It is now desired to scale these results to estimate the power requirements of geometrically similar but larger fans operating over a range of revolutions per second. This information is required to be able to size the electrical motors needed to drive the fan. Derive an appropriate dimensionless group that will provide this information.

Hint: Use Rayleigh's method.

8.5 **Design of a breakwater** A model of a rubble breakwater is constructed of rocks weighing 1 kg each. The rocks have the same specific gravity as those to be used in the prototype. If appreciable damage to the model is observed when the wave height exceeds 0.25 m, what is the minimum weight of each rock to be used in the prototype, if it is to withstand geometrically similar waves 5 m high? *(Hint*: Use the arguments developed in Example 8.5.)

ANSWER 8000 kg

8.6 **Simultaneous linear algebraic equations: numerical calculation of voltages in an electrical network** An electrical network consisting of resistances in parallel and in series is shown in Fig. P8.6. From Ohm's law, the current flowing from node p to node q is given by

$$i_{pq} = \frac{V_p - V_q}{R_{pq}}$$

In addition, Kirchhoff's first rule (see Table 1.2) applies, which states that the sum of currents arriving at and leaving a node must be zero. Thus

$$\sum_{\text{node}} i = 0$$

Application of these two laws at node 1 leads to

$$i_{A1} + i_{21} + i_{61} = \frac{100 - V_1}{3} + \frac{V_2 - V_1}{3} + \frac{V_6 - V_1}{15} = 0$$

or

$$11V_1 - 5V_2 - V_6 = 500$$

Applying the same procedure at the other five nodes, we obtain the set of linear equations given in Table P8.6.

The task is to solve the system using Gaussian elimination. Note that an alternative method relying on matrix multiplication can also be used to solve this problem (see Example 7.9).

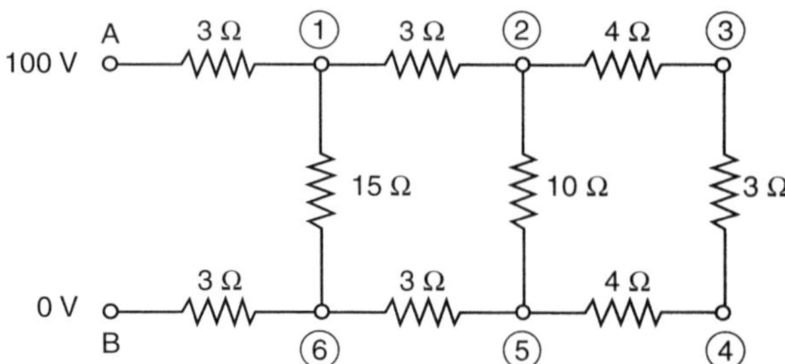

Figure P8.6 Voltages in an electrical network.

Table P8.6 Linear Equation Set for the Unknown Voltages V_g

Node	Equation
1	$11V_1 - 5V_2 - V_6 = 500$
2	$-20V_1 + 41V_2 - 15V_3 - 6V_5 = 0$
3	$-3V_2 + 7V_3 - 4V_4 = 0$
4	$-V_3 + 2V_4 - V_5 = 0$
5	$-3V_2 - 10V_4 + 28V_5 - 15V_6 = 0$
6	$-2V_1 - 15V_5 + 47V_6 = 0$

ANSWER $V_1 = 38.85$ volts, $V_2 = -14.61$ volts, $V_3 = -10.32$ volts, $V_4 = -7.10$ volts, $V_5 = -3.88$ volts, $V_6 = 0.42$ volts

8.7 Adiabatic flame temperature revisited. Solution of a single nonlinear equation by Newton's method In Example 4.8 we acquainted the reader with the notion of an adiabatic flame temperature, the maximum theoretical temperature attainable when a fuel is burned without heat loss to the surroundings. In calculating its value there, we had made the tacit assumption that the heat capacities were constant and independent of temperature. In actual fact, these parameters can vary significantly, particularly over the large temperature range associated with combustion processes.

Consider the combustion of a natural gas with the composition 83 mol% CH_4, 15 mol% C_2H_6, and 2 mol% N_2. The process follows the reaction equations

$$CH_4 + 2O_2 = CO_2 + 2H_2O$$

$$C_2H_6 + 3.5O_2 = 2CO_2 + 3H_2O$$

We wish to calculate the adiabatic flame temperature that results when this mixture is burnt with the theoretical amount of air. The relevant expression to be used, introduced in Example 4.8 [equation (4.100)] and is given now in a generalized form:

$$-\Delta H_c{}^\circ = \int_{298}^{T_a} \sum (n_i C_{pi})_{prod}\, dT$$

The data required to solve this equation for T are given in Table P8.7. Use Newton's method. to solve the enthalpy of combustion equation for T_a. The integral can be evaluated analytically or numerically.

ANSWER $T_a = 1193\,K$

8.8 Flow in a pipe network: numerical solution of simultaneous nonlinear equations by the Newton–Raphson method We consider here the pressures p_i and flow rates Q_{ij} that arise during flow through the pipe network shown in Fig. P8.8. Simple versions of such hydraulic networks had been taken up in Example 7.3 and Practice Problem 7.3.

Table P8.7 Data for Solving Enthalpy of Combustion Equation for T

Compound	C_{pi} (cal/mol K)	n_i
CO_2	$18.036 - 4.474 \times 10^{-5}\,T - 1.581 \times 10^{-2}\,T^{-1/2}$	1.13
H_2O	$8.22 + 1.5 \times 10^{-4}\,T + 1.34 \times 10^{-6}\,T^2$	2.11
N_2	$1.64 + 4.124 \times 10^{-2}\,T - 1.425 \times 10^{-6}\,T^2 + 1.74 \times 10^{-9}\,T^3$	8.24
	$\Delta H_c = -2.325 \times 10^5$ cal/mol gas	

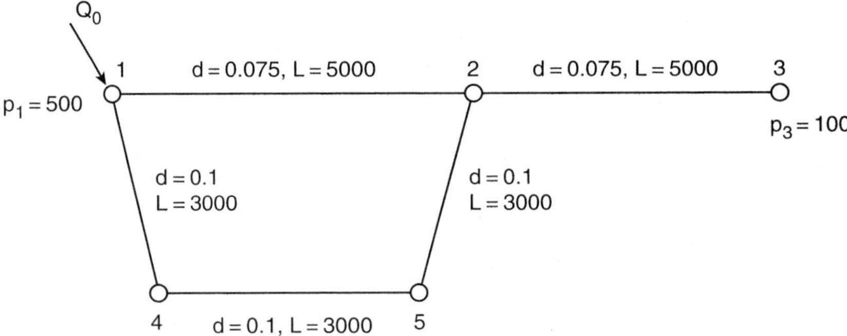

Figure P8.8 Flow in a pipe network: pressures in kilopascals; dimensions in meters.

The case at hand is too complex to be solved analytically, and we proceed instead to use the tools of numerical analysis. To do this, we apply the pressure drop equation (7.22) given in Example 7.3 along each pipe segment and combine these with mass balances around each node. We assume a constant friction factor of 0.001, which is a good average for flow rates in the range $2000 < Re < 10^5$. Substituting numerical values into the pressure drop relation and solving for Q_{ij} results in the following expressions:

$$Q_{12} = 1.22 \times 10^{-5}(5 \times 10^5 - p_2)^{1/2}$$
$$Q_{14} = 3.2 \times 10^{-5}(5 \times 10^5 - p_4)^{1/2}$$
$$Q_{45} = 3.2 \times 10^{-5}(p_4 - p_5)^{1/2}$$
$$Q_{52} = 3.2 \times 10^{-5}(p_5 - p_2)^{1/2}$$
$$Q_{23} = 1.22 \times 10^{-5}(p_2 - 10^5)^{1/2}$$

These expressions are supplemented by the following mass balances:

$$Q_0 = Q_{14} + Q_{12}$$
$$Q_{14} = Q_{45} = Q_{52}$$
$$Q_{12} + Q_{52} = Q_{23}$$

Use a Newton–Raphson routine to substitute the five pressure drop relations just given into the mass balances and solve the resulting four nonlinear equations for Q_0, p_2, p_4, and p_5. Calculate the remaining variables by back-substitution.

Note that typical flow rates in pipes of the specified dimensions are of the order $10^{-3} - 10^{-2}\,\mathrm{m^3/s}$.

ANSWERS $Q_0 = 0.0074\,\mathrm{m^3/s}$, $p_2 = 471.7\,\mathrm{kPa}$

8.9 **Cracking of acetone in a non-isothermal tubular reactor. Solution of simultaneous ODEs by the RK Method** The thermal cracking or decomposition of acetone to ketene is an intermediate step in the manufacture of the important chemical acetic anhydride [$(CH_3COO)_2O$]. The relevant reaction equation is as follows:

$$\underset{\text{Acetone (A)}}{(CH_3)_2CO} = \underset{\text{Methane (B)}}{CH_4} + \underset{\text{Ketene (C)}}{CH_2CO}$$

The reaction is endothermic (requires heat) and is carried out in a long tubular coil heated externally by a furnace.

We shall describe the system by means of a differential mass and a energy balance, which carry the temperature T and the conversion X as dependent variables.

Mass Balance

$$\frac{dX}{dz} - \frac{r\pi d^2}{4n_A^0} = 0$$

Energy Balance

$$n_A^0 C_{pA}\frac{dT}{dz} + n_A^0 \Delta H_r \frac{dX}{dz} + U\pi d(T - T_f) = 0$$

where

n_A^0 = Molar feed rate to the reactor
X = Conversion = $(1 - n_A)/n_A^0$
r = Reaction rate (first order)
C_p = Molar heat capacity of acetone
ΔH_r = Molar heat of reaction
U = Heat transfer coefficient
T_f = Furnace temperature
z = distance variable

After expressing r in terms of temperature and conversion, and substituting numerical values for the parameters, one obtains the following expressions:

$$\frac{dX}{dz} - \frac{7.96(1 - X)\exp[34.34 - 61{,}800/T]}{T(1 + X)} = 0$$

$$\frac{dT}{dz} + 912\frac{dx}{dz} + 1.925 \times 10^{-2}(T - 1895) = 0$$

Here the original English engineering units have been retained, and T is in units of degrees Rankine (°R = °F + 460).

The two equations are to be integrated using a fourth order Runge–Kutta routine and step sizes of 10 ft to a total length of 150 ft. At the inlet, $X = 0$, $T = 1660\,°R$.

ANSWER $X = 0.278$, $T = 1789\,°R$

INDEX